Process Safety

Many chemical engineering graduates leave university with limited practical experience, including in the important field of process safety. Authored by two veterans of industry, this book provides early career engineers with a strong practical background in process safety. It contains basic theory to bolster an understanding of the subject, whilst the bulk of the book gives practical information covering both design and operation of process plants. It offers an interactive approach, with frequent links between chapters and inclusion of numerous case histories to illustrate and support the text.

Key Features:

- Serves as a simple introduction with a bias towards practical and realistic solutions
- Discusses a wide range of safety-related subjects with comprehensive links to more detailed references
- Covers both design and operation of plants
- Features a significant number of brief case histories to offer real-world context, many of which have not been published elsewhere

With its accessible and hands-on approach, this book gives early career chemical engineers a deeper understanding of the issues they will face whilst on the job and practical solutions to address and avoid problems. The book's many case studies will also be of value to more experienced engineers.

Frank Crawley served as a plant commissioning and start-up manager at Olefine Plants in ICI for seven years. He had a further 10 years "special duties" on ICI Nylon Intermediates. For 10 years, he was leader of the safety and loss prevention (now called process safety) engineering team in a major oil company. After retirement, he spent over 10 years as senior lecturer at Strathclyde University in parallel with consultancy activities.

Doug Scott graduated from the University of Bradford with a degree in chemical engineering and management economics. After working for 10 years in a variety of roles for oil and chemical companies, he joined the insurance industry as a risk engineer and over 35 years has visited hundreds of facilities around the world to perform risk surveys and advise on safety-related aspects of design and operation. He has been involved in a number of loss investigations and is the author of numerous technical papers.

Process Safety

A Practical Guide

Frank Crawley and Doug Scott

CRC Press
Taylor & Francis Group
Boca Raton London New York

CRC Press is an imprint of the
Taylor & Francis Group, an **informa** business

Designed cover image: Shutterstock

First edition published 2026
by CRC Press
2385 NW Executive Center Drive, Suite 320, Boca Raton FL 33431

and by CRC Press
4 Park Square, Milton Park, Abingdon, Oxon, OX14 4RN

CRC Press is an imprint of Taylor & Francis Group, LLC

Library of Congress Cataloging-in-Publication Data
Names: Crawley, Frank author | Scott, Doug (Doug S.) author
Title: Process safety : a practical guide / by Frank Crawley and Doug Scott.
Description: First edition. | Boca Raton : CRC Press, 2025. | Includes bibliographical references and index. | Summary: "Many chemical engineering graduates leave university with limited practical experience, including in the important field of Process Safety. Authored by two veterans of industry, this book provides early career engineers with a strong practical background in Process Safety. It contains basic theory to bolster an understanding of the subject, while the bulk of the book gives practical information covering both design and operation of process plants. It offers an interactive approach, with frequent links between chapters and inclusion of numerous case histories to illustrate and support the text. Serves as a simple introduction with a bias towards practical and realistic solutions. Discusses a wide range of safety related subjects with comprehensive links to more detailed references. Covers both design and operation of plants. Features a significant number of brief case histories to offer real-world context, many of which have not been published elsewhere. With its accessible and hands-on approach, this book gives early career chemical engineers a deeper understanding of the issues they will face while on the job and practical solutions to address and avoid problems. The book's many cases studies will also be of value to more experienced engineers"—Provided by publisher.
Identifiers: LCCN 2024061458 (print) | LCCN 2024061459 (ebook) | ISBN 9781032839837 hbk | ISBN 9781032839844 pbk | ISBN 9781003510680 ebk
Subjects: LCSH: Chemical engineering—Safety measures
Classification: LCC TP150.S24 C477 2025 (print) | LCC TP150.S24 (ebook) | DDC 660/.2804—dc23/eng/20250311
LC record available at https://lccn.loc.gov/2024061458
LC ebook record available at https://lccn.loc.gov/2024061459

ISBN: 978-1-032-83983-7 (hbk)
ISBN: 978-1-032-83984-4 (pbk)
ISBN: 978-1-003-51068-0 (ebk)

DOI: 10.1201/9781003510680

Typeset in Times
by Apex CoVantage, LLC

Contents

PART D Aspects of Plant Operation

Preface

The authors, with others, first collaborated on a publication around 30 years ago. Since that time the authors and their contemporaries have approached the end of their working lives. During that time chemical engineering has undergone many changes—in terms of both technology and perception. Between the end of the Second World War and the 1990s, plant throughputs increased by orders of magnitude. Pollution and energy efficiency were often secondary considerations, and safety considerations were basic by today's standards. World events, starting with the mid-1970s oil crisis and then awareness of climate change and pollution, have changed public perceptions of the process industries and forced companies to become "greener".

The process industries are changing. New technologies and new industries are developing. New opportunities for chemical engineers will exist in industries ranging from the production of silicon wafers to pharmaceuticals, as well as in many of today's existing industries. The change to a net zero carbon economy will bring new challenges to sit beside existing safety and environmental concerns. Production and use of coal, oil and gas will be a declining industry, whilst recovery and recycling of all types, plus the development of the hydrogen economy, will be amongst growth areas. Large-scale "green" projects are now planned, and steel produced from hydrogen rather than coke is already available.[1]

Day-to-day practices within the process industries have also changed—in some cases incrementally, in other cases fundamentally. Catalysts have improved, allowing new process routes to be developed and operating temperatures and pressures to be reduced. Automation has resulted in spectacular changes to process control within the working lifetime of today's older operators. Maintenance activities have similarly been revolutionised, by both the greater reliability of equipment and the development of sophisticated condition monitoring and diagnostic techniques. However, the growth of the process industries and the benefits it has brought to society have come at a price. The first major ammonium nitrate explosion occurred in Oppau, Germany, in 1921, and the Beirut explosion, also involving ammonium nitrate, occurred a century later, the interval between being punctuated with other serious explosions. Most chemical engineers will know of the devastating accidents that occurred at Flixborough, Piper Alpha and Bhopal. Numerous similar events have occurred but are less well known only because the death toll was mercifully low.

The job of a chemical engineer has clearly changed in numerous ways over the working lifetime of today's older engineers. They designed heat exchangers and distillation columns using slide rules and graph paper rather than calculators and computer programmes. Maintenance planning and safe control of maintenance were paper-based systems, and fault-finding made use of the "eyeball and ear" method (sometimes connected to the handle of a screwdriver with the point pressed against the equipment casing) rather than today's sophisticated analytical tools.

Although the industry and technology have greatly changed depressingly, safety and environmental incidents continue to occur, often with the same root cause as those of 50 years ago. Occupational safety dates back to the early days of the industrial revolution. Process safety, however, is a relatively new concept, and the two are still frequently confused. In this book we have tried to introduce both sides of the safety coin but with a bias towards process safety which is more relevant to chemical engineering.

We have made a conscious attempt to balance "software" (procedures and management) with "hardware" (design and equipment). Unfortunately, software tends to be less stable than hardware. This is illustrated by the root causes of incidents which reoccur far too frequently.

Whilst every generation needs to relearn the lessons their predecessors learnt, the authors (who have collectively over a century of experience) and others of their generation have concerns that the new tools available to today's engineers may obscure some of the basic practical knowledge included in an engineer's safety toolkit before the age of digitisation. We have attempted to fuse the last generation's experience with today's methods of working—fully aware that this book will become outdated, but perhaps some of the learnings will travel further forward in time.

We have tried to structure the book in a logical way. Following the introductory chapters, we have devoted a chapter to each stage in the life of a project, from conception to demolition, followed by chapters to supplement the operational phase of a process plant's lifetime. Finally, we have attempted to look into the future. Each chapter, indeed each section, could form the basis for more comprehensive guidance, and our intention is to provide an introduction and, where possible, point the reader to more detailed information. There are many excellent textbooks which provide detailed theoretical models for gas dispersion, fire or explosion, but we have made the deliberate decision to limit the number of equations to those where manual calculation is possible. Complex calculations are likely to be performed by computer, but the ability to perform "ranging shots" manually is still considered useful. In this book we have tried to provide a minimum of essential theory with practical experiences drawn from the working lives of ourselves and others.

There is no such thing as "pure" safety. The actions of plant designers, operators and maintenance staff have as much, if not more, impact on safe operation than the activities of the site safety manager. Safety is holistic with its tentacles appearing at any point in a process plant's lifetime. For this reason, many sections of the book in widely separated chapters touch on similar subjects, and there are numerous links between chapters to try to provide useful information without excessive repetition. We have provided examples to illustrate the text through both references and brief case histories; some are well-researched incidents; others are from the authors' own experiences or recounted by their contemporaries. We have not attempted to provide detailed case histories. Many of the classic accidents have been well researched and analysed by others. We have used cases instead to illustrate and amplify issues discussed in the book. These have been kept brief but amplify points made in the text. Where possible, a reference to a more complete description is provided. There are numerous sources of information on all aspects of safety and loss prevention available on the internet, including well-written case histories of minor and major incidents. Of particular use are the UK Institution of Chemical Engineers' Loss Prevention Bulletin and the detailed reports of the US Chemical Safety Board. Both have accessible websites.

The book is aimed at undergraduates in their final year and young engineers approaching chartered status. In the latter case it should be treated as a mix-and-match as each CV or training report will be case specific.

We doubt this book will answer every question the reader may have, but if it creates awareness of safety issues and encourages the reader to go beyond a "push button" or "checklist-based" approach to occupational or process safety, we will have achieved our objective.

Frank Crawley and Doug Scott

DISCLAIMER: As the amount of information available on the internet is constantly increasing, where possible, a web link is provided as well as the original source of information. In a few cases, the reference is direct to a web page. The links were valid at the time of publication.

NOTE

1 International Energy Forum, *Cutting the Carbon Intensity of Steel Using Hydrogen and Renewables*, 23 January 2024, https://www.ief.org/news/cutting-the-carbon-intensity-of-steel-using-hydrogen-and-renewables

Acknowledgements

The production of this book has been interesting, absorbing and, at times, frustrating. It has taken considerable time to put our thoughts in order, with the right level of detail. During this time our families have been, generally, tolerant of the hours we have hidden away in the cycle of typing, editing, retyping and more editing.

We have taken technical comment directly and indirectly from a number of sources, some of which would be embarrassed if named, plus valuable assistance from experts on how to successfully use some of the more obscure (to us) features of Microsoft Word. We should also express our appreciation to the many individuals who provided mentoring, assistance and advice throughout our careers. Most of these individuals have long since died, although their knowledge lives on in the heads of ourselves and others.

Finally, we would like to thank the various individuals and organisations who have allowed us to reproduce tables and figures. Key amongst these are the Institution of Chemical Engineers (IChemE) and a UK website which contains public sector information published by the United Kingdom Health and Safety Executive and licensed under the Open Government Licence. The site can be accessed at www.nationalarchives.gov.uk/doc/open-government-licence/version/3/.

Other organisations which have assisted us include Chartwise UK, Netherlands Organisation for Applied Scientific Research (TNO), the US National Fire Protection Association, the Nobel Institute, RSE Consultants and various US government departments, especially the US Centers for Disease Control and Prevention (CDC) for permission to reproduce Table 10.2. The table, and considerable further information, is available free of charge from the CDC website, and our reproduction of the table does not constitute an endorsement of the publication by CDC, ENRg Consultants and the Keil Centre for permission to reproduce the Safety Culture Maturity Model (Figure 21.3). The Safety Culture Maturity is a registered trademark of the Keil Centre.

We are especially grateful to Tony and Nigel Kletz for permission to use some items from Trevor Kletz's vast volume of work. Finally, we would like to acknowledge the assistance of Allison Shatkin and Ariel Finkle at our publishers, Taylor & Francis.

The author would like to acknowledge the work that Doug's daughter Rosie (Piper) has done to fulfil her father's wishes that this book is completed. This involved checking equations and proof reading. Without her input the second author would would have had difficulties in the completion.

Abbreviations

1-0-0-2 One Out of Two (Protective System Configuration)
2-0-0-3 Two Out of Three (Protective System Configuration)
ACMH Advisory Commission on Major Hazards (UK)
ACVG Alternating Current Voltage Gradient
AFC Approved for Construction
AFD Approved for Design
AI Artificial Intelligence
AIT Auto-Ignition Temperature
ALARP As Low as Reasonably Practical
ANSI American National Standards Institute
APC Advanced Process Control
API American Petroleum Institute
ASA American Standards Association (now superseded by ANSI)
ASL Above Sea Level
ASME American Society of Mechanical Engineers
ASU Air Separation Unit
ATEX Appareils destinés à être utilisés en ATmosphères Explosives—Devices Intended for Use in Explosive Atmospheres (EU)
AWS American Welding Society
BA Breathing Apparatus
BAT Best Available Techniques (effectively the same as BATNEEC)
BATNEEC Best Available Techniques Not Entailing Excessive Costs
BBS Behaviour-Based Safety
BD Bursting Disc
BDV Blow-Down Valve
BLEVE Boiling Liquid Expanding Vapour Explosion
BOD Biological Oxygen Demand
BFW Boiler Feed Water
BP Boiling Point
BPCS Basic Process Control System
CAD Computer-Aided Design
CBA Cost-Benefit Analysis
CCF Common Cause Failure
CCS Carbon Capture and Storage
CCPS Center for Chemical Process Safety
CCTV Closed Circuit Television
CDC Centre for Disease Control and Prevention
CFC Chlorofluorocarbon
CFD Computational Fluid Dynamics
CHAZOP Control Systems Hazard and Operability Study
CIP Cleaning in Place
CIS Close Interval Survey
CISWP Certification Scheme for Weldment Inspection Personnel (now renamed Certification Scheme for Personnel)
CMF Common Mode Failure
COMAH Control of Major Accident Hazards
COD Chemical Oxygen Demand

COP	Code of Practice
CPA	Critical Path Analysis
CPD	Continuing Professional Development
CSTR	Continuous Stirred Tank Reactor
CUI	Corrosion Under Insulation
DCS	Distributed Control System
DDT	Dichlorodiphenyltrichloroethane
DIERS	Design Institute for Emergency Relief Systems
DCVG	Direct Current Voltage Gradient
DLT	Dangerous Toxic Load
ECC	Emergency Coordination Centre
EEMUA	Engineering Equipment Users Manufacturing Association
EHAZOP	Electrical Hazard and Operability Study
EIA	Environmental Impact Assessment
EIS	Environmental Impact Statement
EMF	Electromagnetic Field
EMS	Environmental Management System
EMT	Emergency Response Team
EOH	Equivalent Operating Hours
EPC	Engineer Procure and Construct
ERP	Emergency Response Plan
ESA	Electrical Signature Analysis
ESD	Emergency Shut Down
ESDV	Emergency Shut Down Valve
ESG	Environmental, Social Governance
FAR	Fatal Accident Rate
FDT	Fractional Dead Time
FEED	Front End Engineering Design
F&G	Fire and Gas
FP	Flash Point
HASWA	Safety at Work Etc. Act (UK)
HAVS	Hand-Arm Vibration Syndrome (sometimes known as white finger)
HAZAN	Hazard Analysis
HIPPS	High Integrity Pressure Protective System
HIPS	High Integrity Protective System
HSE	Health & Safety Executive (UK)
HUI	Head-Up Display
HAZOP	Hazard and Operability Study
HEM	Homogenous Equilibrium Method
HVAC	Heating, Ventilating and Cooling
ICI	Imperial Chemical Industries
IEC	International Electrotechnical Commission
IPL	Independent Protective Layers
IR	Infra-Red
IRPA	Individual Risk Per Annum
IS	Intrinsically Safe
JT	Joule Thomson Effect
KPI	Key Performance Indicator
LC_{50}	Lethal Concentration 50%
LD_{50}	Lethal Dose 50%
LFL	Lower Flammable Limit

LIBS	Laser-Induced Breakdown Spectrometer
LPG	Liquefied Petroleum Gas
LME	Liquid Metal Embrittlement
LOPA	Layer of Protection Analysis
LOTO	Lockout Tagout
LOTOTO	Lockout Tagout Tryout
LTA	Lost time Accident
LTEL	Long-Term Exposure Limit
LTIR	Lost Time Incident Rate
MEM	Multi-Energy Method
MFL	Magnetic Flux Leakage
MMC	Motor Control Centre
MMI	Man/Machine Interface
MOC	Management of Change
MSDS	Material Safety Data Sheet
MTBF	Mean Time Between Failures
MTTF	Mean Time to Failure
MTTR	Mean Time to Repair
NATECH	Natural Hazards Triggering Technological Accidents
NDT	Nondestructive Testing
NDT	Non-Destructive Testing
NFPA	National Fire Protection Association
NIMBY	Not in My Backyard
NPSH	Net Positive Suction Head
NRV	Non-Return Valve
O&M	Operation and Maintenance
OSHA	Occupational Health & Safety Administration (USA)
OSHA	Occupational Safety and Health Administration (USA)
OTJ	On-the-Job (Training)
PCB	Polychlorinated Bisphenols
PCV	Pressure Control Valve
PFD	Process Flow Diagram
PFD	Probability of Failure on Demand
PI	Pressure Indicator
P&ID	Process and Instrumentation Diagram
PMI	Positive Materials Identification
POC	Products of Combustion
PPE	Personal Protective Equipment
PRV	Pressure Reducing Valve
PSSR	Pre-start-up Safety Review
PSV	Pressure Safety Valve
PTW	Permit to Work
QA	Quality Assurance
QC	Quality Control
QRA	Qualitative Risk Assessment
REACH	Registration, Evaluation, Authorisation and Restriction of Chemicals Programme (EU)
RIVM	National Institute of Public Health & the Environment (Netherlands)
RPT	Rapid Phase Transformation
RRF	Risk Reduction Factor
RT	Radiographic Testing

RBI	Risk-Based Inspection
RFID	Radio Frequency Identification
RIOV	Remotely Operated Isolation Valve
SADT	Self-Accelerating Decomposition Temperature
S&T	Shell and Tube (Heat Exchangers)
SCC	Stress Corrosion Cracking
SCMM	Safety Culture Maturity Model™ (the SCMM is a trademark of the Keil Centre)
SEP	Surface Emissive Power
SHE	Safety, Health and Environment
SI	Système International
SIL	Safety Integrity Level
SIS	Safety Instrumented System
SLOD	Significant Likelihood of Death
SLOT	Specified Level of Toxicity
SOAP	Spectrum Analysis of Oil Pollutants
SMS	Safety Management System
SOP	Standard Operating Procedure
STEL	Short-Term Exposure Limit
SIMOPS	Simultaneous Operations
TBT	Tool Box Talk
TEL	Tetraethyl Lead
TEMA	Tubular Exchanger Manufacturers Association
TLV-C	Threshold Value-Ceiling
TNO	Netherlands Organisation for Applied Scientific Research
TWA	Time-Weighted Average
TLV-TWA	Threshold Limit Value–Time-Weighted Average
UFL	Upper Flammable Limit
UK	United Kingdom
UPS	Uninterruptable Power Supply
USDA	United States Department of Agriculture
UT	Ultrasonic Testing
UTS	Ultimate Tensile Stress
USA/US	United States of America
UV	Ultraviolet
UXO	Unexploded Ordnance
VCE	Vapour Cloud Explosion
VOC	Volatile Organic Compound
WO	Work Order
WWTP	Wastewater Treatment Plant
XRF	X-Ray Fluorescence

Part A

Introductory Concepts

1 Introduction

1.1 THE PROCESS INDUSTRIES AND CHEMICAL ENGINEERING

Chemical engineering as an engineering discipline is only about 100 years old (the UK Institution of Chemical Engineers was founded in 1922). However, in one form or another, chemical engineering is much older. Brewing, tanning of leather and the processing of metal ores can all be considered as early forms of chemical or biochemical processing. Chemical processing is recorded in early Egyptian and Chinese civilisations.[1,2] The demand for chemicals increased as the Industrial Revolution progressed. Chemical engineering[3,4,5,6] as a distinct discipline evolved in the late 19th century.

The early 20th century saw a rapid development in both the variety and scale of chemical processing, typified by the *Haber* process for industrial scale manufacture of ammonia, used in both fertiliser production and explosives. The expansion of the industry continued in the 1930s and 1940s with increasingly sophisticated processes for the refining of crude oil. Whilst some plastics manufacturing took place prior to the Second World War this was basically of thermosetting resins. The large-scale production of plastics that are familiar today started after the war. Again, the variety and scale of plant increased rapidly between the post-war developments and today.[7]

During the First World War, Britain was at risk of defeat due to a lack of explosives. Kenneth Bingham Quinan and others who would now be called chemical engineers rose to the challenge.[8] As a result, the Institution of Chemical Engineers was formed in 1922. Chemical engineering as a separate engineering discipline was recognised in the USA rather earlier with the formation of the American Institute of Chemical Engineers in 1908. Chemical engineers made numerous positive contributions ranging from the development of special glues used in the construction of the wooden mosquito aircraft[9] to the development of fluidised bed catalytic cracking (FCC) to improve yields from oil refining.[10] Pitt[11] has written widely on the role of chemical engineers in a variety of industries.

Chemical engineering has played a vital role in the growth of the economy and chemical engineering innovations will continue in the future but probably in new directions with the growth of the *net zero* economy and sustainable production. The decline in the use of hydrocarbons and the development of novel feedstocks will lead to new challenges. In the 1960s and 1970s most graduate chemical engineers left university to join a large-scale manufacturing company or a contracting company designing and building new plants. Today's chemical engineering graduates are likely to have a wider range of industries to choose from but require a greater level of specialisation. Industries will, amongst others, include food processing, pharmaceuticals, water treatment and materials recycling.

The chemical industry has brought great benefits to society, but its products can be both beneficial or harmful depending on use. Chlorine, and its derivatives, are very powerful tools for water treatment and will have saved many lives by providing safe drinking water. However, chlorine has also been used as a weapon of war largely in the First World War but more recently in hostilities in the Middle East. Similarly, ammonia is an intermediate in the production of fertiliser, helping to provide food for the world's growing population but is also an intermediate in the manufacture of explosives. Whilst the industry can provide fertilisers, agricultural run-off results in pollution of rivers and run-off from phosphates results in algal blooms, some of which are toxic. The use to which chemicals are put is a function of society and not just the chemical industry.

There have been many unintended incidents resulting in death and destruction during the manufacture, transportation and use of chemicals. Technology and methods of working have evolved over the years—often following a serious incident. Headlines are usually generated by a catastrophic

DOI: 10.1201/9781003510680-2

event with multiple deaths and dramatic photographs of damage. This may be at the expense of less dramatic but equally devastating incidents resulting in widespread loss of life over a period of decades rather than a few seconds. Incidents such as Oppau, Germany (1921), Flixborough, UK (1974), and Texas City, USA (2005) typify the catastrophic incidents that can occur. Industrial diseases such as asbestosis and chemical releases such as the large-scale mercury poisoning in Minamata, Japan[12] (1953–1967), are examples of chronic incidents. Some accidents, such as Bhopal, India (1984), where a leak of methyl isocyanate occurred, cover both categories with over 1,000 immediate fatalities (estimates vary) but a continuing stream of deaths exceeding 20,000 has occurred over the years.[13] The indirect effects of both catastrophic and chronic events are often ignored. The economic effects on the area, the loss of income to bereaved families and permanent injury can have long-term effects on communities and individuals.

Environmental issues are a relatively recent area of concern within the process industries. Whilst standards for emissions of waste have existed since the 19th century, they have not always been rigorously enforced. The location of early process plants has left a toxic inheritance which would not be acceptable in modern facilities. Routine emissions are now more tightly monitored and controlled (with further restrictions possible in the future). The environmental effects of major industrial accidents now form part of any risk assessment. There is a paradox that improved environmental emission standards can result in reduced levels of safety and vice versa. This is explored in Section 4.2.

1.2 THE NEED

Safety is now an integral part of undergraduate chemical engineering courses and several universities run postgraduate courses specialising in process safety. However, there is a difference between theory and practice, and many graduates leave university and enter industry with limited tools to bridge the gap between the academic and industrial worlds. Well-understood academic theories often dissolve in the middle of a night shift when the wind is howling and the rain torrential. The pristine conditions in a university laboratory or lecture hall are very different to the average industrial site. The differences will soon become clear to a fresh graduate, sometimes as a shock when trying to discover why a recalcitrant control loop will not operate in the middle of the night. An understanding of "real life" in the industrial world will be an asset when applying for professional status.

A great deal of information is available on the safe design and operation of process plant. Much of it tends to be in the form of standards, codes of practice (COP) or technical documentation dealing with specific equipment or particular industries. Frequently this information is in a complex technical form. More recently human factors have become an important consideration during plant design and operation. Many bookshelves could be filled with the literature covering design and operation. There is also a mass of information available on occupational safety, sometimes known, dismissively, as "hard hat and boots".

The authors feel, however, there is still a need for an introductory text to assist younger engineers to acquire a basic understanding of the principles of process safety and developing the *duty of care* needed for safe and efficient design and operation. This book aims to fulfil that need.

Chemical engineering technology has changed greatly over the last few decades. Originally "big was beautiful", effluents were discharged to sea or rivers, thermal efficiency was less important and *recycling* was a word not yet in the engineering vocabulary. The *oil crisis* of the mid-1970s focused attention on thermal efficiency but *global warming* was still a concept for the future. Now there is an urgent need to decarbonise industry. Plastic waste is polluting the sea with microplastics, and there is a need to recycle and reuse as much as possible. A casual review of industry publications will show that new processes continue to be produced and developed from the laboratory to pilot plant, to demonstration plant then to full-scale production but a depressing number of accidents (often the same accidents in different circumstances) continue to happen. The principles of process safety (sometimes known as loss prevention) laid down over many years are still valid and remain a valuable tool in the "safety armoury". However, there is a danger that they will be overlooked as shiny new toys (often computer based) become available.

This book is an attempt to present a "mix-and-match" of techniques which have been developed over many years. Some may not apply to a particular process but are still relevant in the overall context of safe operation and run the risk of being overlooked. The costs in terms of death/injury, damage to property, environmental damage, property damage, loss of profit and damage to company reputation can be significant—perhaps to the point of company survival.

If any reinforcement of this requirement is needed a few examples will dramatically illustrate the point.

Case 1

In the mid-1960s there was a push for innovation and higher efficiency and all-round better plants. Two plants with capital costs (2023) combined of approximately £180 million failed to reach beneficial production and were demolished.[14]

Case 2

Piper Alpha 1988[15] was an exact equivalent to an onshore event which occurred in 1965. In the case of Piper Alpha, the costs included 167 lives and billions of pounds in replacement/legal costs. A more recent incident in 2023 also shows similarities with Piper Alpha.[16]

Case 3

The Chernobyl Nuclear Power Station reactor[17] was inherently unstable in certain areas of the operating envelope. The reactor efficiency was a little higher at the expense of increased risk of reactor *runaways*. In 1986, a reactor runaway did occur during a safety-related experiment due to poor design, poor operating, poor procedures and possibly—equally importantly—poor assessment of management of change (MOC). Several hundred premature deaths are predicted to result from the incident.[18,19]

Case 4

A Mexico City LPG storage area[20] erupted in a major fire, with at least half the storage vessels rupturing from Boiling Liquid Expanding Vapour Explosions (BLEVEs). The cause was a mixture of poor design, poor inspection, poor operation and poor site location. Over 500 deaths occurred.

In all of these events the cost to society in the form of loss of life, pollution and financial damage was considerable. Alarmingly, some of the design and operational features which gave rise to these events are reappearing in new projects and modifications to existing plants. An incident in 2020 at Visakhapatnam, India, leading to the release of styrene monomer has similarities to Bhopal but, mercifully, with fewer casualties.[21]

1.3 OCCUPATIONAL AND PROCESS SAFETY

Until the 1970s, safety was largely considered in terms of occupational safety. Model reviews based on scale models of the proposed plant would ensure safe access to equipment and escape routes. Exposure to danger in the form of toxic chemicals, excessive heat, noise and vibration would be assessed based on codes and standards. Hazard identification was largely absent from the design process. Protection against poor design would be based on the use of codes and standards. However, these alone were not adequate for the increased size and complexity of facilities being built. Improved means of identifying and quantifying hazard and risk were required not only for the process industries but also for other industries such as aerospace and nuclear power. Within the chemical and process industries, the use of hazard and risk identification techniques plus improved safety technology and methods of working have led to the term *process safety* to differentiate the subject from traditional occupational or "technical" safety.

In general, the outcome of accidents can be broken down under three headings:

Process/Occupational: Process hazards are those which directly relate to the chemistry, the process materials, and the manner in which they are processed. Occupational hazards are those which are general such as falls, cuts, impacts and will also include asphyxiation due to poor equipment preparation. The general test is, "Can I remove the hazard by changing the process route?"; if "Yes", it is a process hazard.

Simple/Multiple: This is self-explanatory.

Catastrophic/Chronic: Again, this is self-explanatory, but the time definition of chronic is that the results are there for all time and cannot be rectified. *Catastrophic* generally refers to fatality.

These three aspects can be broken down into a three-dimensional representation. *The accident cube* is shown in Figure 1.1.

Examples of the eight corners of the diagram are as follows:

A1	=	Single chronic occupational—broken leg or permanent eye damage
A2	=	Single catastrophic occupational—death from nitrogen asphyxiation
A3	=	Single chronic process—exposure to toxic gas or acid burn
A4	=	Single catastrophic process—small fire
B1	=	Multiple chronic occupational—post-traumatic stress syndrome
B2	=	Multiple catastrophic technical—structural collapse
B3	=	Multiple process chronic—Minamata
B4	=	Multiple process catastrophic—Piper Alpha

Some incidents can fit more than one corner of the cube such as Bhopal and Chernobyl which have both process chronic and process catastrophic aspects.

Safety and loss prevention must be considered throughout any project's lifetime from initial concept through to, and including, demolition. Even experienced process designers make mistakes, and with new process engineers constantly joining the profession as older staff leave, there is obvious scope for error as knowledge of a particular problem fades. This is known as *corporate* or *design*

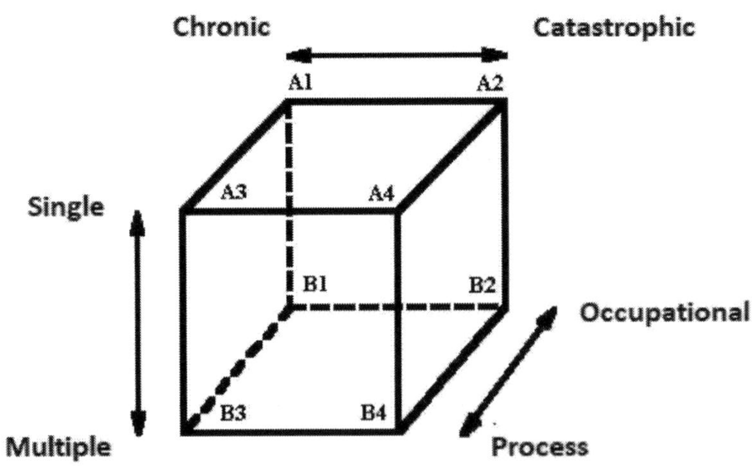

FIGURE 1.1 Classification of Accident Types

memory fade. It may become exacerbated as company mergers, rationalisations and downsizing reduce the availability of experienced personnel who have a wealth of knowledge which they take with them when they leave employment.

1.4 PROJECT LIFE CYCLE

A decision to invest in new plant is not taken lightly. The evolution of a project consists of several stages. Ideally each is completed and reviewed before the next begins, but in practice there is often a degree of overlap. Whilst the number of stages and definitions vary between organisations, eight phases in the project life cycle are defined here (and used throughout this book):

0 Inherent safety
1 Conceptual design
2 Front end engineering design (FEED)
3 Detailed engineering
4 Construction
5 Commissioning
6 Operation
7 Mothballing and demolition

Chapter 5 gives an overview of hazard identification techniques appropriate to each stage, and a chapter is devoted to each phase later, but brief descriptions are given in the following sections.

1.4.1 Inherent Safety

The concept of *inherent safety*, like many pioneering process safety concepts, was developed by Imperial Chemical Industries (ICI) in the 1960s and 1970s but is still not widely used. The concept is based on reducing the risk by the following:

1 Replacing hazardous materials with less hazardous ones
2 Reducing the inventory of hazardous materials
3 Reducing the amount of energy in the system by reducing operating temperatures and pressures
4 Attenuation and intensification
5 Simplification
6 Change
7 Getting it right first time
8 Ease of recovery

Whilst some progress has been made, the drivers for change have been financial or technical rather than safety related. However, the development of novel technologies could significantly change this situation. This is discussed further in Chapter 11.

1.4.2 Conceptual Selection and Development

Key questions to be defined at this stage include plant size, process technology and location. Before proceeding, the economics of the plant must be checked and found to be viable.

In some cases, choices may be limited—locations will generally be dictated by the source of raw materials or markets. Plant capacity will be determined by market size and economics. However, there are still important choices to be made, for example the choice of process and the use of proven technology instead of new, potentially, more economic but untried technology.

Whilst process chemistry may be limited to commercially available processes, inventories, types of chemicals used, energy/utility requirements and pollution prevention are important factors in determining the process route. Safety and environmental issues must also be addressed. During this phase the potential of inherent safety concepts should be reviewed. Once a concept has been selected there may still be fundamental safety and environmental issues to be resolved; for example utility supply and waste disposal must be clearly defined before the project is considered viable. At the end of this phase risk levels will have been determined and set against corporate/national criteria. Only if they pass these criteria will the project proceed to FEED and detailed design.

It is during this concept phase that the big decisions are made. Major errors will make the project unviable; minor errors will increase costs and schedule at a later stage.

Preliminary work to gain permission for plant construction will also take place at this stage and will include agreements on a variety of topics including discharges to land, air and water, utility requirements, provision of local infrastructure and, in the developing world, the provision of facilities such as schools and medical facilities for the local community.

This is discussed further in Chapter 12.

1.4.3 Front End Engineering Design

Once the concept design has been finalised and the critical parameters of the new plant agreed, a detailed design takes place; this specifies equipment parameters, detailed process flow diagrams (PFDs), and process and instrumentation diagrams (P&IDs). The FEED adds sufficient detailed to the framework of the conceptual design to allow accurate cost and scheduling information to be produced. The output forms the basis for the detailed design. Whilst some risk assessment will have been undertaken at this stage they may be limited and inherent hazards in the design may be present. This is discussed in Chapter 13.

1.4.4 Detailed Design

Detailed design builds on FEED to produce drawings and procedures for use during construction. Documents are generally marked "for construction". Whilst earlier stages of the project will have established the basis for the project and FEED produced a basis for detailed design, the design process should virtually start from scratch. Detailed hazard studies, generally hazard and operability (HAZOP) studies, should be undertaken. At the conclusion of detailed design all the documentation, specifications and other information needed to turn the design into reality should be available, including detailed general arrangement drawings of piping and foundations, cabling layouts, instrument loop diagram and control system architecture. At this stage not only should the final design of the plant for operation be defined but also the practicality of construction and, ultimately, demolition. For example, are the roads adequate for equipment to be brought to the site, or will part-fabricated equipment need to be provided with final assembly on-site? Quality assurance/quality control (QA/QC) control requirements will also need to be determined at this stage.

Design progress will be measured, and changes made to budgets to reflect the modifications, large and small, which inevitably occur. At this stage purchase orders will be issued for equipment and agreements made with main and subcontractors involved at the construction stage.

Having obtained outline planning permission at the conceptual design stage, detailed permission is required; local and government authorities may require safety and environmental studies to obtain a licence for construction and then operation (see Section 1.6). This is also discussed in Chapter 14.

1.4.5 Construction

Preparatory work will often have been started during the detailed design phase. Long delivery items such as large compressors and major process vessels may have been ordered early in the design and changes requested where necessary. It is not unusual for ground preparation to start before detailed design is complete, which will shorten the project schedule.

Other preliminary work before major construction activity starts will include setting up the site infrastructure including construction offices, workshops, storage areas and in remote areas, accommodation and recreation facilities. A schedule for the project will be in place with progress recorded and detailed descriptions of work activities.

Frequently design changes are made during construction to reflect construction problems, changing economics and other factors such as new legislative requirements. These changes must all be reviewed from a safety viewpoint and all documentation updated (see Section 1.7).

QA/QC will be performed throughout the construction progress. Initially this will concentrate on activities such as testing of concrete mixes but, as time progresses, will cover weld integrity, application of corrosion protection systems and many other aspects. Checks will generally not cover just the final product but also the method of working. Pre-commissioning tests will ensure integrity of pressurised systems, correct rotation of electric motors and calibration of instruments, including pressure safety valves (PSVs).

Typically, the plant will be split into systems and subsystems which are inspected by a multidiscipline team who record any incomplete/substandard construction in a *punch list*.

This is discussed in Chapter 15.

1.4.6 Commissioning

Once the integrity of the plant has been confirmed, commissioning activities will begin. This is often the most hazardous period during the life of a plant and is performed in a staged manner. Some safety systems may need to be defeated as a temporary measure.

Following a successful startup, a proving run will take place to ensure that the plant can operate at full capacity for extended periods. This is discussed in Chapter 15.

1.4.7 Operation

Although the least dramatic, this should be the longest of the life cycle phases. It is regulated by procedures and instructions for safe operation and maintenance which are amended with experience.

Modifications will be made and must be assessed using appropriate risk assessment techniques. It is important that design stage operating and maintenance assumptions are adhered to. If the assumptions become invalid the assessment must be updated and the results may indicate that changes to equipment or procedures, or both, are necessary.

Many older plants still in operation have incomplete records: piping and cabling routes may not be known, materials of construction may be suspect—particularly where design standards have changed since construction. Maintenance may often be difficult due to the lack of availability of spare parts and equipment becoming worn out rather than suffering from random failures. This is discussed in Chapter 16.

1.4.8 Mothballing and Demolition

Whilst final shut down of a plant is probably the least of the plant designers' concerns, provision for final disposal should be considered at an early stage. Plants designed in line with current-day practice should minimise the potential for environmental contamination. However, disposal of by-products, intermediates, catalysts and other contaminants may be difficult and expensive. In some cases, for example, mercury contamination and asbestos removal, special precautions are required to prevent contamination of the environment or injury to personnel. Many of the plants currently in operation have been designed and constructed without regard to their eventual decommissioning or closure. Demolition and site restoration therefore become difficult. In future, the choice of the process to be adopted may be influenced by the need to return the site to future beneficial use.

An increasingly common practice is the *mothballing* of plant and equipment which is not immediately required but may be returned to service in the future. Depending on the expected length of shut down, process fluids may be left in place with limited measures to prevent corrosion. If a lengthy shut down is anticipated, more comprehensive methods are necessary, including emptying of pipework and equipment then refilling with non-corrosive materials (often nitrogen) and removal of sensitive instruments to a controlled environment. An active maintenance programme is still required to ensure that rotating equipment is not damaged. Restarting the plant will require a detailed inspection to ensure the continued integrity of plant and equipment.

This is discussed in Chapter 17.

1.5 ACCEPTABILITY OF RISK

Determination of the acceptability of risk has been a topic of discussion at least since the 1970s when the UK Council for Science and Society published "The Acceptability of Risk".[22] Determining the acceptability of risk involves balancing the risks against the potential benefits and determining if the risk is *As Low As Is Reasonably Practical* (ALARP)[23] (see Figure 1.2). This is almost the same wording as the UK's Health and Safety at Work Etc. Act (HASWA): "so far as is reasonably practicable".[24]

There are a number of sources which discuss the acceptability of risk and criteria of acceptability in detail. This section can only be an introduction so that the reader may appreciate the complexity of the subject, as well as its boundaries. The development of criteria in the process industries started in the 1970s with the first report of the UK Advisory Committee on Major Hazards (ACMH)[25] which states in rather vague terms:

> *If, for instance, such tentative conclusions indicated that with reasonable confidence that in a particular plant a serious accident was unlikely to occur more than once in 10,000 years (or—to put it another way—1 in 10,000 chance in any one year) this might perhaps be regarded as just on the borderline of acceptability bearing in mind the known background risks faced every day by the general public.*

Whilst not being specific it pointed to the "risk" of over-topping Dutch sea defences as once in 1,000 years.

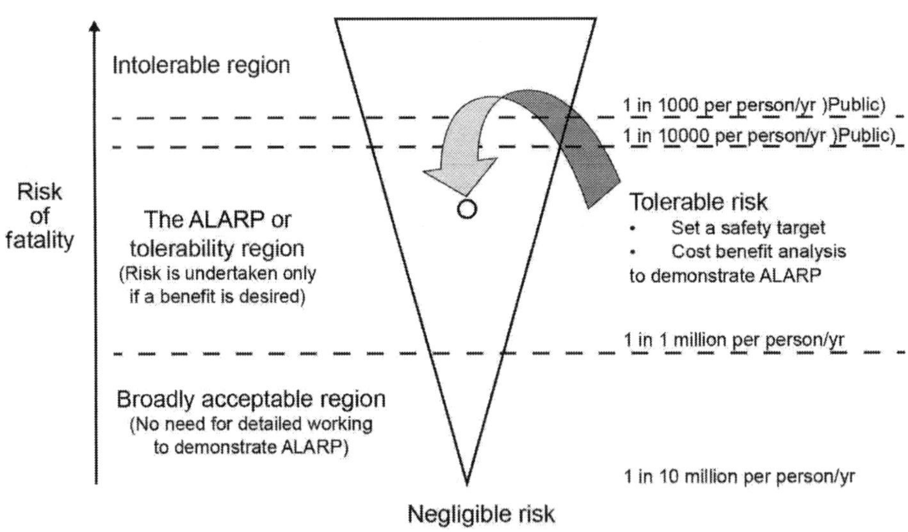

FIGURE 1.2 ALARP

If the assessment shows the risk to be too high, changes must be made to lower the risk to a tolerable level. If further risk reduction is possible without endangering the project economics, this should be implemented. A concept known as *best available techniques not entailing excessive costs* (BATNEEC) relates risk reduction to cost. This unwieldy acronym is now commonly shortened to BAT for *best available techniques.* Note that "techniques" might be procedures or methods of working rather than hardware.

A 1989 UK Health & Safety Executive (HSE) discussion document, "Risk Criteria for Land-use Planning in the Vicinity of Major Industrial Hazards",[26] gave a listing of "everyday risks". This document also introduces the idea that tolerability of risk should consider the concept of *societal risk.* It lists four typical locations and populations and provides indicative criteria for each: for low populations, 10^{-5} per person per year, and for larger populations, 10^{-6} per person per year. This is totally consistent with *land use planning* where farms may be allowed near a hazardous site but other developments with higher population densities are not. Land is at a premium in most of Europe and, increasingly, in other parts of the world. A major review of land use planning[27] in the UK and elsewhere was undertaken as part of the investigations into the major UK fire at Buncefield in 2005. The most recent UK land use planning guidance is available from the HSE.[28]

Criteria are available from a number of sources. In the UK, this includes criteria for the nuclear industry[29,30] and the offshore industry.[31]

Societal risk is often illustrated by the use of F/N (frequency/number) curves. These illustrate the correlation between frequency and the number of people suffering from a specified level of harm within a population from the occurrence of a specific hazard.[32]

1.6 REGULATORY ASPECTS

Legislation is dynamic, constantly changing with time. In general, it tends to be reactive and not proactive. Historically legislation is enacted as a result of an event or events which have occurred with the aim of preventing a future reoccurrence. Legislation is not usually designed to predict some specific undesired event that has not yet occurred and then prevent it from occurring, although there are moves in this direction.

Legal systems vary around the world. In the developed world there tends to be a greater level of regulation of high hazard process industries. Regulation in some countries is basic and not much advanced beyond basic occupational safety plus inspection of pressure vessels.

Where regulation specific to the process industry exists, it tends to be either

- goal setting or
- rule based (prescriptive).

Goal setting is based on the concept of continuous improvements and requires risk assessment to determine that the overall risk at the facility has been reduced to a level that is *as low As reasonably practical* (ALARP)[33] (see Section 1.5 and Chapter 6). The scope of reasonable practicability should be clear to professional engineers and is generally easy to achieve with the tools available—provided common sense is used. Many installations achieve this requirement at startup, but over a period of time standards slip; audits and monitoring are essential safety tools to identify and control them (see Section 21.7). The duty of the employer and the employee must always be to satisfy themselves that safe working conditions are maintained.

A goal-setting approach is used in most of Europe. The framework within the European Union (EU) is defined by the *Seveso Directive* (Seveso was the result of a serious process industry accident in Italy in 1976).[34] Each country within the EU incorporates the directive into national law. Within the UK, the Control of Major Accident Hazards (COMAH)[35] regulations specify what constitutes a major hazard within the process industries, plus the methods used to demonstrate that adequate levels of safety are maintained. This information is contained in a *safety case* demonstrating how safe conditions of work are maintained. Under the Seveso Directive[36] hazardous locations are split into *upper* and *lower tier* sites depending on the substances handled and their amounts. A greater level of assessment is required at upper tier sites. Dangerous substances are classified according to their hazardous properties in Part 1, Annex 1 of the regulations, and specific substances are listed in Part 2 of appendix 1. The annex also defines the amounts necessary to meet upper and lower tier requirements. The annex, and indeed the whole of the directive, is likely to be modified from time to time.

Other countries, notably the USA, use a rule-based approach (OSHA CFR 1910.119). This outlines 14 elements of process safety which must be met:[37]

1	Process safety information
2	Process hazard analysis
3	Operating procedures
4	Employee participation
5	Training
8	Contractors
7	Pre-start-up safety review
8	Mechanical integrity
9	Hot work permit
10	Management of change
11	Incident investigation
12	Emergency planning and response
13	Compliance audits
14	Trade secrets

Some multinational companies use a goal-based approach for all their operations throughout the world, regardless of the legal requirements.

1.6.1 Goal Setting versus Prescriptive Approach

Two major difficulties exist with prescriptive regulation: In a democratic society, legislation is often the result of the desires of government modified by lobbying from industry and pressure groups which tend to dilute the intent somewhat. In some cases, this may result in the necessary level of regulation being compromised. Sometimes impractical legislation may be enacted, imposing requirements which may be difficult—if not impossible—to achieve. Alternatively, if the requirements are achievable, they may not be updated in a timely manner to take account of new and developing technology. The need for regulation was identified by George Stephenson as far back as the 1840s, who reported to the UK Parliamentary select committees.[38]

> *I am quite sure that some interference on the part of this government is much wanted. I see many young engineers, and such a variety of notions, I am convinced that some system should be laid down, to prevent wild and visionary schemes, being tried, at great danger of injury or loss of life to the public.*

A goal-setting approach is less defeated by changes in technology or conditions. The general duties are laid down, and it becomes relatively simple to determine if, in general terms, they are being complied with. Prescriptive approaches tend to be more difficult to enforce and can, from a safety viewpoint, be counterproductive. There may be attempts to obey the letter of the law, but not its intent, in the misguided belief that compliance with the law produces a safe working environment.

There are advantages and disadvantages to both approaches. Goal setting requires continuing improvement as new techniques become available. However, unless approached carefully, there may be confusion over exactly what is required. The prescriptive approach defines exactly what needs to be done to meet legal requirements; however it is more difficult to keep up to date with changes in technology. There is also the danger that an over-prescriptive approach could lead to conflicting requirements which cannot be met. A bigger issue is that minimum standards are defined, and the potential for significant further improvements is ignored. Some companies develop a dangerous mindset that compliance with rules ensures safety. Whilst this may be a defence in law, it traps the level of safety to that at the time the regulations were made and discourages improvements.

1.6.2 Other Regulation

In addition to regulation related to overall level of risk, most countries will have a variety of regulations dealing with specific aspects of safety, for example:

- Safe design using appropriate standards and COP
- Machinery guards and other personnel protective features
- Exposure to noise (which will also apply outside the site)
- Design standards (for example, earthquake resistance)
- Handling and exposure to hazardous chemicals
- Aesthetic considerations (such as landscaping)

Different criteria often extend outside the site, for example:

- Access, particularly during construction when large numbers of contractors' vehicles and personnel may be present—special arrangements may be necessary for dealing with the arrival of large pieces of equipment or the removal of mud or other contaminants from public roadways
- Discharge levels of effluents (including noise, light and flaring)
- Emergency planning for dealing with off-site incidents

1.6.3 Other Interested Parties

Although governments and local authorities are the most obvious bodies to impose conditions on plant operation, they are by no means the only ones. Other organisations who may have specific requirements include the following:

- Banks and other providers of capital, who may insist on process performance guarantees and insurance cover in addition to safety and environmental standards
- Trade unions who may have concerns over the safety of the workforce
- Insurance companies who may make recommendations for the design and operation of plant, particularly in the provision of fire and explosion protection (provision of adequate protective features at the design stage and good operating procedures will influence the cost and availability of insurance)
- Equipment vendors who may specify process limitations, ancillary equipment or process trips, without which warranties may be withdrawn.

1.7 DOCUMENTATION

A librarian once earnestly remarked that the information explosion was as serious as the population explosion but not as much fun. Each phase of the project life cycle will generate documentation, and by the time the plant reaches the end of its life, the cumulative documentation will be substantial. This is essential information for the *Safety Case,* which details the original design intent which may be corrupted by subsequent changes to both plant design and operation. For older facilities, most of the documentation may still be in paper format with poor indexing. Some documentation may also be missing or outdated.

Newer plants are likely to have documentation in an electronic format with paper documentation (acceptance certificates, commissioning reports and so on) being scanned. In this format searching for a particular document is simplified and the evolution of documents through the design and operational phases is easier to trace. Regardless of format, documents must be capable of reproduction if the original is lost. Nowadays with the use of remote servers and back-ups, it is usually possible to recover data. The same is not true of paper records where the loss of the original information might require a lengthy trawl through memos and transmittals to track copies. This is time-consuming and often only partially successful.

For both computer servers and document libraries the provision of fixed fire protection (generally a gaseous extinguishing system) is desirable. Loss of data, as has occurred at a few sites, creates major problems.

Almost as big a problem is ensuring documentation is up to date. Where physical changes take place, drawings, procedures, training manuals and so on must be updated. This should be a requirement of the *Management of Change* (MOC) procedure (see Section 16.9). Even so, it is common practice to require a site survey before a modification is made rather than relying on drawings which might be out of date.

Most people at one time or another have printed or saved a document for future reference only to find it is out of date or superseded by events. Plant staff are no exception to this rule and often have individual copies of key procedures. Use of outdated documents can lead to errors, for example the use of outdated P&IDs to determine isolation points for maintenance.

Ensuring that only up-to-date documents are used is a significant challenge and a disciplined approach is necessary. When documentation was paper based, many companies maintained a *master copy* of each document which would be reproduced and distributed to a controlled circulation list. Each copy would be marked *controlled copy* with a red stamp (which would show up black if photocopied) and circulated to a list of authorised staff. Often the individual copies were numbered. Outdated copies would be returned to the distributor for destruction. Only an archived copy of outdated documents and drawings would be retained. If copies were needed by people not on the circulation list, the document would be marked *uncontrolled copy*. If there was a delay in updating the master document or drawing, a "red line" alteration would be shown. When updating a document, the revision date and number would be noted and often, a brief description of the change. Any document which is not controlled should be treated as outdated.

Nowadays documents/drawings can often be downloaded or printed from a server, and it is common practice for a watermark to show that this is an uncontrolled copy. Computer terminals are increasingly installed in control rooms and other locations so that the master copy can be viewed—but not edited. This can cause difficulties if documents need to be consulted in a plant emergency, particularly if the power supply is lost and a controlled copy of P&IDs, operating instructions and other relevant documents should be available.

Preparation of a safety case (see Section 1.6) will require the preparation and referencing of significant amounts of information which will need to be reviewed at intervals, typically every five years, or when a significant change is made.

The use of quality systems, such as the ISO 9000 series, will help to ensure that documentation is correct and up to date. However, the temptation to update documentation just prior to an audit should be resisted. It is far better to ensure that information is up to date on a day-to-day basis.

1.8 CHANGE AND ITS MANAGEMENT

Almost any facility, organisation or method of working will evolve over time. Almost from the point of completion, change is possible either as "erosion" from its original form or as a deliberate act. The former can be checked by a good system of technical auditing, but where change is deliberate, safeguards must be in place to ensure changes are adequately assessed to ensure that new hazards are not introduced, or the overall level of risk does not increase. Any change must be adequately documented and any change to the design intent clearly identified. Whilst some forms of change, such as a physical change to the plant, are relatively obvious, others may be missed. For example:

- Process changes (such as increases in flow-rate, temperature or pressure)
- Change of chemical suppliers
- Change in organisational structure which might leave responsibility for some activities unclear or cause a conflict of interest.

Management of change is considered in more detail in Section 16.9.

1.9 DEFINITIONS

Each branch of technology has developed its own language with words or phrases which have an exact definition; the same is true of process safety. Throughout this text the following words or phrases will be found. Care has been taken to ensure that the words or phrases have been used in a consistent manner, and in some cases, the definitions have been expanded to assist in an immediate understanding.

Table 1.1 gives a glossary of definitions that were largely developed by the IChemE.[39] Where necessary additional definitions will be provided in the chapter concerned:

TABLE 1.1
Common Definitions

Term	Definition
Loss Prevention	A systematic approach to preventing accidents or minimising their effects. The activities may be associated with financial loss or safety issues.
Hazard	A physical situation with a potential for human injury, damage to property, damage to the environment or some combination of these.
Risk	The likelihood of a specified undesired event occurring within a specified period or in specified circumstances. It may be either a frequency (the number of specified events occurring in unit time) or a probability (the probability of a specified event following a prior event), depending on the circumstances.
Individual Risk	The frequency at which an individual may be expected to sustain a given level of harm from the realisation of specified hazards.
Societal Risk	The relationship between frequency and the number of people suffering from a specified level of harming a given population from the realisation of specified hazards.
Residual Risk	The remaining risk after all proposed improvements to the facility under study have been made.
Risk Analysis	An imprecise term which infers the quantified calculation of probabilities and risks without taking any judgements about their relevance.
Process Safety (often known as **Loss Prevention)**	A systematic approach to preventing accidents or minimising their effects. The activities may be associated with financial loss or safety issues. Note that loss prevention has a different meaning in the transport industry.

Note that the word *risk* is used to mean different things to different industries. In the insurance industry it refers to a site or installation, but in the "safety industry" it is as defined earlier. It is important that the correct meaning is verified to avoid confusion. The world of finance also has a variety of risk definitions which will differ from those given earlier.

REFERENCES

1 Williams T J, *The Chemical Industry*, Pelican, 1953

2 Pitt M, 100 Centuries of Chemical Engineering, *The Chemical Engineer*, 27 January 2022, www.thechemicalengineer.com/features/100-centuries-of-chemical-engineering/

3 Pitt M, Pee Purple Potions and Pills, *The Chemical Engineer*, 1 June 2022, www.thechemicalengineer.com/features/pee-purple-potions-and-pills/

4 Pitt M, The Story of Oil 1022–1922, *The Chemical Engineer*, 14 July 2022, www.thechemicalengineer.com/features/the-story-of-oil

5 Pitt M. A Wasted Life, *The Chemical Engineer*, 15 December 2022, www.thechemicalengineer.com/features/a-wasted-life/

6 Pitt M, The Forgotten Founder of ICHEME, *The Chemical Engineer*, 28 April 2022, www.thechemicalengineer.com/features/the-forgotten-founder-of-icheme/

7 Reuben B G & Burstall M, *The Chemical Economy*, Longman, 1973

8 Evans T, An Explosive Start, *The Chemical Engineer*, May 2015

9 Kinloch A J, Adhesives in Engineering, *Proceedings of the Institution of Mechanical Engineers*, Vol 211, 1997, https://journals.sagepub.com/doi/pdf/10.1243/0954410971532703

10 Fletcher R P, *The History of Fluidized Catalytic Cracking: A History of Innovation: 1942–2008*, Innovations in Industry and Engineering Chemistry, ACS Symposium Series, 2008

11 www.thechemicalengineer.com/authors/martin-pitt/

12 Tucker A, *The Toxic Metals*, Pan Ballentine, 1972

13 Howard S, Bhopal's Legacy: Three Decades on and Residents Are Still Being Poisoned, *British Medical Journal*, 11 December 2014, www.bmj.com/content/349/bmj.g7602.full

14 Personal communication

15 Hon Lord Cullen, The Public Inquiry into the Piper Alpha Disaster, *HMSO*, November 1990, www.hse.gov.uk/offshore/piper-alpha-public-inquiry-volume1.pdf, www.hse.gov.uk/offshore/piper-alpha-public-inquiry-volume2.pdf

16 Personal communication

17 Hall A N, Hall S F & Nixon W, The Chernobyl Accident, *Loss Prevention Bulletin No. 076*, ICHEME, www.icheme.org/media/5930/lpb_issue076p022.pdf

18 Gale R P, Chernobyl at 35 Years: An Oncologist's Perspective, *The ASCO Post*, 25 May 2021, https://ascopost.com/issues/may-25-2021/chernobyl-at-35-years-an-oncologist-s-perspective/

19 Ritchie H, *What Was the Death Toll from Chernobyl and Fukushima, Our World in Data*, https://ourworldindata.org/what-was-the-death-toll-from-chernobyl-and-fukushima

20 Pietersen C M, Analysis of the LPG Disaster in Mexico City, *Loss Prevention Bulletin No. 064*, ICHEME, www.icheme.org/media/5838/lpb_issue064p001.pdf

21 Hailwood M, Release of Hazardous Vapours at LG Polymers Chemical Plant in Visakhapatnam Andhra Pradesh India, *Loss Prevention Bulletin No. 274*, ICHEME, www.icheme.org/media/14335/lpb274_pg09.pdf

22 Council of Science and Society, *The Acceptability of Risk*, Barry Rose (Publishers) Ltd, 1977.

23 M Goose, *Model Code for the Management of Risk Issues*, ICHEME, 2007, www.icheme.org/media/10670/mcmri1.pdf

24 UK Health and Safety Executive, *Health and Safety at Work Etc. Act 1974*, www.hse.gov.uk/legislation/hswa.htm

25 Advisory Committee on Major Hazards, *First Report*, June 1976, HMSO

26 Risk Criteria for Land-Use Planning in the Vicinity of Major Industrial Hazards (Appendix 2), HMSO, 1989

27 Buncefield Major Incident Investigation Board, Recommendations on Land Use Planning and the Control of Societal Risk around Major Hazard Sites, *HMSO*, July 2008, www.icheme.org/media/10694/recommendations-on-land-use-planning.pdf

28 UK Health and Safety Executive, www.hse.gov.uk/landuseplanning/methodology.htm#princiles

29 UK Health and Safety Executive, *The Tolerability of Risk from Nuclear Power Stations*, HMSO, 1992

30 UK Government, www.gov.uk/government/publications/how-we-regulate-radiological-and-civil-nuclear-safety-in-the-uk/how-we-regulate-radiological-and-civil-nuclear-safety-in-the-uk-webpage

31 UK Government, Offshore Installations (Safety Case) Regulations (SI 2005/3117), 2005

32 UK Government, Offshore Installations (Safety Case) Regulations (SI 2005/3117), 2005
33 M Goose, *Model Code for the Management of Risk Issues*, IChemE, 2007, www.icheme.org/media/10670/mcmri1.pdf
34 Marshall, The Seveso Disaster: An Appraisal of Its Causes and Circumstances, *Loss Prevention Bulletin No. 104*, ICHEME, April 1992, www.icheme.org/media/6208/lpb_issue104p014.pdf
35 UK Health and Safety Executive, www.hse.gov.uk/comah/background/comah15.htm
36 European Chemicals Agency, https://echa.europa.eu/regulations/clp/understanding-seveso#:~:text=The%20Seveso%20III%20Directive%202012,human%20health%20and%20the%20environment
37 US Occupational Safety and Hygiene Administration, www.osha.gov/laws-regs/regulations/standardnumber/1910/1910.119
38 Letter from George Stephenson to Henry Labouchere, 31 March 1841, www.railwaysarchive.co.uk/docsummary.php?docID=160
39 Jones D, *Nomenclature for Hazard and Risk Assessment in the Process Industries* (2nd edition), ICHEME, 1992

2 Hazards and Precautionary Measures

2.1 INTRODUCTION

This chapter introduces the types of hazards that may be encountered in process plant and the measures to make the process safe *as far as reasonably practical,* a phrase first used in the UK Health and Safety at Work Ect. Act (1974).[1] The term safety, in the present context, has a broad meaning and covers not only injury to personnel but damage to the facilities, releases to the atmosphere plus damage to the business and loss of reputation for the site owners.

There is a very subtle difference between safety and environment. In safety the objective is to prevent or avoid all incidents. In environmental terms there is a need to minimise releases—accepting that some will routinely occur such as the release of carbon dioxide or aqueous discharges agreed by the regulatory authorities in the form of *consent limits*. By comparison, safety has clear boundaries.

It is impossible to eliminate all hazards to personnel and property, regardless of how much effort is put into the task. There will always be the possibility that a hazardous event will occur. Hazards have a complex interplay of causes, and no universal rules can be laid down. In general, the effects of hazards can be divided as follows:

- Pollution (including visual, noise, dust, odour, gas and liquids)
- Chemical reactions and reactivity (in particular exothermic reactions)
- Toxicity (including asphyxiation and long-term effects)
- Mechanical failure
- Corrosion
- Nuclear radiation
- Fire
- Explosion
- Domino effect (small event leading to a larger event)

Occupational safety covering "slips, trips and falls", as well as human ergonomics issues such as lighting and noise, was covered briefly in Section 1.3 and is further explored in Chapter 3.

Hazards may affect the following:

- The environment (visual, noise, land, water, air)
- Company employees within or the public outside the site; the hazard may produce single or multiple fatalities and may result in "delayed fatalities" (see Figure 1.1).
- Plant equipment, storage facilities, offices, warehouses, laboratories and so on
- Property outside the site
- Company cash flow (by loss of revenue, replacement of damaged equipment and/or payment of claims for damages).
- In exceptional cases, an incident may threaten the survival of the company concerned or result in changes to legislation. The 2020 ammonium nitrate explosion in Beirut resulted in the fall of the Lebanese government.[2]

DOI: 10.1201/9781003510680-3

The effects of common hazards and their mitigation are covered in Chapters 7 to 10. Hazards are commonly controlled by

- elimination
- containment
- reducing frequency
- reducing effect or magnitude
- "First aid" or mitigation measures.

In some cases, hazard can be addressed by engineering (or *hardware*) solutions and in others by a management (or *software*) procedure. Hardware solutions are generally used during the design stages of a project and software procedures during the start-up and operating stages. The relative costs, ease of implementation and robustness of the solution will affect the final choice.

As accidents cannot be totally eliminated, the aim is to reduce them to an acceptably low level. However, it should be recognised that reducing one risk may increase another and the final result may be a compromise. A solution which reduces human hazard may increase the environmental hazard and the designer must balance both sides of the equation. The total risk to the environment, humans, plant fabric and cash flow must be acceptable both to the company and to the regulatory authorities.

2.2 ENGINEERING STRATEGIES FOR HAZARD REDUCTION

There are a number of high-level strategies which should be considered during design. These are written in general terms and should not be treated as unique solutions—a combination of solutions may be the best strategy. Common strategies are shown in Table 2.1.

Table 2.1 represents something of a "wish list" which cannot always be achieved. Engineering judgement and risk assessment techniques will help to determine which strategies offer the greatest improvement and where hazards need to be accepted and mitigated. Sometimes the application of one risk reduction strategy will throw up problems in other areas. More detail on specific issues is to be found in Chapters 11–17.

2.3 MINIMISING ENVIRONMENTAL DISCHARGES

Any new plant will need to consider disposal of pollutants to land, sea and water.

Gaseous effluents are discharged to the atmosphere and diluted to a low level. Their subsequent history is less clear. Particulates are breathed in by humans and animals, with possible health effects. Toxic gases may affect all forms of organic life. Acceptable levels of exposure frequently change and the latest standards must be applied. It is sometimes necessary to modify or retrofit existing equipment to comply with changes in standards.

Many pollutants can be captured at source by means of absorption, refrigeration/condensation or scrubbers/filters. Modern cars now have a catalytic converter to remove nitrous oxides (NO_x) and volatile organic compounds (VOCs). In the case of solid waste, the materials could be encapsulated, combusted, neutralised or converted into a less harmful form.

Safety and environmental concerns sometimes run counter to each other; a reduction in emissions might increase the overall risk associated with a plant or process. In this situation engineering judgement based on a balance of safety/environmental risk and hazard will be necessary. This is further discussed in Section 4.2.

Aqueous effluents can be cleaned by a variety of chemical, physical or biological processes ranging including flocculation, neutralisation or other forms of chemical treatment. Oil can be removed

TABLE 2.1
Common Risk Reduction Strategies

1 **Process Chemistry**	A process which is stable and does not involve toxic or flammable chemicals must be one of the objectives of the conceptual design. This is seldom possible, and there may also be hazardous intermediate or by-products. An example is the production of methyl acetylene in an olefin plant.[3] Process chemistry must be carefully assessed to minimise the hazards of feedstocks, products, intermediates and by-products.
2 **Reduce the Inventory (Intensify)**	A smaller inventory reduces the absolute potential for an emergency arising from fire, explosion or toxic release; this may require process intensification[4] or operating at higher pressure and temperature.
3 **Reduce the Pressure**	A lower pressure will reduce the energy release from a catastrophic vessel rupture. Similarly, the release from a lower pressure source will have a lower discharge rate than from a high-pressure source for a given hole size.
4 **Reduce the Temperature**	The lower the temperature, the lower the proportion of vapour generated from a volatile liquid. Therefore, the spread of vapour will be reduced. There may also be other benefits such as reduced corrosion and lower cyclic stressing. However, working at low (especially below 0°C) temperatures can be challenging, with the need for special materials of construction and design features.
5 **Reduce Process Complexity**	This is self-evident. A simple and easy-to-understand process is inherently safer and less prone to failure from equipment failure or operator error. However, there will always be a need to look at thermal and chemical efficiency, which often runs counter to simplification.
6 **Increase the Distance between Equipment**	Increasing the distance between equipment has two benefits. First, there is a reduced risk of an incident escalating and leading to a larger event *(the Domino effect)*. Second, ventilation and dispersion of leaks are improved, thereby reducing the concentration of toxic gases or the potential for a vapour cloud explosion, whilst also minimising the effects of an explosion. Whilst this is a useful strategy, it is important to recognise that there are cost implications due to extra piping lengths and supports. Increased pipe runs will increase pressure drop, possibly requiring larger pumps with increased power requirements. A secondary advantage is improved access for maintenance staff and emergency responders.
7 **Containment**	In some cases, processes are considered sufficiently hazardous that some form of barrier is provided, the most common being blast resistant enclosures or fire proofing. In the case of toxic gases, personnel may be prevented from entering specific areas of the plant. In some processes double wall piping is used, with the annulus between the inner and outer walls acting as a containment.
8 **Reduced Operator Exposure**	Design the plant so that operators are remote from hazardous areas. Buildings should be designed to safeguard the occupants. Providing blast resistance to prevent explosion damage or pressurising the building to prevent the ingress of toxic or flammable gas is common practice. Windows should never overlook hazardous areas.
9 **Equipment Layout**	Incompatible systems should be separated from each other: humans from toxic fluids, or corrosive chemicals from low-grade steel pipework and equipment. The following are some examples: Large volumes of fluids should be separated from each other and from sources of ignition. Utilities should be separated from process units. Pumps and other sources of liquid leakage should, where possible, not be located below other equipment to minimise damage from a pool fire—this is particularly important with fin-fan coolers, where air movement will increase flame size. Cabling on pipe racks should be located above piping carrying process fluids to avoid leaking pipes dripping onto cabling and causing damage. Cabling should also be routed away from likely leak sources such as pump seals.
10 **Minimise Sources of Ignition**	This is again self-evident, particularly where flammable materials are being handled. Typical examples include locating electrical switch rooms away from the process area, using steam heating rather than fired heaters where possible and avoiding the use of belt-driven machines.

by skimming and then biologically digested to form a monocellular protein. A useful overview of waste-water treatment processes has been produced by Kemera.[5]

In addition to the generation, control and disposal of waste from normal operations, the effects of a significant accident must be considered. Watercourses may be polluted and toxic materials and smoke discharged. Clean-up following a major incident can also be an expensive and time-consuming incident; the locations, type and amount of hazardous chemicals may not be known, and hazardous materials of construction, such as asbestos, may be present.

Previous industrial activities have often resulted in contamination by industrial waste; for example, coal tips, steel slagheaps and chrome waste have destroyed significant areas of land, and, in some cases, the *leachate* has entered aquifers and rivers. Leachate from landfill and aqueous effluent has left rivers sterile and the continuous pollution can result in silting. Methane from landfill sites has been implicated in explosions in buildings. Oils float on water and, in polluted areas, can resurface on the ground following heavy rain. Layers of silt on a river or seabed may also be contaminated with toxic materials, which may be disturbed by fishing or dredging.

Environmental issues are covered in more detail in Chapter 4.

2.4 WHAT CAN GO WRONG?

Something, sometime, somewhere will be wrong with the design or operation of the facilities. At the design stage, control will be largely by calculation checks and hazard/risk assessments. During construction, quality control/quality assurance (QA/QC) is essential. During operation, plant should be walked each day following a different route and, if possible, at a different time of the day. All equipment should be inspected regularly to ensure no degradation, for example fouling of heat exchangers. Plant records should be critically examined on a regular basis. Regular safety audits (see Section 21.7) will be necessary to ensure that standards do not slip and a robust MOC system will need to be in place (see Section 16.9).

The price of safety is eternal vigilance.

REFERENCES

1 UK Health and Safety Executive, *Health and Safety at Work Etc. Act 1974*, www.hse.gov.uk/legislation/hswa.htm

2 The Guardian, 10 August 2020, www.theguardian.com/world/2020/aug/10/lebanese-pm-to-resign-after-more-than-a-third-of-cabinet-quits

3 MAPD Stability and Management in Ethylene Plants, *AIChE*, April/May 2013, www.aiche.org/academy/videos/conference-presentations/mapd-stability-and-management-ethylene-plants

4 Stankiewicz A J & Moulijn J A, Process Intensification: Transforming Chemical Engineering, *Chemical Engineering Progress*, January 2000, www.aiche.org/sites/default/files/docs/news/010022_cep_stankiewicz.pdf

5 Hemira's Water Treatment Handbook, www.kemira.com/insights/water-handbook-2020/?utm_source=google&utm_medium=ad&utm_campaign=iw-chemicals&utm_content=handbook&gclid=EAIaIQobChMImKnh-LzW_AIV1u7tCh3A4Q6OEAAYASAAEgLspvD_BwE

3 Occupational Health and Safety

3.1 INTRODUCTION

Occupational health and safety is a complex topic, and this chapter provides only a basic overview. Many of the areas covered are highly specialised, and expert guidance should be sought for specific problems. Health and safety guidance and regulations are frequently reviewed and updated, particularly toxicity data. The latest version should be consulted for guidance and decision-making.

There is no universal safety system which will work in every location or in every situation; the best practical solution is one which addresses health and safety issues on a unified, operation-specific basis. In one case, it may be appropriate to design out the problem; in another, it may be appropriate to protect the employee, and in yet another, it may be appropriate to use a strictly enforced procedure.

Whilst there are some systems which are designed for totally automated or remote operation, particularly in the nuclear industry, most facilities have an element of human interface with equipment. Different strategies will be necessary in different areas of the plant. Issues to be addressed within control rooms, for example, will generally relate to comfort and efficient working without skeletal damage or eyestrain. Within plant areas, there will be greater concerns regarding noise, vibration and access to equipment.

From an occupational health and safety viewpoint, the hierarchy of safety controls[1] is often expressed as follows:

1 Elimination—physically remove the hazard
2 Substitute—replace the hazard
3 Engineering controls—isolate people from the hazard
4 Administrative controls—control the way people work
5 Protection—use of personal protective equipment (PPE)

3.2 THE HUMAN "SPECIFICATION"

As a means of understanding human frailty, consider the following as a description of a human as observed by a visiting alien:

> *The object is poorly designed; it has some interesting features but is vulnerable to a number of simple physical effects. The main assets seem to be the "manipulators" and the small in-built computer which is able to execute complex logical processes without being told what to do. The manipulators are relatively weak but are exceedingly manoeuvrable and contain very sensitive pads. The computer takes around 20 years to program but then has a strange ability to update itself from data acquired through its sensors. The object is inherently unstable and requires a continuous feedback control loop. It will fall over if subject to an acceleration of less than 1 m/s or a force of a few 10s of Newtons.*
>
> *The main structure has an appendage which contains its computer and three sensors. The visual senses are stereoscopic but only perform in strong light, the visual focus is limited. The aural sensors are stereophonic but are affected by high ambient noise levels. The third sensor seems to be used to detect a few chemicals; its function is still being studied. The positioning of all the main detectors and the computer in close proximity produces the potential for common mode failure. The computer*

DOI: 10.1201/9781003510680-4

malfunctions under conditions of stress, boredom or stimulation by noise, movement, light or smell outside its normal working range.

The main structure is very weak and members are broken by falls of approximately one metre, it can also be damaged by a few kilograms falling from a height of one metre. The external coating is soft and can be damaged by sharp objects.

The power source occupies about half the volume of the object. It takes fuel in the form of a mixture of organic molecules three times a day and also ingests water. Yet, strangely, the object does not function underwater as it requires oxygen to keep the power source operating. If the oxygen partial pressure is less than 10 kPA the object falls over and does not function. The same effect is noted if it is fed 250 cc of ethanol and a number of other gaseous, liquid or solid chemicals. The whole unit can also be put out of operation by fire, heat, dusts, acids/alkalis or small quantities of many other chemicals.

It seems to be affected by small invisible organic systems which appear to be named as "viruses"

We are still studying its strange capability to renew parts of itself and its capacity to produce mini objects which grow up to become full sized Homo Sapiens.

When reduced to a simple specification it is clear that humans need to be carefully protected. Apart from the brain and manual dexterity the human has little to offer that a robot could not do equally well. The human advantages are narrowing with time as robotics and *artificial intelligence* (AI) improve.

Humans survive within an "envelope" of maximum and minimum temperature, humidity and other factors such as low oxygen levels. This is sometimes known as "the Goldilocks zone". However, there will be some whose tolerance is lower than average and some with greater tolerance to extremes. Similarly, there will be variations in tolerance to burns, exposure to blast, toxic chemicals and so on. The old and young will have lower tolerances than the bulk of the population, but there will be many exceptions—healthy young men could survive a significant wound with severe loss of blood, but even a minor cut could be fatal to a haemophiliac. Most people would expect to survive a fall of several metres—possibly with broken limbs—but a fall of only a metre or less could be fatal if the victim landed on their head. At the other extreme, there are a few recorded cases of people falling thousands of metres from an aircraft and surviving.[2]

Further information on the effects of thermal radiation and blast waves is given in Chapters 9 and 10.

A smaller area within the survivability envelope defines the conditions where a human can operate efficiently. This is partly a matter of training but concentration and the ability to make correct decisions will be influenced by levels of noise, vibration, temperature and other factors.

In view of the physical and mental attributes necessary for many positions, pre-employment medical tests are often required, together with periodic retests. The depth of a medical examination will depend on the type of work being undertaken. For military and emergency service personnel, physical fitness, stamina and dexterity will be more important than for an office worker. Personnel who are likely to work in remote environments (offshore or the desert) will be required to have a higher level of general health than a worker in a town or city where there is good access to medical facilities. Specific tests might be necessary for some roles, for example to ensure firefighters (who may be perched at the top of a ladder) do not suffer from vertigo and control room operators are not colour-blind. Periodic health checks are required in many industries to ensure long-term health problems, such as hearing loss or cumulative exposure to chemicals, are identified. Tests on workers in the nuclear industry to check for exposure to radiation are one example, but there are many others.

3.3 ERGONOMICS

Humans have many useful attributes but also have physical and psychological weaknesses. Limitations range from strength, reach, stretch, focal vision, balance, attention span plus unpredictable reactions under stress and in unforeseen circumstances. These limitations impose serious problems for the plant designer, manager and supervisor.

The importance of the physical health of staff is self-evident and checked by regular medical examinations. An unfit process operator will be unable to function efficiently and will be more likely to make mistakes. Physical weakness might inhibit the operator from some activities such as tightly closing valves. The level of physical fitness required will vary with job function. An office worker will not require the same level of fitness as a process operator who, in turn, will not require the same levels of strength and endurance as a firefighter. It would be common to expect all staff to have basic periodic health checks such as blood pressure, electrocardiogram and blood tests. Enhanced levels of testing will be necessary in many cases, and many industries will have specific tests where occupational diseases may occur. Whilst alcohol or drug dependency is not a common problem, it can occur, particularly on the night shift, where there is time for a drink or two before the start of the shift. Staff working offshore or at remote locations might also "binge" before starting their turn of duty. For this reason, many companies have introduced random drink and drug tests.

The study of humans under stress is complex and, whilst useful, it is impossible to accurately predict behaviour under all circumstances. The best solution is to avoid putting personnel into stressful situations. However, upset conditions and emergencies can occur and must be planned for. The design of vehicle controls illustrates this point, with direction indicators and windscreen wipers mounted on the steering column for ease of use and the speedometer mounted on the facia behind the steering wheel where it is clearly visible. Modern military aircraft (and some vehicles) utilise *head-up displays* (HUD), which allow the pilot to gather data with limited movement of the head. In the design of analogue control panels, now increasingly uncommon, the most important controls should be nearest to the operator.

3.3.1 Physical Access

Access to equipment is an obvious requirement but sometimes forgotten. It is common in new plants to find scaffolding installed to allow access to a valve which was not foreseen by the designer. There is no excuse for fitting a valve close to an elevated walkway so that an operator has either to crouch or, even worse, to stretch over or through the handrails. Both situations involve the operator being in a poor state of balance and unable to provide adequate leverage. Another example of difficult access is the use of chain-operated valves; these are mechanically inefficient and difficult to reach for a short operator—they are also a potential "lasso" for a tall operator. There are other obvious forms of poor access, such as the inclined valve, which does not allow uniform leverage, the rising stem valve which juts out into a passageway and, of course, the vent or drain valve which always seems to be just out of reach.

Access ways should have a height clearance of at least 2.2 m. Too often it is forgotten that a safety helmet will add 5 cm to the height of an individual and thermal insulation on piping can remove another 5 cm of clearance. Boots add 2–3 cm and the peak of the helmet obscures upward vision—consequently, collisions occur, the helmet usually preventing injury.

Stairways should have adequate tread depth, rise and stair angle. This arrangement should ensure that a toe is not caught on the tread whilst ascending or a heel caught on the tread whilst descending. Where it is necessary to install vertical ladders, for example on distillation columns, landings should be installed every 25 m for rest. There are a number of other important safety features associated with stairways and ladders:

- Where stairs or ladders have to have a short step, it is better for this to be at the bottom of the flight and not at the top—it is better to stumble at the bottom rather than the top where there is further to fall.
- Ladders should have a safety cage above 2 m in height. Whatever the height of the ladder, there should be a self-closing safety gate at the top.
- Handrails should be at least 1 m tall, and all overhead walkways should have a toe or kick-board to prevent objects being knocked onto the unwary below.

- All emergency exits should be duplicated, preferably at opposite ends of the walkway or structure. Escape stairs should be at least wide enough for two persons to pass in comfort, and, if used regularly, they should be a minimum of 1.5 m wide. For smaller walkways or platforms, a stairwell at one end with a vertical ladder at the opposite end may be adequate. The use of rope ladders or other makeshift means of escape is not acceptable in most circumstances.
- All stairs and landings should be sized so that stretchers can be manoeuvred without lifting them over handrails.

3.3.2 Manual Handling

Physical handling of heavy objects requires skill and training. The *straight-back lift* can be achieved with compact systems, but it is not always possible with sacks or when lifting equipment within a crowded area. Figure 3.1 gives an indication of the correct posture to prevent excessive stress on the lower back.

Similarly, Figure 3.2 shows good and bad shovelling techniques.

The top row illustrates bad techniques; the techniques shown in bottom row are good.

FIGURE 3.1 Safe Lifting

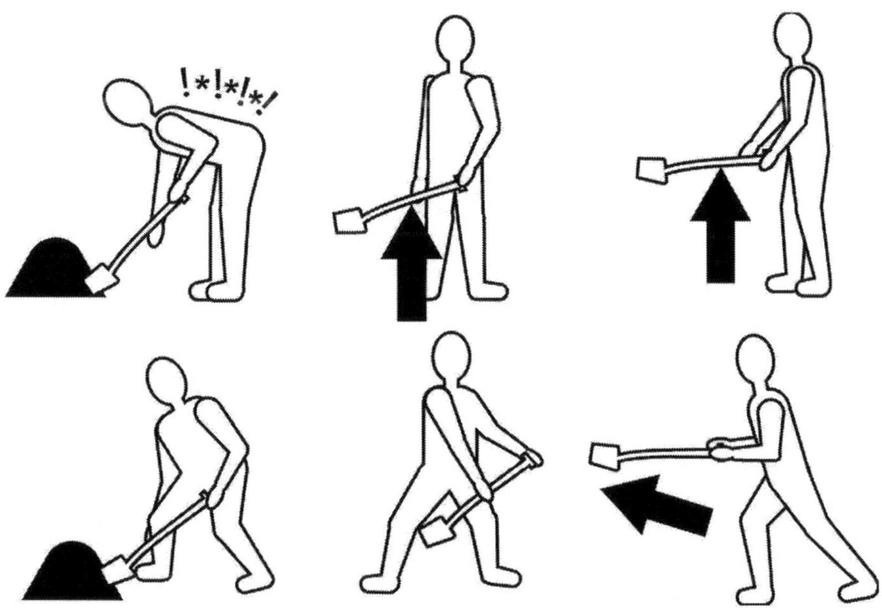

FIGURE 3.2 Shovelling

Site layouts should be reviewed at the design stage by operations and maintenance personnel as well as by mechanical handling experts to ensure suitable lifting equipment is available. Simple lifting frames and block and tackle arrangements supported from overhead runway beams will greatly ease mechanical handling for maintenance. Layout is further discussed in Section 13.8.

3.3.3 Control Rooms

The design of control rooms also has a significant ergonomic element—is the temperature/humidity controlled to prevent drowsiness? Is the lighting level adequate (with no reflection on computer screens)? Are audible alarms loud enough to be heard but not loud enough to cause panic where the operator "freezes" and cannot make decisions? These and similar questions should form an integral part of the control room design.

Human factors in the control room have been the subject of much recent research, and comprehensive guidance[3,4] is available. Traditional analogue control panels have now been largely superseded by screen-based systems which often still retain push-button *emergency shut down* (ESD) buttons. These should be easily accessible but protected against accidental activation.

Whilst analogue control panels are largely obsolete, some older facilities, particularly in the developing world, still use analogue control as do some smaller equipment items. However, these are becoming fewer and are being replaced by more efficient *distributed control systems* (DCSs). The few remaining major plants with analogue control usually have large panels stretching the length of the control room, displaying key parameters. Whilst the human eye has a limited cone of sharp vision, peripheral vision is useful for detecting movement and flashing images. Reading data is most easily done at eye level, but if head movement is required, it is easier to look up than to look down. Where large vertical display panels are installed only the top half is of real use. An audible alarm is often the first indication that a control parameter is out of the specified range.

Spare parts for analogue panels are likely to become increasingly hard to source, and as older operators (who sometimes have difficulty adapting to DCS) retire, the advantages of newer systems become overwhelming. A DCS will generally consist of a series of interchangeable computer screens, usually with one or more screens plus keypad dedicated to each major plant section.

The architecture of each "page" or display on the screen must be carefully developed. The design of the graphics for the screens is a skilled activity with a requirement to ensure that there is not too much, or too little, information on each screen. Although the layout on the screen is an essential safety feature, no less important is the order with which the pages are accessed, and the data displayed on each page. In traditional analogue instrument panels, all data was visual, and the operator could "scan" different panels rapidly; however, only a limited amount of data can be stored on each page of a computer screen. The analysis of the plant using a hazard and operability study (HAZOP) (see Section 5.3) will help the control system designers devise the best and most informative displays. Key process parameters should be instantly accessible regardless of which page is shown on the screen. This can be achieved by dedicating a screen to key parameters. It is increasingly common practice to install a large screen on the control room wall showing key plant information.

The use of screen-based systems means that operators are spending more time sitting in front of a screen, and a workstation risk assessment[5] should be conducted to ensure there is no long-term damage to their health, back and eye issues being the most common.

As well as displaying the plant, section by section, the DCS will contain other information, such as alarm lists which will alert the operator to alarms which are not shown on each page but which are still active.

The *man/machine interface* (MMC) also needs consideration. The complexity of the screen pages can inhibit comprehension by the operator and the use of red and green to indicate open or closed valves can cause problems for colour-blind operators. Problems can also occur with the accidental rubbing of touch-sensitive screens.

A reactor was being started up. It was filled with the contents of another reactor which was already on line, and the panel operator started to add fresh feed, gradually increasing the flow whilst he watched the temperature on a recorder conveniently situated at eye level. He intended to start a flow of cooling water to the reaction cooler as soon as the temperature started to rise—the usual method. Unfortunately, there was a fault on the temperature recorder, and although the temperature actually rose, this was not indicated. Result: A runaway reaction.

The rise in temperature was, however, indicated on a six-point temperature recorder at a lower level on the panel, but the operator did not notice this. Fortunately, the runaway was not serious because a high temperature alarm on the six-point recorder alerted the operator before the temperature became dangerously high.

3.4 HUMAN FACTORS

Contemplative novelists often write of the "human condition", the complex mixture of experience and emotion that shapes our lives. The same complex mixture, known in an industrial setting as *human factors*, shapes the reactions of us all and is important in determining the behaviour of personnel during both normal operation and upset conditions. The science of human factors has grown in importance in recent years, and there is a vast and expanding literature on the subject. A particularly useful, if basic, guide has been developed by the UK HSE.[6] The importance of human factors is demonstrated in Table 3.1 which is based on experience and data from various sources.

It is important to note the difference between *human error* and *human factors*. Human error is a catch-all phrase which, in the 1970s and earlier, was used to explain any human behaviour which caused, or contributed to, an accident. The term is now considered too broad and unhelpful in determining accident cause (and preventing a recurrence). Human factors involve detailed assessment of human behaviour to determine how behaviour can lead to accidents and are also useful to identify how human activity helps to cause accidents. This system can be used to prevent recurrences of previous incidents. The critical aspects of human factors are discussed in the subsequent sections.

TABLE 3.1
Basic Human Error Probabilities

Task	Probability of Error
Totally unfamiliar, perform at speed, no idea of outcome	0.55
Restore system to new or original state on a single attempt without supervision or procedure check	0.26
Complex task requiring a high level of comprehension and skill	0.16
Fairly simple task performed rapidly or given scant attention	0.09
Routine, highly practised, rapid task involving a relatively low level of skill	0.02
Restore system to a new state following procedural checks	0.003
Totally familiar task, performed several times per hour, well-motivated, highly trained staff, time to correct errors.	0.0004
Respond correctly when there is an augmented supervisory system providing interpretation	0.00002
Miscellaneous task—no description available	0.03

Source: Smith D J, Reliability Maintainability and Risk (5th edition), Butterworth-Heinemann, 1997

3.4.1 Personality

Personality can affect performance in a number of ways. The successful operation of a team will depend on the strengths and weaknesses of the individuals. An individual may feel another member or supervisor of the team is "picking on" them and as a result they become reclusive and depressed. This will dull decision-making skills, performance and reaction to unusual situations.

A strong-minded individual may believe, rightly or wrongly, that they "know best", and develop a corresponding *mindset.* Such a person is likely to deviate from a fixed procedure, believing that he/she is performing the correct operation. This is the psychological make-up of the person and may override any training or knowledge base (or maybe the knowledge base is in error).

A variation on this may be the domineering manager or supervisor who challenges the competence of an employee or who is unable to accept advice and adopts a superior attitude. The end point is usually a comment such as "come on, make up your mind" or "is this the best you can come up with after ** days?". The employee will rapidly learn that the best solution is to "give the manager what he wants rather than take the correct action".

Some people cannot make accurate decisions when confronted with a mass of complex information (often called *information overload*); they freeze and do nothing. This is sometimes called panic, but it is more realistic to put it down to an individual's make-up or lack of training. A variation of this is a *blame culture* (see Section 21.2) where everyone is in fear of making a mistake because of potential blame or recriminations. As a result, the employee overlooks situations or avoids taking the correct decision such as highlighting a possible hazardous situation or procedure.

3.4.2 Training

Adequate training, and refresher training, is essential for an operator to perform well. All staff require training to become proficient in a role. At one time training would be informal, with a new recruit being thrown in at the "deep end" and instructed by experienced staff. It is now standard practice at operator level for a period of classroom training for new recruits prior to *on-the-job* training under the supervision of an experienced operator. Before being allowed to work unsupervised, the trainee must demonstrate competence in the role. For all but the simplest plants, there will be a number of operational roles. A new operator will generally become competent in one role before moving into other roles in turn. Process simulators are often provided to allow training which would not be possible on a live plant such as ESD. The provision of clear, well-written operating instructions will assist training and comprehension (see Section 18.5).

Traditionally it is assumed that it takes about six months for a new recruit to be able to fill an operator's role fully. For a further two years the recruit develops and improves the role in a creative and constructive manner. Then the enthusiasm falls off and, for around six months, performance, though good, is not always as good as it had been previously. To keep up enthusiasm and interest refresher training and training in other roles, possibly to aid promotion opportunities, is desirable.

Refresher training is often required to reinforce good working habits and discourage poor ones. Refresher training for infrequent activities such as plant shut down, startup or emergency situations is particularly necessary. The latter is critical but often performed poorly, with emergency scenarios being discussed in classroom sessions during the night shift. A better alternative is to conduct drills on the plant, with field operators placing tags to show the locations of valves they would operate and panel operators writing down the actions they would take. This allows a check to ensure the shut down procedures are being followed correctly.

Training and retraining is an essential requirement not only for operators and technicians; professional engineers also require continuing professional development (CPD) to maintain their chartership in the UK.

3.4.3 Experience/Confidence

A competent operator with a depth of experience is a valuable asset. Additionally, as operators generally work in shift teams, a good *team spirit* is essential. Support and guidance from plant management are vital in this regard. The shift leader plays a key role. He/she must know the strengths and weaknesses of individual members when assigning work activities whilst, at the same time, providing training and instilling confidence in the team.

It is now common practice to maintain a *skills and training matrix* for each employee. Increasingly, these matrices include "soft" skills such as team building, leadership and communication. They are used to ensure that each member of staff has maintained the skill levels necessary for his job and has suitable training/experience for future promotion.

3.4.4 Stress and Unpredictable Behaviour

Human performance declines with boredom and information overload. The peak performance is reached when there is just enough activity to keep the person alert and "on their toes". Human error tables are produced for normal situations and many protective systems are designed on the basis of a certain level of "operator error".

Operator performance depends on a number of factors; there is an absolute limit to the ability to handle objects—beyond that limit information gets "dropped". If too much data is available to the operator, he or she will become confused and make irrational decisions. This is a greater issue with digital control systems where numerous alarms are installed and many may be activated during an upset condition. The noise from audible alarms may be distracting and the number of alarms generated can make it difficult to identify the source of the problem. This is known as *alarm flooding*[7,8] and most modern control systems have an *alarm management system* to prioritise alarms in the event of a major upset (see Section 13.9).

Information overload can be dealt with by two strategies. At one level, the operators have sufficient resources to handle all of the activity and, at the other, the information is filtered and presented in a clear and unambiguous manner. On a small plant, the supervisor may be able to handle more data as there is less time used in moving between screens to analyse information. This will be less likely in a large plant. The presentation of clear, unambiguous data with the appropriate diagnostics is fundamentally important. The human can only accept a limited amount of information at any one moment and clarity is key.

Stress and panic are also major issues during emergencies. Adequate training, especially for upset conditions, will increase the speed and efficiency of a response and reduce the potential for emergency responders to be injured.

Staff may also suffer psychological stress, for a number of non-work-related reasons including

- financial worries
- health worries
- family worries

Stress is an increasing concern in all walks of life and needs to be considered in relation to both control room and field operators. A good manager should be alert to the warning signs of stress in his staff.

In summary good operating instructions, good management, good training, good plant design and good control room layout can go a long way to overcoming stress. Ultimately the final protective system should be the ESD system which will put the plant into a safe condition and allow the operator to restart in an orderly manner.

3.4.5 System-Induced Errors

Modern control systems are generally DCSs. These have nearly every critical parameter displayed and available for visual access. DCSs utilise a limited data display on every "page" and many "pages" or screens exist for each plant. With the old analogue "long panel" the control room operator could scan the parameters and identify any deviations from a normal condition—this can still be done with a DCS but it requires one screen for "scanning" and inevitably interrupts some of the data display. The control room operator may wait until an alarm sounds before a page is interrogated—by which time the problem may be well advanced.

System-induced errors are legion and a subject in themselves. A few examples are given here:

During startup cold fluids were fed to a distillation column with metal temperatures notionally at ambient conditions but whose operating temperature was –50°C. No base level was detected after one hour, and there was a mass imbalance. Feed was stopped, and a faulty level measurement was corrected. No recommendations were needed.

During startup, cold fluids were fed to a distillation column notionally at ambient conditions but whose operating temperature was –50°C. No base level was detected and feed continued for some time—eventually cold fluids spilled out of the column via the pressure control system. The fluids overfilled the flare knock-out drum and entered the flare stack. The flare stack was not designed for low temperatures and failed in brittle mode. A major fire resulted.

A distillation column developed a high-pressure drop during a process upset. The mass balance of feed rate minus product rates after the upset was zero. The integrated balance over two hours showed a 10% discrepancy—namely flooding—and the feed was stopped whilst corrective action was taken.

No recommendations were needed, the event was managed correctly and mass balance techniques were used to assess the diagnosis of the event.

3.4.6 Avoiding Information Overload and Mindset

The advent of digital control makes it easy to install large numbers of alarms. This is a temptation that should be resisted. The problems of alarm flooding have already been discussed and alarms should be calibrated so that set points are at reasonable levels outside the normal operating envelope. The worst situations occur when alarms frequently sound in the control room. Operators then become engrossed in the task of control and further alarms are ignored. Alternatively, operators try to respond to alarms and forget the main task of control. The opposite can be true on stable plant; operators may ignore changes in variables where experience shows there is no effect on plant operation. Parameters may rise or fall with no action being taken, even to the point of alarms being ignored.

Control instruments should have an amount of supplementary *check indicators* or other diagnostics. These should be simple, easy to read and readily accessible to the operator. The simple expediency of recording control values and actual values on log sheets will show if instruments are "drifting" or giving false readings. It may be necessary to install additional diagnostics to assist the operator. The need for this will usually be identified during a hazard and operability study. Modern control systems usually include data logging which will record parameters for a period of around 30 days. Most sites download data giving parameters over an extended period. Field operators generally record readings on field instruments at regular intervals. This will assist in identifying instruments which are drifting and will also require them to regularly patrol around the plant.

The problem of *cognitive dissonance* (sometimes called *mindset*) can sometimes occur, particularly during periods of stress. Cognition is a set of mental activities including judgement and reasoning which can sometimes break down (dissonance). When this happens, the "mind user" forms a mental model based on training and experience and accepts information which fits in with their mental model. Information which does not support the mental model is ignored.

An example would be the case of a catalyst change where the original catalyst was a *bulk flow* (high thermal inertia) type, with the replacement having to be controlled carefully to avoid a *runaway* reaction.

3.4.7 Ageing

Ageing of equipment is a well-known and documented issue[9] (see Section 23.3); however, humans also age and the effects of ageing should not be ignored. Older, and unfit, operators will be less able (or inclined) to climb steps or ladders. This may lead to operators taking "short cuts" particularly during hot, cold, wet or windy weather. Their ability to escape in an emergency will also be impaired. Staff who require a higher level of physical fitness, such as emergency responders, are unlikely to maintain the levels of fitness required until retirement age.

Mental attributes may also suffer as age increases; reaction times will increase, and the ability to adopt to new technology may be limited.

In the past there may have been a temptation to "side-line" older employees into dead end jobs, which is demotivating. Currently employers tend to be less benevolent and more likely to force poorly performing employees to retire early. In either case, their many years of experience will be wasted. They can usefully be fitted into other roles that make use of their experience, such as training and as operator representatives in HAZOP studies. In some cases, retired operators have been re-employed as temporary staff to make up for a shortage of in-house trained staff after overzealous redundancy programmes.

3.5 PERSONNEL PROTECTIVE EQUIPMENT

Good design and comprehensive operating and maintenance procedures help ensure safe operation, but the potential for injury will always exist. For the same reason that cars are fitted with seat belts and motorcyclists wear helmets, the use of PPE is now considered essential throughout industry. Additional requirements will be needed for specific activities (such as handling chemicals) or certain areas of the plant (for example, high noise areas). Some industries have specific requirements, for example medical facilities, biohazard areas, clean rooms for the manufacture of computer chips, the nuclear industry and explosives handling. These highly specialised activities are not considered further.

In most cases, PPE will be required even though other protective measures are in place. For example, exposed moving parts on machinery should be fitted with guards. However, ties, cuffs and long hair may slip between the guard and rotating parts; management will normally supply coveralls and require the use of hair nets where necessary. The most vulnerable parts of the human body, starting at the top, are listed in Table 3.2.

Various forms of PPE are available, the specification depending on the activities that the user may perform—bearing in mind that sometimes more than one hazard may be present. For example, an operator unloading tankers of corrosive chemicals will require clothing, including boots which will protect against splashes and spills. The boots will also need toe protection in case the end of a hose or other item is dropped. This requirement to protect against more than one hazard simultaneously is sometimes forgotten.

PPE is discussed in general terms later, but the exact specification will depend on the materials being handled, ambient conditions and local legislation. More detailed guidance on personnel protection is available from a number of sources.[10,11]

3.5.1 Head

A safety helmet is designed to arrest or deflect dropped objects and to deflect the head away from beams, fittings, brackets and other obstructions. A helmet will not protect the head from sharp falling objects but should protect from serious injury from, say, a dropped spanner. The wearer may, however, be knocked out by the impact.

Safety helmets are a necessary protection but are not particularly comfortable; they are hot and "sweaty", and the peak obscures upward vision. As a result workers cannot be relied on to wear

TABLE 3.2
Areas of Human Vulnerability

Area	Sources of Vulnerable
Head	Impact by falling objects or when passing through low structures with limited height.
Eyes	• Light (low or high—ultra violet) • Dust • Acid or alkali • Projectiles
Brain (see Section 3.3)	• Information overload • Stress • Confusion • Panic
Balance	• Wind over 50 m/s • Acceleration of 0.1 g laterally
Ears (see Section 3.2)	Noise above 120 dB(A) causes pain and above 90 dB(A) for eight hours causes hearing loss
Nose/lungs	• Toxic materials • Dust • Low oxygen concentration • Hot/dry air • Nuclear radiation
Skin	• Chemicals • Nuclear radiation • Heat/cold • Thermal radiation • Projectiles • Sharp objects • Trips/falls
Bone	• Dropped objects • Projectiles • Trips/falls
Muscle and Tendons	Damage due to strain or lifting heavy loads

them under all circumstances. Particular problems arise in arc welding where a standard safety helmet and welder's mask cannot be worn simultaneously, although combined helmets/masks are available and widely used. Where combined helmet/masks are not available, welding areas should therefore be isolated with warning tape and "No Entry" signs displayed. Where height is restricted black and yellow warning strips should be installed and warning signs fitted. If necessary, padding should be installed around low horizontal beams.

3.5.2 Eyes

Many types of eye protection are available (glasses, goggles, visors). Some protect the eyes from wind-borne dust or chemicals and others from high levels of light or other forms of radiation. In most cases standard safety glasses with side shields, often with darkened lenses, are satisfactory. For people not used to wearing glasses they can be a nuisance in rain. In a few cases, face visors are used, particularly in laboratories where a clear field of view is required whilst performing experiments. Goggles are less common but sometimes necessary; however, they can become uncomfortable, sweaty or mist up. Goggles, by nature of their design, restrict peripheral vision and can make the operator's task more difficult. The fit of the goggle round the face and particularly round spectacle frames is seldom "tight" and drops of chemicals may still penetrate.

3.5.3 Ears

Ideally noise reduction at source should be attempted, a highly specialised subject. Noise levels in many process areas are sufficiently high that hearing protection is routinely required. Where necessary warning notices must be posted on entry routes. Noisy machinery can be incorporated within enclosed, acoustically sealed areas. However, in other cases, particularly compressor houses, the risk of a flammable gas leak then explosion in a confined space will override the need for an acoustic enclosure. Where access within high-noise areas is required, hearing protection is essential.

High noise levels—over 85 dB(A) for eight hours' exposure per day and sharp impactive noises (as in sheet metal work, bottling or pile driving) can lead to hearing damage. Noise-induced hearing loss is not obvious until damage is quite advanced—only then will it be noticed that conversation becomes difficult to follow in noisy rooms. Ears can be protected by means of ear plugs which cut out the most destructive high frequencies, but to be effective, they must be a good fit. Earmuffs or ear defenders can be used but these must be fitted correctly and have a good seal around the edge. This can create problems with facial or long hair and for wearers of glasses. Earmuffs can become uncomfortable and "sweaty" after prolonged use and are, again, not popular with operators. Earplugs are more comfortable but less efficient.

Ear protection is not an ideal solution to high noise levels as sound energy can still be transmitted to the inner ear through body tissues. Additionally, ear defenders may "leak" sound and communication is made difficult, if not impossible. If design and mitigation efforts cannot reduce the noise it may be necessary to use sound compensating microphones and speakers. Some designs of helmets incorporate these features but are not in common use.

Noise is further discussed in Section 3.8.

3.5.4 Lungs

Lungs can be severely affected by dusts or toxic gases. The simplest protective device is a dust mask which fits over the nose and mouth and filters out solids. These can be useful for applications such as grinding; the filter efficiency is influenced by the particle size distribution. The seal around the face is important; beards greatly reduce efficiency. Ideally beards should be shaved off, but this may be difficult in some parts of the world for cultural reasons. Dust masks have a limited life and need frequent replacement of the filter.

Where toxic gas may be present a compressed air supply feeding to the face mask, commonly called *breathing apparatus* (BA), represents the best protection. The air supply can be from a compressed air cylinder or from a static airline. Where a static airline is used, the connections to the masks must be taken from the top of the header to prevent slugs of liquid entering the masks. There must also be drain points in the header. These systems are operated via either a *demand valve*, which supplies air when the breather requires it, or a *supply valve* which maintains a positive pressure within the mask. The positive pressure will prevent the entry of toxic gas into the mask. Cylinder-based BA gives the wearer greater mobility—not being limited by the length of the hose—and avoids the possibility of the hose snagging or rupturing. However, the bulky cylinders can make access through manholes difficult.

Activated charcoal (or other media depending on the toxic materials being handled) is widely used for masks, but the filter element can only be used once and then requires regeneration/replacement. Masks with activated charcoal filters were widely used during both world wars and are still commonly available but should only be used as emergency escape masks and not for deliberate exposure for an extended period.

Work should not routinely be performed in a toxic environment but masks may be necessary in some instances, for example where a gas escape may occur (such as the unloading of chemicals or during maintenance).

3.5.5 Hands

Hands can be injured by sharp objects, hot surfaces, chemicals and of course dropped objects. The usual protection is gloves, either "heavy duty" or gauntlets. They should be appropriate to the protection required. They are not popular with operators who lose their "feel". Gloves are often clammy or sweaty. The primary objective must be to protect the operator at source, but in some cases, this will not be possible. Furnaces always have hot surfaces and sheet metal has sharp edges ("rags").

3.5.6 Feet

Feet are vulnerable to impact through scuffs and dropped objects as well as hot or hazardous fluids. The typical industrial shoe or boot protects the foot against glancing impact on sharp objects and small dropped objects. They cannot be fully protected against heavy dropped objects, and a higher level of protection may be required. Many boots have ankle supports for protection in case of slips, trips or falls.

Specialist footwear may be required in some areas, for example to provide electrical insulation or resistance to chemicals.

3.5.7 General Body Protection

In addition to the protections described earlier, site workers generally wear long-sleeved flame-retardant coveralls where there is a risk of fire. Generally, these are of a bright or luminous material so that workers can be clearly identified.

Specialist protection against hazardous chemicals, particularly acids and alkalis which attack the skin, may be required. In a well-run plant significant leaks should not happen, but small discharges from pump or valve glands may occur, as can leaks from couplings used during loading/unloading. In a few instances, process materials may be so hazardous that full-body protection is required at all times in case a leak suddenly occurs. Hydrogen fluoride handling is a common example.

The face can be protected by a full visor or a headpiece and visor, the body by a totally enclosed suit made of impervious fabric. The seal between the parts of the suit and, of course, boots and gauntlets requires careful consideration; the top half must overlap the lower half. Similarly, the overlap between body protection and boots and gauntlets must be tight.

3.5.8 Other Types of Protective Equipment

There are numerous other types of protective equipment available, many for specific industries or applications. Common examples are shown in Table 3.3.

3.5.9 Training

Whilst the use of some kinds of PPE is self-evident, training will be required for more complex equipment such as escape masks, life jackets and breathing apparatus which are likely to be required in an emergency and must be fitted quickly and correctly. In some cases, such as personnel working in environments exposed to hydrogen sulphide, training is mandatory and a certificate issued. Repeat training is required at intervals.

TABLE 3.3
Other Types of Protective Equipment

Item	Industry
High Visibility Jackets	Warehouse and truck movements
Hair Nets and Paper Suits	Clean rooms (electronic industry), food processing and pharmaceuticals
Life Jackets	Offshore and dockside work
Harness	Working at heights, underground locations and also with some confined spaces (columns and the like), a lifting davit and a "stand-by operator" will also be necessary.
Bunker Gear and Aluminised Suits	Used by firefighters. The suits are designed to protect the firefighters from high levels of thermal radiation and drenching from water during a fire. Aluminised suits allow a closer approach to a burning fire. In addition, breathing air sets will probably be required.

3.5.10 Maintenance of PPE

Protective equipment has a limited lifetime. Some items are single use, such as some types of ear plug and dust masks; others may have a longer lifetime but will still require replacement or repair at some point. Repeated washing of fire-retardant overalls will remove the fire-retardant chemicals. Filters in chemical escape masks will have an expiry date, as will breathing masks and cylinders. Damaged equipment must be replaced without delay. Common failures are damage to the soles of safety boots or the harness of a safety helmet. Other forms of subtle damage can occur; solvent spillages on plastic helmets, for example, can result in a softening of the plastic (this has occasionally happened due to stickers being fixed to helmets).

3.5.11 Control of the Working Environment (Airborne Materials)

Some plants—particularly enclosed plants which handle dust or toxic fumes—require ventilation; this can be achieved in several ways:

- Containment of materials in an enclosed area to avoid affecting the employee (such as a laboratory fume cupboard).
- Dilution of the contaminant to acceptable levels by mechanical ventilation
- Under-pressurised ventilation and discharge through a scrubber system.

In the first case, equipment is either in an enclosed environment or located under an exterior hood. In all cases air is drawn from the hooded area and discharged into a safe location where the material may be recovered in dust cyclones, scrubbers or discharged to the atmosphere to dilute by normal dispersion mechanisms.

Where VOCs may be released, such as degreasing operations, more sophisticated ventilation, possibly involving the use of refrigeration, may be required.

3.6 TOXICOLOGY

Toxicology refers to the study of harmful effects caused by the presence of chemicals. Many chemicals, including some occurring naturally, are potentially harmful to humans and other life forms. The degree of harm depends on the form of the substance, its concentration, its method of ingestion and a number of other factors. Different animal or plant species may react to the same substance in different ways and at different concentrations.

The main routes for toxic materials to enter the human body are ingestion by nose/mouth or absorption through the skin. The concentrations, durations and effects are specific to the chemical concerned and the effects may be specific to the site of entry to the body. There is often difficulty in determining safe levels of exposure to chemicals as the effects are difficult to measure and the response will vary from one individual to another. Much of the available data has been obtained from animal tests where the animal's metabolism differs from that of humans. There can, therefore, be difficulty in translating the results of laboratory tests into useful guidance. Similarly, the same level of exposure can have very different effects on different individuals. Some materials are especially harmful to women of childbearing age and may cause abnormalities in their children. Toxicology is therefore a very complex and specialised subject.

The basic "vocabulary" of the subject is shown in Table 3.4.

TABLE 3.4
Common Terms in Toxicology

Term	Definition
Acute	Immediate, short-term. Relating to exposure: conditions which develop rapidly and may cause harm within a short time. Relating to effects: effects which appear promptly after exposure.
Asphyxiation	Endangering life by causing a deficiency of oxygen
Carcinogen	A substance which produces cancer
Chronic	Persistent, prolonged and repeated. Relating to exposure: frequent or repeated or continuous exposure to substances. Relating to effects: when physiological effects appear slowly and persist for a long period or with frequent recurrences.
Corrosive	In the context of toxic substances, a corrosive substance is one which may, on contact with living tissues, destroy them.
Dose	Used as a synonym for exposure.
Exposure	Amount of toxic substance to which an individual is exposed. This may represent the amount ingested, absorbed or inhaled or it may refer to the integral of concentration with time in the immediate environment. Where ambiguity may arise the basis used to define the exposure should be specified.
Irritant	A non-corrosive material which may, through immediate prolonged or repeated contact with the skin or mucous membrane, cause pain, discomfort or minor injury. Such reactions may appear as a precursor to more serious injury or an allergic reaction.
Mutagen	A compound causing genetic damage
Sensitiser	Compounds which generate an immune response (trigger allergies)
Teratogen	A compound that causes birth defects when the developing foetus is exposed to it
Toxic	The property of substances which, when introduced into or absorbed by a living organism, destroys life or injures health
Long-Term Exposure Limit (LTEL)	A time-weighted average concentration, usually averaged over 8 hours, which is appropriate for protecting against the effects of long-term exposure
Short-Term Exposure Limit (STEL)	A time-weighted average concentration, usually averaged over 10 minutes, aimed at avoiding acute effects.
Threshold Limit Value-Time-Weighted Average (TLV-TWA)	The time-weighted average concentration for a normal 8-hour workday or 40-hour workweek to which nearly all workers may be exposed, day after day, without adverse effect (to be superseded by the term "control limit").
Threshold Value-Ceiling (TLV-C)	The concentration which should not be exceeded, even instantaneously.
Lethal Dose (LD_{50})	The quantity of material administered orally or by skin absorption which results in the death of 50% of the test group within a 14-day observation period.
Lethal Concentration (LC50)	The concentration of airborne material, the four-hour inhalation of which results in the death of 50% of the test group within a 14-day observation period.

Other commonly used terms are as follows:

- Dangerous toxic load (DTL)
- Specified level of toxicity (SLOT)
- Significant likelihood of death (SLOD)

These are defined in mathematical terms in the subsequent sections.

3.6.1 Calculation of Exposure Levels

The list of definitions mentioned earlier may appear bewildering; however, they are all necessary and specific even to a basic understanding of the subject. They help to differentiate the effects of different chemicals, the means of access, the consequence and immediacy of the effect.

Toxic effects and concentrations are often derived by experimentation on animals with a very limited amount of inferred data from industrial incidents. The animals tend to be rabbits, rats and mice. Whilst experimentation in vivo is unpopular with the public, it has been an unpleasant necessity. There are potential hazards with experimentation as the physiology and biochemistry of the subject must be a close match to that of homo sapiens. There are a number of readable summaries concerning toxicology and the limitations of this form of experimentation. Bridges[12] gives the following noteworthy warning:

> *Assessment of the lethal effects of chemicals is usually conducted only in rodents yet findings in a rodent may be a poor indicator of toxic hazard to man. Large interspaces differences are also common. Since the Seveso Directive is based on an incident in which Dioxin (2, 3, 7, 8—tetrachlorodibenzo-p-dioxin—TCDD) was released into the atmosphere it is appropriate to use this as an illustration. Ingested Dioxin is about 100 times less acutely toxic to mice than to guinea pigs, and the Syrian hamster is about 600 times less susceptible than the guinea pig. Differences of this magnitude between three species make extrapolation from animals to man somewhat problematical. In an attempt to identify whether one species is more predictive of lethal properties of chemicals in man than another, Krasovskii [14] analysed the scientific literature on the acute toxicity of some 260 chemicals in man and other mammals. Krasovskii concluded that man was usually more sensitive than the commonly used tested species to the acute toxic effects of chemicals.*
>
> *If no data exists to identify, for a particular chemical, which species is likely to be the most appropriate representative of the human response the findings from the most sensitive species must be adopted for hazard assessment purposes.*
>
> *Differences in LC_{50} and LD_{50} also commonly occur between animals of different ages, sex, and between strains of animals of the same species. Other factors include timing of observation, housing conditions and diet may also contribute to variability of results*
>
> (Reproduced courtesy of IChemE).

Successful tests on animals are therefore not an indication that there will be no adverse effects on humans. The best-known example is Thalidomide which was tested on animals with no side effects noted but caused widespread deformities to the children of pregnant women who took the drug.[13]

LC_{50} and LD_{50} refer to an average healthy human. There may be significant differences between healthy and less healthy subjects, for example, those suffering from bronchitis or other respiratory problems. Values quoted must be treated with caution and the data source checked carefully for applicability and authenticity. The history of Thalidomide is a warning against the blind acceptance of experimental data derived from animals.[14] A particular problem is found with carcinogens which have long induction periods and where the history of exposure may be limited, particularly in a migrant working population.

The hazardous effects of chemicals can be found in a number of sources, the best-known being Lewis[15] and Bretherick.[16] These are massive and expensive books and information may be available from *material safety data sheets* (MSDS) which are freely available from chemical suppliers or the internet. Data for most common materials are available from a UK HSE publication[17] which also gives details of safety precautions, hazards and treatment in the event of exposure. Most chemicals are labelled with warning notices and classified by hazard potential. The UN numbering system[18,19] is a common way of identifying the contents of containers or tanks if no other information is available. In an attempt to provide a comprehensive, consistent and accessible chemical safety information the European Union (EU) has launched the Registration, Evaluation, Authorisation and Restriction of Chemicals (REACH) initiative[20] to identify the hazardous property of materials used within the EU.

Depending on the chemical and concentration, injury may occur after only a single exposure. In other cases, repeated exposure at lower concentrations may be necessary before the ill effects become apparent. Toxicology and industrial hygiene of chemicals were not adequately explored at the earlier stages of industrialisation and accepted levels of exposure were capable of causing injury. Widespread exposure within an industry resulted in *industrial disease*. Some examples are well known, such as asbestosis,[21] a chronic lung condition caused by exposure to concentrations of asbestos fibres in the air. There are numerous other examples including pneumoconiosis,[22] another lung disease affecting coalminers. Exposure to wood dust gives an increased risk of lung cancer[23] and exposure to cutting oils[24] used in machinery can cause scrotal cancer. Many other examples exist and increasingly sophisticated methods of assessment are improving the ability to identify them, even at relatively low levels of exposure.

Not all chemical or biological illnesses involve a specific industry. Legionella[25] could occur anywhere if the water supply is not correctly treated and tetanus,[26] now fortunately controlled by vaccines or treatment in the early stages of exposure, could affect anyone who is exposed.

3.6.2 Toxic Dose

The toxic dose for chemicals in common use is generally available from a number of sources and is defined as

$$C^nT \tag{3.1}$$

where C = Concentration (ppm)
n = Chemical Dependent Factor of 1.6 to 2.8
T = Time (s)

Equation 3.1 must be treated with care as the concentration in a gas cloud is not constant and may vary by a factor of ±2 or 3 throughout the cloud. Most dispersion programmes give *time-weighted average* (TWA).[27]

TWA concentrations allow for variations in exposure limits throughout the working day. A more recent development by the UK HSE for *land use planning* is the concept of *dangerous toxic load* (DTL)[28,29] using the same relationship as defined earlier.

The *specified level of toxicity* (SLOT) is defined as causing the following:

- Severe distress to almost everyone in the area
- Substantial fraction of exposed population requiring medical attention
- Some people seriously injured, requiring prolonged treatment
- Highly susceptible people possibly being killed

Significant likelihood of death (SLOD) is the exposure conditions which are predicted to cause mortality of 50% of an exposed population. Based on information for toxicity of methyl isocyanate (the material released at Bhopal), the CT relationship is shown in Table 3.5.

For methyl isocyanate the SLOT DTL is 750 ppm. min (150 × 5, 25 × 30 . . .)

For methyl isocyanate the value for n is 1. For other toxic materials the value for n may differ. For example, for ammonia, chlorine and sulphur dioxide the value for n is 2. The time/concentration relationship is shown in Table 3.6.

TABLE 3.5
SLOT Values for Methyl Isocyanate (reproduced courtesy of HSE)

T (min)	5	10	30	60	120
C (ppm)	150	78	25	12	6

TABLE 3.6
SLOT Values for Sulphur Dioxide (reproduced courtesy of HSE)

T (min)	5	10	30	60	120
C (ppm)	965	682	394	279	197

For sulphur dioxide the SLOT DTL is 4.6×10^6 ppm^2.min ($965^2 \times 5$ or $394^2 \times 30$).

Values for SLOT and SLOD are available from UK HSE.[30]

Note that these values will provide steady-state concentrations which may differ significantly from the stochastic concentrations which occur in reality. This is discussed further in Section 8.1.

Alternatively, toxic effects may be described by *Probit Functions*[31] which describe the relationship between the concentration of a substance, the exposure time and the effect on humans. Probit functions are derived by testing and converted into exposure/response relationships for humans. Probit functions to estimate the effects of thermal radiation are also *available.*

Probit values have been formally determined for a number of substances by the Netherlands National Institute of Public Health and the Environment (RIVM).[32] For substances for which no probit values are available, a procedure to derive, evaluate and formalise the function has been developed.

$$Pr = k_1 + k_2 ln \Sigma C_i^n T_i Pr = k_1 + k_2 ln \Sigma C_i^n T_i \quad (3.2)$$

where Pr = probit (relationship to the probability of fatality)
k_1, k_2 and n = Constants
T = Time of exposure (s)

Probits have also been developed for thermal and overpressure injuries. There are considerable areas of uncertainty in these equations, and they may not accurately predict the effects of long-term, low-level exposure.

Whilst discussions of toxicity often relate to hazardous chemicals, there are many other materials which are overlooked but still represent significant hazards:

Dusts: There are many examples of dusts with harmful effects. The best-known cases being silica leading to pneumoconiosis and asbestos leading to asbestosis and mesothelioma. Wood dust (and wood smoke) can also lead to damage to the lungs. The general principles of dust exposure are summarised in a UK HSE publication.[33]

Dust explosions are also a significant explosion hazard and are discussed in Section 10.2.

Metals: Many metals are toxic.[34] These include cadmium, chromium, mercury and lead, all of which can gain access to the lung in the vapour phase. Low-level concentration, for example ingested in food, can also cause serious health problems.

3.7 LABELLING

Both the UK and USA have developed warning systems to identify hazardous materials and provide guidance on emergency response. The UK system is known as HAZCHEM, with chemical containers having a warning panel as shown in Figure 3.3.

The top left gives emergency response information (see Table 3.7) with the UN number below plus a telephone number of emergency contact.

1	JETS
2	FOG
3	FOAM
4	DRY AGENT

FIGURE 3.3 HAZCHEM Label (reproduced courtesy of Chartwise UK Limited)

TABLE 3.7
HAZCHEM Emergency Response Information

PR	V	FULL	
S		BA	
S	V	BA for FIRE only	
T		BA	
T		BA for FIRE only	DILUTE
WX	V	FULL	CONTAIN
Y	V	BA	
Y		BA for FIRE only	
Z		BA	
Z		BA for FIRE only	
E		CONSIDER EVACUATION	

Notes for guidance:

FOG: In the absence of fog equipment, a fire spray may be used.

DRY AGENT: Water must not be allowed to come into contact with the substances at risk.

V: Can be violently or even explosively reactive.

FULL: Full body protective clothing with BA.

BA: Breathing apparatus plus protective gloves.

DILUTE: May be washed to drain with large quantities of water.

CONTAIN: Prevent by any means available the spillage from entering drains or watercourse.

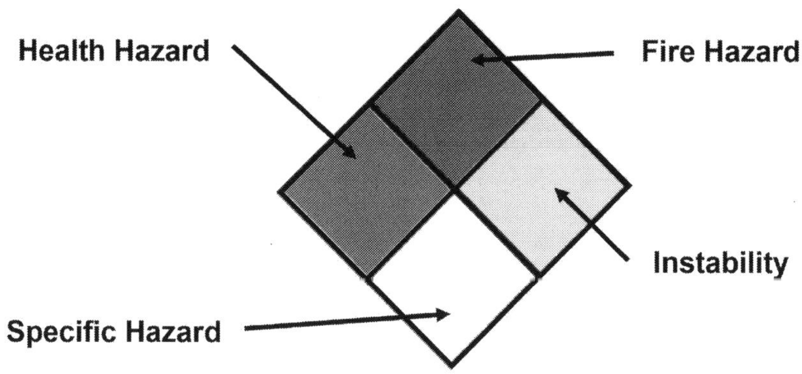

Health Hazard [blue]	Fire Hazard [red]	Specific Hazard [white]	Instability [yellow]
0–Normal	0–Will not burn	OX–Oxidiser	0 – Stable
1–Slightly Hazardous	1–Flashpoint above 93°C (200°F)	ACID–Acid	1–Unstable if Heated
2–Hazardous	2–Flashpoint above 37.8°C (100°F), not exceeding 93°C (200°F)	ALK–Alkali	2–Violent chemical change
3–Extreme danger	3–Flashpoint below 37.8°C (100°F)	COR–Corrosive	3–Shock and heat may detonate
4–Deadly	4–Flashpoint below 22.8°C (73°F)	W–Use no water	4–May detonate
		☢–Radioactive	
		P–Polymerizes	

FIGURE 3.4 NFPA Hazardous Chemical Classification

The US system is based on the National Fire Protection Association's (NFPA's) Standard System for the Identification of the Hazards of Materials for Emergency Response.[35] The classification is shown in Figure 3.4.

Well-developed systems are in place for the transport of hazardous materials.[36]

3.8 NOISE

All process plant has the potential to produce noise; the frequency and strength will be a complex mix from many sources. Frequencies produced are generally equipment specific: electric motors will produce frequencies which are multiples of 25 Hz or 50 Hz; gearboxes emit frequencies at shaft speed and multiples of shaft speed, as well as the gear mesh frequency (usually 2–4 kHz). Noise due to the flow of fluids is typically in the range 1 kHz to 20 kHz, with leaks producing even higher frequencies (detection of high frequency sound is now a recognised method of leak detection)[37] and ultrasound has many uses including flaw detection in welds and wall thickness measurement.[38] Turbines, control valves and steam ejectors also produce noise at high frequencies. Noise and vibration can be considered as synonymous. Low frequency noise has relatively low energy and is less damaging to human hearing, but noise above 1 kHz will potentially affect the inner ear.

By convention, noise is defined in decibels (dB) as

$$20 \log 10\, P_2 P_1 \tag{3.3}$$

where: P_2 = pressure of the signal (N/m^2)
P_1 = reference pressure of 2×10^{-5} N/m^2

As noise is measured on a logarithmic scale, a doubling of the sound level is equivalent to an increase of 3 dB. Some frequencies are more damaging, particularly the range 1–8 kHz, so the various frequency bands are given a loading factor to average out the damage potential. Lower frequencies have a lower loading factor than the higher frequencies. The result of this "normalising" is the A scale, dB(A).

A UK HSE publication[39] gives a brief overview of noise issues and a more comprehensive discussion of noise by Crocker[40] is available.

Exposure to noise is measured by a dose relationship, with the unit of measurement called the Leq. Within the UK, the Control of Noise at Work Regulations[41] apply and the current action level of exposure is 85 dB(A) for eight hours. Above this level, hearing protection must be provided. Within the UK, for noise levels above 80 dB(A), employers must assess the risk to the health of workers and provide information and training. Additionally, there is an exposure limit of 87 dB(A) above which workers must not be exposed, after taking account of hearing protection measures. Different standards may apply outside the UK.

Where there are high noise levels, it is necessary to establish each employee's baseline hearing level and then regularly monitor for hearing loss. Apart from the use of Leqs, there are overriding limits for noise in some industries. Special considerations may be necessary during construction and demolition activities, for example, during pile driving or steam flushing of pipelines.

Whilst hearing protection will prevent damage to hearing, it can prevent employees hearing audible alarms and alternative means of warning are necessary. For example, the use of visual alarms in permanently noisy areas, the use of a "buddy touch" system and/or flashing beacons where noisy machinery is used in a normally quiet area (such as maintenance on rail lines).

Table 3.8 gives typical examples of noise levels.

TABLE 3.8
Typical Noise Levels

Approximate Sound Pressure Level dB(A)	Source/Location
140	Jet aircraft at take-off
130	Pneumatic drill
100	Very noisy factory or compressor house
90	Tractor
	Power tools
80	Alarm clock
70	Traffic
40	Quiet office/library
20	Leaves rustling

TABLE 3.9
Noise Sources and Possible Solutions

Noise Source	Design Solution
Piping	Install heavy wall piping
	Install dense insulation
	Reduce the flow velocity
Control Values	Install silent Trims
	Install silencers
	Enclose in acoustically insulated cubicles
Rotating Equipment	Install silencers
	Specify low noise Fans
Compressors/Turbines	Enclose in acoustically insulated cubicles. Reduce blade tip velocity
Gearboxes	By design—low mesh frequency or velocity
	Enclose in acoustically insulated cubicles
Intakes/Exhausts	Install silencers
	Install diffusers
Flare and Vent Stacks	By design

Noise from several sources, such as pumps, motors, control valves and piping, can add up rapidly to a significant overall noise level. As stated previously, noise is measured on a logarithmic scale, noise from two sources is not numerically additive. The planning consent for a new greenfield site may impose strict noise limitations and, of course, the addition of new plant on an existing site may require extensive noise abatement across the site. The public may justifiably complain if excessive noise levels occur. Noise reduction is a specialist activity which often makes use of the techniques listed in Table 3.9.

3.9 VIBRATION

Equipment vibration was briefly covered earlier, and specific aspects of vibration are further discussed in future chapters. Vibration, generally at low frequency, can cause a variety of occupational health problems including injury to bones, joints and tissue, collectively known as hand-arm vibration syndrome (HAVS). The most well-known vibration problem, *white finger*,[42] can be a problem for personnel working with machinery such as pneumatic drills and riveting guns. It usually takes years for damage to become apparent, and it is usually irreversible.

Within the UK, hazards from vibration are covered by the Control of Vibration at Work Regulations.[43] Other countries will have similar regulations.

3.10 LIGHTING

Industrial facilities must be adequately illuminated to ensure instruments can easily be read and hazards such as poorly lit stairways avoided. A simple guide to lighting requirements has been developed by the UK HSE[44] and more comprehensive publications are readily available. The site should be reviewed regularly to ensure that any failed lights are repaired and newly installed equipment does not block lighting and result in shadowy areas. Failed lights create their own localised issues of shadow until repairs are made. Access ladders and maintenance staff may reduce the level of illumination on a temporary basis and temporary lighting might be required until repairs are completed. Temporary lighting must conform to the same standards as permanently installed

TABLE 3.10
Illumination Levels (reproduced courtesy of HSE)

Activity	Typical Locations/Type of Work	Average Illumination Lux	Minimum Measured Illumination Lux
Movement of People and Vehicles [(a)]	Lorry park, corridors, circulation routes	20	5
Movement of People, Machines and in Hazardous Areas (Rough Work Not Requiring Perception of Detail)	Construction site clearance, excavation and soil work, loading bays, bottling and canning plant	50	20
Work Requiring Limited Perception of Detail[(b)]	Kitchens, factories, assembling large components, potteries	100	50
Work Requiring Perception of Detail[(c)]	Offices, sheet metal work, bookbinding	200	100
Work Requiring Perception of Fine Detail[(d)]	Drawing offices, factories assembling electronic components, textile production	500	200

lighting. A common failure is to install temporary lighting which doesn't meet the required hazardous area classification (ATEX) rating for its location (see Section 14.6).

In addition to the main lighting systems, separate emergency lighting must be installed and capable of operation in the event of a power failure. Quellette and Rea[45] provide an overview of emergency lighting; however, requirements are often specified by regulation.

Light from a source is measured in *lumens* with *lux* being the unit of illuminance, measuring luminous flux per unit area. Table 3.10, taken from the UK HSE guide "Lighting at Work",[46] gives an indication of desired lighting levels. However, there will be instances where more detailed guidance will be necessary.

(a) Only safety has been considered because no perception of detail is needed and visual fatigue is unlikely. However, where it is necessary to see detail to recognise a hazard or where error in performing the task could put someone else at risk, for safety purposes as well as to avoid visual fatigue, the figure needs to be increased to that for work requiring the perception of detail. The CIBSE Code for lighting[47] gives more information and recommendations based on scientific knowledge, practical experience, technical feasibility and economic reality.
(b) The purpose is to avoid visual fatigue; the illuminances will be adequate for safety purposes.
(c) The purpose is to avoid visual fatigue; the illuminances will be adequate for safety purposes.
(d) The purpose is to avoid visual fatigue; the illuminances will be adequate for safety purposes.

Lighting has its own health hazards; flood lights can cause eye damage if stared at and, if touched, will cause burns. Site-run cables for temporary lighting can create a tripping hazard and may cause electric shocks if the insulation is damaged.

Lighting may be considered an environmental nuisance if close to residential or rural areas. Flares can also be considered a source of light pollution, and a special case is the flaring of hydrogen which burns with a near-invisible flame during the day but can be clearly seen at night.

3.11 IONISING RADIATION

Ionising radiation exists naturally. It is present in air, rocks, food, cosmic rays and from atomic weapons tests. Everyday uses of radiation, in controlled circumstances, include medical investigations/treatment and for non-destructive testing (NDT). Most countries have strict controls for the use of ionising radiation.

Radioactive devices in process plant are used for a variety of applications ranging from level measurement to radiographic inspection of welds and non-intrusive scanning of equipment with source strengths powerful enough to penetrate fluids and metal.

3.11.1 Types of Radiation

Four basic types of radiation exist; their effects and penetration ability differ significantly:

Alpha (α) Particles	Essentially charged helium atoms—they have relatively little energy and can be stopped by a layer of dead skin. Damage potential for humans from external exposure is low, but if they enter the body damage will be high.
Beta (β) Particles	Essentially charged electrons. They have sufficient energy to penetrate approximately one centimetre under the skin.
Gamma (λ) Rays	Electromagnetic waves such as X-rays. They are highly penetrating and very difficult to arrest. Dense layers of lead and/or concrete are needed.
Neutrons	Uncharged. They are highly penetrating but can be arrested.

In summary: Alpha particles are relatively safe when received externally but dangerous if received internally. Beta particles are potentially dangerous if received externally, but neutrons and gamma rays are very dangerous if received externally or internally.

3.11.2 Units

The rate at which radiation is produced is called the Becquerel (Bq) and is the equivalent to the decay of one radionuclide per second.

3.11.3 Dose

The absorbed dose is called a Gray (Gy) and is defined as the absorption of one joule of radiation energy per kilogram of matter (J/kg). The damage potential for the four types of radiation differs and the damage potential is normalised into a dose equivalent, called a Sievert (Sv). A Sievert is defined as metre squared per second squared (m^2/s^2).

Within the UK, the levels of exposure are given in the latest version of the ionising radiations regulations.[48] Other countries will have similar regulations. The limits vary depending on age, exposure over time and other factors.

3.11.4 Working with Radioactive Materials

Personnel who use radioactive substances have specialised training; in the UK they are designated *as classified workers* by law and are subject to

- periodic health checks
- radiation dose monitoring (film badges)
- detailed record-keeping.

3.11.5 Uses of Radioactive Sources

Radioisotopes or radioactive sources have found increasing use in the process industries. Obviously, there are special precautions needed in their application, but under proper control and supervision their use both in control and as a diagnostic tool is a potent addition to the safety armoury. Table 3.11 gives some of the more common uses.

The use of radioactive sources in the process industries is reducing as alternative techniques such as ultrasonic inspection are used (see Chapter 20).

3.11.6 Safe Use of Radioactive Sources

Radioactive sources must be used under strictly controlled conditions. When they are used the surrounding area must be restricted and cordoned off. The cordon must cover all three dimensions, extending over and under the radioactive source as well as horizontally. Clear, visible warning signs must be displayed at all edges of the cordon. The area around the source must be monitored for radiation leakage. Radiographic work (such as weld inspection) is often undertaken at night when fewer people are on-site. However, this introduces hazards associated with night-time working and will require good levels of illumination.

Regulations control the use of radioactive sources. Generally, these must be registered with the regulatory authority and held in a safe at the site when not in use. When they are out of the safe a record of their movement must be kept. The use of radioactive materials is generally undertaken by specialist companies.

TABLE 3.11
Some Uses of Radioisotopes

Application	Technique
Flow Measurement in Pipework	Pulsed isotope injection
Flow Measurement in Thermosyphon Reboilers	
Leaks in Heat Exchangers	Drip feed injection
Tracing Underground Pipes or Drains	
Density Measurement	Neutron absorption*
Level Management/Control	Absorption*
Interface Measurement/Control	
Scanning Vessels for Solids Build Up	λ ray scans*
Scanning Distillation Columns for Tray Performance	
Tray Damage	
Tray Blockage	
Downcomer Blockage	
Jet Flooding	
Inspecting Welds and Castings for Defects	λ ray/X-rays*
Inspecting Filters, Online, for Build-up of Debris	
Inspecting Non-return Valves for Damage/Performance in Flowing Conditions	
Monitoring Equipment for Corrosion	
Alloy Identification	X-ray*

* Non-intrusive techniques

3.12 WHAT CAN GO WRONG?

Regardless of the amount of effort expended to develop an effective safety system for occupational safety, something is likely to be undetected such as the following:

- Change of a chemical supplier with a different formulation
- A change in the work programme—where some items of PPE were not considered necessary when the work was planned
- PPE to an incorrect specification is supplied
- PPE has been used beyond its expiry date
- Materials supplied (or disposed of) were incorrectly specified
- Panic during normal operation or an emergency

Some examples may appear trivial but have severe consequences. It is therefore important that all chemicals that could be encountered and the methods of work are reviewed by a risk assessment before any activity takes place. A simple example for glass fibre-based insulation to be installed in a roof space is given in Table 3.12.

Despite stringent, in some cases excessive, precautions incidents can still occur, as the following real-life example demonstrate.

> A plant supervisor in full PPE had just returned from the process area and was observing operations in the control room when he leaned forward for a closer view. His helmet became detached and fell onto a keyboard associated with the plant DCS, causing a plant shut down.
>
> The site concerned now requires helmets to be removed and stored outside the control room, but this is not widespread practice (6).

TABLE 3.12
Simple Risk Assessment for Glass Fibre Insulation installation

What is glass fibre and what are the issues?	Glass fibre (and mineral wool) are fine fibres of glass (or spun minerals)
What parts of the body might be affected due to its nature?	• Fibres in the eye • Fibres in the lung • Fibres penetrating clothing a creating irritation
What protection should be supplied?	• The eyes should be protected by tight-fitting goggles • The arms should be protected by full-length (elbow or higher) plastic gloves and a full work boiler suit, Preferably a PVC suit • The lungs should be protected by a dense fibre mask (well fitting). Preferably a breathing air set which will also protect the face. • Finally, a long shower to remove strands of fibre after the task is complete

Note this assessment covers the material only—slips, trips, falls and other hazards should also be considered and will be case-specific

REFERENCES

1 The Food, *Farm Hotels and More Global Union*, www.iuf.org/news/a-trade-union-guide-to-fighting-back-against-behavior-based-safety/
2 Alverez J, *LBV Magazine*, June 2019, www.labrujulaverde.com/en/2019/06/the-world-war-ii-airmen-who-survived-falls-from-thousands-of-feet-high/
3 UK Health and Safety Executive, www.hse.gov.uk/humanfactors/topics/control-room.htm
4 Edmonds J, *Applying Human Factors Engineering to Control Room Upgrade Projects and to the Design of New Build Control Rooms*, ICHEME, Hazards XXV, 2015, www.icheme.org/media/8464/xxv-paper-18.pdf
5 UK Health and Safety Executive, *Display Screen Equipment (DSE) Workstation Checklist*, UK HSE, 2013, https://www.hse.gov.uk/pubns/ck1.pdf
6 UK Health and Safety Executive, Reducing Error and Influencing Behaviour (HSG 48), *The Stationary Office*, 1999, www.hse.gov.uk/pubns/books/hsg48.htm
7 UK Health and Safety Executive, *Better Alarm Handling (Chemicals Sheet 6)*, HSE, www.hse.gov.uk/pubns/chis6.pdf
8 UK Health and Safety Executive, *Human Factors: Alarm Management*, www.hse.gov.uk/humanfactors/topics/alarm-management.htm
9 UK Health and Safety Executive, Plant Aging: Management of Equipment Containing Hazardous Fluids or Pressure (RR509), *HSE Books*, 2006, www.hse.gov.uk/Research/rrpdf/rr509.pdf
10 UK Health and Safety Executive, *Personnel Protective Equipment (PPE) at Work*, www.hse.gov.uk/ppe/index.htm
11 Health and Safety Executive of Northern Ireland, *Personnel Protective Equipment (PPE)*, www.hseni.gov.uk/articles/personal-protective-equipment-ppe
12 Bridges J W, *The Assessment of Toxic Hazards, Hazards VIII 22–24 April 1985*, ICHEME, www.icheme.org/media/12177/viii-paper-26.pdf
13 *The Thalidomide Trust*, www.thalidomidetrust.org/about-us/about-thalidomide/
14 Kim J H & Scialli A R, Thalidomide: The Tragedy of Birth Defects and the Effective Treatment of Disease, *Toxicological Sciences*, 122(1), 2011, www.gvsu.edu/cms4/asset/F51281F0-00AF-E25A-5BF632E8D4A243C7/thalidomide._kim_scialli.pdf
15 Lewis R J, *Sax's Dangerous Properties of Industrial Materials* (12th edition), Wiley, 2 October 2012
16 Urben P, *Bretherick's Handbook of Reactive Chemical Hazards* (7th edition), Academic Press, January 2008
17 UK Health and Safety Executive, Workplace Exposure Limits (EH40/2005), *The Stationary Office*, 2005, www.hse.gov.uk/pubns/priced/eh40.pdf
18 Chemsafetypro, *UN Dangerous Goods List*, www.chemsafetypro.com/Topics/TDG/Dangerous_Goods_List.html
19 Chemsafetypro, *How to Determine Correct UN Number and Proper Shipping Name*, www.chemsafetypro.com/Topics/TDG/Propper_Shipping_Name.html
20 European Commission, *REACH Regulation*, https://ec.europa.eu/environment/chemicals/reach/reach_en.htm
21 Asbestosis, www.asbestos.com/asbestosis/
22 American Lung Association, *Coal Workers Pneumoconiosis (BlackLung Disease)*, www.lung.org/lung-health-diseases/lung-disease-lookup/black-lung
23 Matrat M, et al., Occupational Exposure to Wood Dust and Risk of Lung Cancer: The ICARE Study, *British Medical Journal Occupational & Environmental Medicine*, 2019, https://oem.bmj.com/content/76/12/901
24 US Occupational Safety and Health Administration, *Metalworking Fluids: Safety and Health Best Practices Manual*, www.osha.gov/metalworking-fluids/manual
25 UK Health and Safety Executive, *Legionella and Legionnaires' Disease*, www.hse.gov.uk/legionnaires/
26 World Health Organisation, *Tetanus*, www.who.int/health-topics/tetanus#tab=tab_1
27 A time-weighted average is the average concentration over a period of time (usually eight hours)
28 UK Health and Safety Executive, *Toxicity Levels of Chemicals*, www.hse.gov.uk/chemicals/haztox.htm
29 Fearnley J & Nair S R N, Using Predictive Risk Assessments to Develop User Friendly Tools for On and Off Site Emergency Planning, *Hazards XXI*, IChemE symposium Series, 2009, www.icheme.org/media/9570/xxi-paper-073.pdf
30 UK Health and Safety Executive, *Toxicity Levels of Chemicals*, www.hse.gov.uk/chemicals/haztox.htm

31 Imbrechts K, The Use of Probit Functions for Toxic Substances in Quantitative Risk Analysis, *Chemical Engineering Progress* (Vol. 77), AIChemE, 2019, www.aidic.it/cet/19/77/005.pdf
32 Netherlands National Institute of Public Health and the Environment (RIVM), *Reference Manual Bevi Risk Assessments*, July 2009, www.rivm.nl/documenten/reference-manual-bevi-risk-assessments-version-32
33 UK Health and Safety Executive, *Controlling Airborne Contaminants at Work: A Guide to Local Exhaust Ventilation (LEV)*, www.hse.gov.uk/pubns/books/hsg258.htm
34 Järup L, Hazards of Heavy Metal Contamination, *British Medical Bulletin*, 68(1), December 2003,
35 https://academic.oup.com/bmb/article/68/1/167/421303
36 UK Health and Safety Executive, www.hse.gov.uk/cdg/manual/classification.htm
37 Moon C, Brown W C, Mellen S, Frenz E & Pickering D J, *Ultrasound Techniques for Leak Detection (09NVC-0269)*, www.bksv.com/media/doc/bn1831.pdf
38 Iowa State University, www.nde-ed.org/NDETechniques/Ultrasonics/index.xhtml
39 UK Health and Safety Executive, *Noise at Work: A Brief Guide to Controlling the Risks (INDG362)*, www.hse.gov.uk/pubns/indg362.pdf
40 Crocker M J (ed), *Handbook of Noise and Vibration Control*, Wiley, 2007
41 UK Health and Safety Executive, *Control of Noise at Work Regulations*, 2005, www.hse.gov.uk/noise/regulations.htm
42 UK Health and Safety Executive, *Hand-Arm Vibration*, www.hse.gov.uk/mvr/topics/vibration.htm
43 UK Health and Safety Executive, *Hand-arm Vibration the Control of Vibration at Work Regulations*, 2005, www.hse.gov.uk/pubns/priced/l140.pdf
44 UK Health and Safety Executive, *Lighting at Work* (2nd edition), HSG38, 1997, www.hse.gov.uk/pubns/priced/hsg38.pdf
45 Quellette M J & Rea M S, Illuminance Requirements for Emergency Lighting, *Journal of the Illuminating Engineering Society*, 18, September 2013, www.researchgate.net/publication/44054881_Illuminance_Requirements_for_Emergency_Lighting
46 UK Health and Safety Executive, *Lighting at Work*, HSG38, 1997, www.hse.gov.uk/pubns/books/hsg38.htm
47 The Society of Light and Lighting, *SLL Lighting Handbook*, Chartered Institution of Building Services Engineers (CIBSE), 2018, www.cibse.org/knowledge/knowledge-items/detail?id=a0q0O00000F4MeJQAV
48 UK Health and Safety Executive, *Radiation Legal Base*, www.hse.gov.uk/radiation/ionising/legalbase.htm

4 The Environment

4.1 INTRODUCTION

Any activity will have some environmental effect, but for millennia human activity was relatively limited. Until recently, natural events have caused greater worldwide changes than human activity, the most famous example being the meteorite which hit the earth approximately 66 million years ago, causing the demise of the dinosaurs. For most of history, human impact on the environment was generally limited, in part due to the relatively small number of humans and the low level of industrialisation. This started to change during the industrial revolution with the widespread use of coal to power industry and the uncontrolled release of industrial pollutants. Environmental protection was relatively slow to develop with the first legal restrictions in the UK being the Alkali Act 1863[1] to moderate the discharge to air of muriatic acid gas (gaseous hydrochloric acid) from Leblanc alkali works. Some sources[2] claim the act was passed for the protection of landowners rather than as a health measure benefiting the general public.

Environmental standards have increased tremendously since the initial and often half-hearted attempts at pollution control. A common trick used to be burning waste or discharging effluent at night to make pollution hard to detect. Hazardous waste was sometimes buried with limited record-keeping of what had been buried and exactly where. The most famous example being Love Canal in the USA.[3] Rather than controlling discharges a "dilution is the solution to pollution" approach was often adopted. Since those days, environmental controls have tightened worldwide but, regrettably, are not always enforced. It would be prudent to assume environmental regulation will become increasingly strict; however there are signs that standards might be eased in some parts of the world for ideological or political reasons. Regardless, it would be unwise and immoral to assume standards will actually decline and not be enforced.

Modern regulations cover routine discharges to air, land and water and may require special measures for the control of hazardous materials—particularly from *heritage sites* where environmental control might have been limited and not always enforced. A typical example on elderly hydrocarbon locations is the sheen of oil after rainfall which "lifts" oil from previous spillages out of the ground.

The effects of pollution can be indirect: rivers can be *killed* by oxygen depletion. Agricultural run-off (nitrogen and phosphates) can produce toxic algal blooms. The decomposition of organics/chemicals requires oxygen and oxygen depletion produces sterile rivers. The longer-term effects of poisoning a river are uncertain—there could be an accumulation of debris, leading to the silting up of watercourses. Even now, it is clear that the recovery of floodplains for human use in many countries is a potential time bomb. Concerns are being expressed for the potential for future flooding along the River Thames in the UK as sea levels rise. Changes to improve navigation on the Rhine[4] have allowed flood waters to reach the lower reaches of the river due to bypassing or the elimination of *flood plains*. This has resulted in inundation in areas which previously had not experienced flooding.

The immediate safety effects of a major incident will be relatively local but the environmental consequences of continuous discharge of effluent from a major release can be widespread, affecting multiple countries. In some cases, such as the release of radiation from the atmospheric testing of nuclear weapons, the effects are worldwide.

The world ecosystem is finely balanced and small changes can have a major long-term cumulative effect. Whilst global warming, resulting in more extreme weather events, is the major—and still

DOI: 10.1201/9781003510680-5

uncontrolled—environmental hazard, it is by no means the only one.[5] The following examples are well known but are only the tip of the pollution iceberg.

- The *dust bowl* due to the removal of vegetation in the *Prairie Corn Belt* of the Central USA in the 20th century.
- The effects of dichlorodiphenyltrichloroethane (DDT) on bird life (in Africa, DDT did produce beneficial effects in the suppression of malaria by controlling the mosquito population).[6]
- The recovery of land near rivers for human habitation reduced floodplains and increased downstream flooding. The effects in the Rhine valley have already been mentioned.
- The production of *acid rain*[7] due to the burning of coal containing sulphur compounds for power generation. Oxides of sulphur would discharge into the atmosphere, creating *acid rain*, which would often be precipitated hundreds of miles from the point of release and create structural damage to both steel and stone, as well as causing acidification in lakes.
- The use of *chlorofluorocarbons* (CFCs) from refrigeration and firefighting accumulated in the ozone layer[8] affecting its ability to adsorb ultraviolet radiation. In a good example of international cooperation, the Montréal Protocol was agreed and the use of CFCs phased out. The ozone layer is now recovering.[9]

Of course, environmental disasters will occur regardless of human actions. Earthquakes and lightning have long caused casualties and damage. Meteorites could have had a significant but infrequent environmental impact. Drought, high winds and rainfall have caused death and destruction throughout the ages. However, there is growing evidence that global warming is exacerbating the effects. Occasional bizarre events occur such as the release of a huge bubble of carbon dioxide from Lake Nyos[10] in 1986, which resulted in approximately 1,700 deaths.

Even if the majority of commercial organisations wish to act responsibly, one or two irresponsible players can sabotage successful action for improvement. This situation is neatly described by the illustration of the *Tragedy of the Commons*. Imagine an area of common land where a number of farmers each keep a single cow. Grass on the common grows at the rate at which the cows eat it. One of the farmers now gets a second cow so the grass is consumed more quickly than it grows. What should the other farmers do? If they do nothing the common will slowly deteriorate. On the other hand, if they each get a second cow they will have a larger share of the grass on the common but increase the speed at which it deteriorates.[11]

Whilst the most obvious concern is pollution, other factors need to be considered:

- The noise from industrial facilities produces a low-frequency "throb"; the higher-frequency sources are more attenuated by distance than low-frequency sources. The increase in noise levels in previously "rural" areas can produce a significant environmental impact. The effects of noise from increased traffic levels also need to be considered.
- Visual impact can be seen in many locations; different people will have a very different reaction to visual changes. However, changes to the landscape and their potential effects must not be ignored but may be masked by trees or landscaping.
- Increase in traffic on local roads—particularly at the start and end of the working day—must be considered, especially with regard to other road users such as schools, hospitals and farmers who use the roads for moving livestock between fields. Noise has been mentioned earlier, but an increase in traffic fumes will also occur and access to emergency service could be impaired. Improvements to roads could attract additional traffic unconnected with the development. Construction traffic, particularly during civil works, may result deposit mud on public roads.

- Effects on fauna and flora at the site must be considered:
    - Fumes, noise or light could affect vegetation.
    - Animal habitats might be affected.
    - Migration or nesting patterns for birds might be affected.
    - Changes in drainage caused by changes to ground level and permeability of the site. Concrete is virtually impermeable. This will result in more 'run off' in soil surfaces, with less rainwater soaking into and being captured by the soil.

Most developments require an *environmental impact assessment* (EIA), which may require a baseline measurement of flora and fauna, an assessment of the restitution of the site, and the routine monitoring of the near- and far-field environment for changes in flora and fauna.

4.2 THE SAFETY/ENVIRONMENTAL CONFLICT

There is frequently conflict between environmental and safety issues. Some of the benefits of technology have been achieved at the expense of the environment. Refrigeration provides a simple illustration: hydrocarbon-based refrigerants often resulted in fires and explosions. ammonia—previously a commonly used refrigerant in the food industry—was both flammable and toxic. The development of *chlorofluorocarbons* (CFCs) offered a "safer" alternative, but the impact on the *ozone layer* was not predicted.[12] Whilst some new, more environmentally friendly refrigerants have been developed, many industrial refrigeration systems still use ammonia or hydrocarbons.

An older example is chlorine—widely used in water treatment chemicals—which has saved countless lives by ensuring supplies of clean drinking water. However, the gas itself is toxic and was used as a weapon in the First World War[13] and, to a lesser extent, in more recent conflicts.[14] Kletz,[15] Crawley[16,17] and others have attempted to integrate both topics within overall *safety, health and environment* (SHE) criteria. A more recent example is sulphur hexafluoride (SF_6). It is a non-toxic, odourless, non-flammable gas, often used as a dielectric medium in electrical switchgear where it is more efficient than air or nitrogen. It also has a number of other significant uses. Unfortunately, SF_6 is also a major greenhouse gas, thousands of times worse than carbon dioxide.[18]

There are many similar examples and there is now greater care in the assessment of chemicals before widespread use, The EU REACH[19] regulations being one example, but there is no guarantee that future mistakes will not be made and there is still a huge legacy of pollution from earlier times.

Whilst few people would argue against high standards for safety or for environmental protection there are times when conflicts occur. An increase in safety standards may be detrimental to the environment and vice versa. The following are a few examples:

- Disposal of hydrogen sulphide (H_2S) is often achieved by incineration and discharge of the products of combustion, including oxides of sulphur (SO_x) to the atmosphere. As noted earlier, SO_x can form acid rain but hydrogen sulphide (H_2S) is toxic. Dispersion of SO_x is therefore considered preferable to dispersion of H_2S—even though acid rain will be precipitated at locations which are generally well away from the point of release.
- To prevent the build-up of flammable gases, compressor houses are generally open-sided to allow natural ventilation of any gas releases. However, noise (and light) levels around an open compressor house will be higher than around an acoustically enclosed structure.
- Ammonia is a toxic and flammable gas but is used for NO_x removal from combustion flue gases[20] to limit pollution from combustion processes. The use of ammonia may drastically increase if it is used as a means of transporting hydrogen as a step towards the *net zero economy*.
- Screening of production facilities with trees will reduce noise and light pollution as well as hide the outline of a facility which many would find ugly. However, trees can act to obstruct the flame front in the event of a gas cloud igniting. This may increase the flame speed with the potential to turn a flash fire into an explosion.[21] Explosions are further discussed in Chapter 10.

Fauna and flora can also have unexpected effects:

- Birds' nests or bee hives can block atmospheric vents or pressure safety valve (PSV) discharge lines.
- Birds roosting in structures can lead to corrosion of steelwork[22] and slipping hazards.
- Mice and rats in electrical areas may eat through the insulation of instrument or electrical cabling
- Cats walking between uninsulated phases of electrical busbars can cause a major short circuit and power outage
- Moss (the growth of which is a good indication of low pollution levels) can cause slips, trips and falls
- Process plants, in some parts of the world, can provide an unwelcome home for wildlife, including venomous snakes on steps and walkways and alligators in cooling/firewater ponds

There can sometimes be positive effects from releases, a common example being the use of waste heat for fish farming, district heating or other purposes.

Often, safety and environmental concerns are in full alignment. If there is no leak, there is no safety or environmental issue, and the use of lower pressures will reduce the size of release for a given hole size. As often occurs in real life, a balance will need to be made between safety and environmental concerns. The safety/environment dilemma may become more pronounced as net zero and the hydrogen economy approaches (see Chapter 23).

On 1 August 1991,[23] an explosion occurred at the Coode Island terminal, Melbourne, Australia. This was a modern terminal with a number of safety and environmental features, including the blanketing of tanks with nitrogen to prevent the formation of a flammable atmosphere and the inclusion of impervious mats beneath the storage tank to contain spillages and prevent them from entering the ground.

The initial explosion occurred in a tank of acrylonitrile and was followed by fire. The cause of the explosion is disputed, but the official report gives an atmospheric electrical discharge—St Elmo's Fire. The initial fire propagated to a number of other tanks. Some were nearby and could be expected to be damaged by direct thermal radiation. However, the terminal had a vent system connecting the tanks to an incinerator which burnt waste gases before discharge to the atmosphere. The vent/incineration system was installed for good environmental reasons—to prevent the discharge of VOCs—but its presence also provided a pathway for greater destruction than might otherwise have occurred.

4.3 ROUTINE RELEASES

Almost any industrial site will have routine releases to the atmosphere. Noise and light pollution can be controlled by a variety of measures, and once control measures are in place, there should not be a long-term problem. The main issues relate to releases of gas, liquids, solids and possibly heat. National and local regulations will limit the concentration and possibly the volume of releases. Regular, possibly continuous, monitoring will be necessary to ensure that discharge levels are not exceeded.

Most countries place a *duty of care* on the plant operator. This requires that waste materials are disposed of in an environmentally acceptable manner. The operator must ensure that waste removed from the site is disposed of responsibly and not passed to a contractor with no thought to its ultimate disposal. A document trail will be required to show that safe disposal has occurred. Undocumented and unauthorised disposal of waste is a relatively common criminal activity.

In November 1983,[24] surplus medical equipment was picked up by truck in Juarez, Mexico. The mostly steel waste included a capsule containing approximately 1,000 pellets of Cobalt 60 (CO–60). During transit to a scrapyard, the capsule ruptured, spreading the pellets throughout the truck. Some of the pellets were offloaded with the scrap, but some remained on the truck which was then parked for several weeks in a residential area.

The pellets at the scrapyard were scattered as the equipment was moved around before being sent to two local foundries. One foundry produced rebar, which became contaminated with Co–60. The contamination was discovered by chance when a truck carrying rebar made a wrong turn and set off a radiation alarm. Two workers at the scrapyard became sterile. Whilst there were no immediate deaths there was concern about an increased risk of cancer amongst the numerous people who had been contaminated.

There may be times when allowable disposal limits are exceeded. This should only occur during emergency situations (see Section 4.4). They must be reported to the authorities and corrective action taken. There may be a temptation for the plant management to hide or minimise such events, but the long-term consequences of this approach will be very negative. At a minimum, fines are likely to be imposed, but the effect on the company's reputation could be severe and, in extreme situations, could result in plant closure.

4.3.1 Gases and Vapours

The most obvious gas discharged in most plants is carbon dioxide. Whilst this is not the most damaging gas to the environment, the sheer volume released makes it the major contributor to the greenhouse effect. *Carbon capture and storage* (CCS) technologies[25] have been developed but are not yet widely installed. This is likely to change in the future.

A wide variety of gases and vapours may be released during operations, and routine venting of surplus gases often takes place. Small but frequent releases may take place due to sampling. Fugitive emissions from pump seals, valve glands, joints and other sources are also common. In recent years sampling techniques have improved as have sealing methods for pumps and compressors. Overall, there is now much more attention to fugitive emissions than in the early years of the process industry. Releases due to opening flanges during maintenance can still be a problem and detailed control of this activity is necessary (Section 16.6).

Whilst some gases and vapours may be discharged directly to atmosphere, others will need treatment prior to release. Combustion in a flare or incinerator is one common method and *gas scrubbing* is used to remove hazardous vapours such as chlorine in chloralkali plants. Increasingly restrictive discharge limits will spur the development of new techniques.

Some common methods for cleaning contaminated gas streams prior to discharge are in the following list:

• Volatile Organic Compounds	Control by improved piping connections, valve glands and pump seals. Control by ventilation and disposal (by adsorption or combustion) or recovery of vapours (at vehicle fuel dispensers)
• Oxides of Sulphur (SO_x)	Hydrodesulphurisation or scrubbing processes using lime
• Oxides of Nitrogen (NO_x)	Catalytic conversion or scrubbing with ammonia
• Particulates	Cyclones, electrostatic precipitation or filtration

Tank *breathing* can also result in intermittent discharge of vapours to the atmosphere. In fixed-roof storage tanks, the vapour space will be a mixture of air and vapour from the tank contents. When liquid is removed from the tank the liquid level is reduced and air enters the tank. When the level is raised, air and chemical vapours are discharged. In some instances, vapour discharges from the tanks are manifolded together and lead to a vapour recovery or disposal systems.

The modelling of gas releases is a complex task and is further discussed in Chapter 8.

4.3.2 Liquids

Many routine releases will be liquids, with the largest volume being water, often with contaminants. These will generally flow into a watercourse of some kind or to the sea. Some process waste such as *boiler blowdown* can be discharged directly to the effluent discharge system; others with a high concentration of process materials will need specific treatment, recycling to the process or some form of separate disposal.

The extent of contamination is often measured in terms of the following:

• Biological Oxygen Demand (BOD)	The amount of oxygen consumed by bacteria and other microorganisms whilst they decompose organic matter under aerobic conditions[26]
• Chemical Oxygen Demand (COD)	A measurement of the oxygen required to oxidise soluble and particulate organic matter in water

The comments on fugitive emissions made earlier are relevant to spillages entering the site drains system. Surface drains will also collect rainwater, dirt and rubbish. If solids are allowed to build up blockages will occur and regular drain cleaning is necessary. Flows of waste water can vary by orders of magnitude. Much of the time flow will be limited but will be much larger during heavy rainfall and during fires, if large volumes of firewater are used. This latter point is often missed during design. Variations in flowrate can be covered by design for the largest anticipated flowrate and by the provision of a buffer storage facility for waste water awaiting treatment.

Most sites will have some form of *waste-water treatment plant* (WWTP). On simple facilities such as an oil terminal, treatment may be limited to a skimmer to remove any hydrocarbons that are floating on top of water flowing through the system. This may be followed by biological treatment. Depending on the processes being undertaken, more complex facilities may be required. The following table shows common water treatment processes:

• Skimmer	To remove oil/grease floating on top of water. The most common type is a simple weir with liquids flowing underneath and oil/grease flowing over the weir. Variations on the basic principle are available, such as a tilting plate separator designed for use in confined spaces such as ships and offshore platforms.
• Neutralisation	For pH control
• Flocculation Followed by Skimming	For the removal of solids
• Oxygen Demand	Oxygenation
• Biological Treatment	There are a variety of methods depending on the chemicals to be removed.
• Chemical treatment	Precipitation and filtering.
• Incineration	Incineration can be used to dispose of hazardous chemicals but can generate hazardous gases such as dioxin if chlorine/chlorides are present.[27]
• Recycling	Some materials, such as waste oil and used solvents, can be recycled for reuse.
• Storage	Some industries, particularly mining, produce large quantities of waste which are stored in tailings dams. These require continuous inspection and repair, possibly for long periods after the site has closed. Several serious dam failures have occurred.

In January 2019,[28] the tailings dam at the Córrego do Feijão iron ore mine near Brumadinho, Brazil, failed. The slurry released by the failure caused widespread damage and over 250 deaths.

Toxic waste is sometimes stored in a supposedly secure repository. Perhaps the best known example in the UK is the Pitsea site. This is not a "store and forget" solution. There are cases of records being lost or the containment being ineffective or breached. Love Canal (Section 23.2) provides a chilling example.

- Evaporation — In desert environments, waste water is often directed to shallow lagoons where it evaporates, with the deposited solids being further treated or used as landfill.

Many sites will connect to local authority facilities for disposal of sewage and kitchen waste, but in remote locations and offshore facilities for treating these wastes will also be required.

4.3.3 Solids

Solid wastes vary enormously from one site to another. In some simple cases, it may be limited to normal domestic waste plus packaging materials, used components from machinery repairs and similar items. Where possible, materials should be recycled and others may be disposed of as normal commercial waste.

However, many sites will have significant amounts of waste ranging from out-of-specification polymer to ash from combustion of coal or solid waste. Some processes produce an organic slurry which is treated in an anaerobic digester[29] to produce biogas (largely methane) which, after treatment, can be used as a fuel.

Incineration, possibly with heat recovery for process heating or power generation, is a commonly used method of disposal of combustible solids. The combustion process must be closely controlled to ensure that there is no discharge of ash and no toxic materials of combustion are generated.

Some wastes may only be generated on a periodic basis, such as a plant shut down for maintenance or during demolition. Spent catalyst and depleted nuclear waste are examples. There is a market for waste with a significant commercial value (such as used catalyst containing platinum or other valuable metals). For high-hazard wastes, such as radioactive scale or mercury-contaminated pipework (from hydrocarbon production) or asbestos, specialist disposal is necessary.

4.4 ACCIDENTAL RELEASES

Whilst most releases will be within the discharge consent limits, there may be occasions where they exceed either the allowed level of contaminants or volume of release. In most cases the deviation from allowable levels is likely to be limited and short term. There are however a few cases where catastrophic releases could occur. The examples of Bhopal[30] and Deepwater Horizon[31] are well known and documented, but there are many others covering production of hydrocarbons, processing and transportation. Examples include the following:

- Large-scale releases from oil tankers (Torrey Canyon, Amoco Cadiz, Exxon Valdez and Braer)
- Pipeline ruptures (Nordstream 1 and 2 resulted in large methane releases but were caused by sabotage. Ageing infrastructure has resulted in a number of large-scale leaks from pipelines.)
- Rupture of storage tanks, in many cases secondary containment, should prevent spillage, but this is not always the case, either because secondary containment was not provided (increasingly unusual) or the containment was inadequate or failed.

- Oil well blowouts.[32]
- Tailing dams[33] containing industrial waste.

4.4.1 Gases and Vapours

Where process upsets release a large quantity of gas a control mechanism should be in place to prevent or limit environmental damage. This is often a vent or a flare and, in the case of toxic gas, a scrubber. Whatever system is used, the design should allow for the effects of a process failure (such as loss of the power supply) and ensure that the concentration of hazardous materials will be reduced to a safe level before reaching inhabited areas (see Chapters 8).

Major fires can often release large quantities of smoke with toxic components. In these situations, the offsite emergency plan (see Section 22.3) should prevent access to the area and advise people within the affected zone to stay indoors with the windows shut. Clean-up will be necessary once particulate matter falls to earth.

4.4.2 Liquids

Spillages of hazardous materials are depressingly common. Spillages are often the result of human error but also occur as a result of piping or equipment rupture following, for example, an earthquake, explosion or fire. Process and utility piping are the most likely source of leaks but more hazardous chemicals may be released such as polychlorinated bisphenols (PCBs) from elderly transformers and electrical switchgear. Examples of large spills were outlined at the start of this section and are generally beyond the ability to control easily. The first priority is to rescue people and livestock and, as far as possible, limit the spread of the spillage. Collecting resources for clean-up might take time, with resources needing to be marshalled and logistics planned.

Oil spills usually involve the use of booms to contain oil on water, plus the collection and incineration of contaminated earth. Oil booms are deployed to contain floating spills and are routinely installed as a precaution during marine loading/unloading operations. Biological processes for oil removal have also been developed; incineration of polluted earth or in situ injection of air/oxygen may be possible, but sometimes landfill is still used.[34]

A particular issue occurs with water-soluble chemicals which can enter watercourses and contaminate drinking water supplies. Ironically firefighting foam has been a source of contamination and new formulations are being developed.[35]

In some cases, fire water is so contaminated that it cannot be discharged directly from the site. In this situation, containment must be provided or the fire allowed to burn in a controlled manner.

> On 1 November[36] a fire started at the Sandoz site in Basel, Switzerland. The warehouse contained a number of powerfully toxic and harmful substances, including over 800 tonnes of organophosphorus insecticides and, of even more concern, 11 tonnes of water-soluble mercurial fungicides.
>
> The firefighting response involved tug boats on the Rhine as well as ground-based fire water application. Fire water runoff exceeded the capacity of the drainage system and contaminated water was released into the Rhine and also penetrated the ground to a depth of 14 m—potentially affecting ground water. There were delays in raising the alarm, preventing downstream mitigation measures from being fully effective. Widespread damage to the marine environment occurred. Dead eels were noted 200 km downstream from the incident and contamination was recorded up to the point where the Rhine entered the North Sea.

4.4.3 Solids

The situation with solids is similar to that described in Section 4.3, but in many cases it may not be clear what the waste is and how it needs to be disposed of. Materials that are generated in a fire may have totally or partially combusted solids and contain a variety of different components, some with harmful properties. Hazardous materials of construction may also be found following a fire, particularly in older plant where asbestos may have become exposed.

Ideally, the plant owners should be aware of the condition of the plant, the materials of construction and any chemicals, residues and contamination that exist. This is often not the case and a cautious approach to post-incident clean-up should be adopted.

Explosions can send projectiles considerable distances—possibly outside the site boundary.

4.5 ENVIRONMENTAL IMPACT ASSESSMENT

Most countries now require some form of *Environmental Impact Assessment* (EIA) before a new plant starts operation along with periodic updates. It is also common for an EIA to be required to cover the construction phase of a project as the environmental impacts will differ from those found during routine operation. The EIA must consider the significance of impacts from the plant on the surrounding area. This goes beyond the assessment of potential pollutants but considers their spread and the effect they will have on the surrounding fauna and flora. For example: will discharge of heat into a river affect the marine life? will light, noise or vibration affect the behaviour of animals, birds and fish in the area? Similarly, will noise, light and road traffic affect the animal or human population in the area? Whilst guides and checklists are available on the internet, they mostly relate to national planning regulations and are of limited value. There are numerous questions to be asked and there is no simple, universal checklist available. Detailed assessment of all issues must take place. Like most major activities, EIA is best considered in a number of stages[37]:

1 Screening	The EIA process runs from the start of a project and addresses two questions: • What will be the effects of this development on the environment? • Are those effects significant?
2 Scoping	Scoping will define the issues that need to be addressed so that the available resources can be deployed on the most critical issues.
3 Baseline Study	This involves the collection of information on the current environmental conditions
4 Impact Prediction	Once the baseline study information is available, the important task of impact prediction can begin. Impact prediction involves forecasting the likely changes in the environment that will occur as a result of the development.
5 Impact Assessment	The effects of the potential impacts of the new plant can be determined.
6 Mitigation	Having determined the potential impact of the development, it may be possible to implement measures to reduce the impact of the plant.

Developing the EIA will be an iterative process and will likely become more onerous as more detailed understanding of the effect of chemicals and different forms of energy on the environment are determined.

The EIA will often be used to develop an *environmental impact statement* (EIS) which will be used in the application for plant construction and operation. Updates of the EIS will be required at intervals to determine if changes to the environment match those predicted.

Archaeological surveys of construction areas are becoming increasingly common and, if relics of historical importance are found, this can cause a delay in construction.

4.6 NATURAL HAZARDS

Natural hazards can represent a significant exposure to process plants and a number of serious accidents have occurred:

- On floating oil/gas production units extreme weather conditions can cause shut down and, in some cases, moving off station.
- Vessel unloading can be stopped during high winds, hurricanes, typhoons and other extreme conditions—leaks in loading arms and hoses can occur if safe working conditions are exceeded. Marine construction activities such as dredging and piling can be curtailed for the same reason.
- Low temperatures can cause a range of problems, including freezing and condensation. Often the most serious problems occur in utility systems such as cooling water or in instrument air systems when the air has been inadequately dried. For very low temperatures such as in the Arctic and Siberia special materials of construction may be necessary to avoid brittle fracture.
- High winds, subzero temperatures and high humidity can disrupt power supplies.
- High winds can "push in" the side of large storage tanks (if coupled with heavy rain, and a build-up of water in a tank bund, an almost empty tank can float, rupture pipework and release a significant amount of liquid).[38]
- Deforestation can result in ground instability, particularly during or following heavy rainfall.
- Wildfires can affect process plants in a number of ways, ranging from high thermal radiation levels in the plant, restrictions on access where roads pass through wildfire areas, and fumes which may affect plant operations.
- Earthquakes can result in direct damage to plant and equipment, but greater damage can be caused by rupture of pipework releasing toxic or flammable materials. The same earthquake might also damage fire water mains. A tsunami from an offshore earthquake can generate tidal waves.

Events of this type are commonly referred to as Natural Hazards Triggering Technological Accidents (NATECH).[39] Consequences may range from minor but repeated issues to more complex situations. In the former category are the problems faced by a port in Asia where the boundary fence is frequently breached by marauding elephants. The more serious end of the scale is demonstrated by the Fukushima tsunami,[40] which caused widespread damage to nuclear power facilities in Japan. A feature of many incidents initiated by a natural peril is the way the incident develops. The initial event might be relatively minor, such as a pipe rupture, power loss, or instrument freezing, but the resulting 'knock-on' or domino effects can be far more severe.

In principle, the use of appropriate design codes, supplemented by procedures, should safeguard the plant against natural perils. However, note that some design criteria are frequency-based criteria such as the 100-year storm; there is an acceptance that the event will occur but sufficiently infrequently that it can be ignored—until it happens. There are cases where the design underestimated the frequency and magnitude of natural hazards. Global warming may result in design assumptions being exceeded. Degradation may defeat the intended protection. For example, over time, insulation in cold-weather zones may become damaged and ineffective.

On 31 August 2017,[41] A series of fires started at the Arkema plant at Crosby, Texas, when refrigerated organic peroxides decomposed. The peroxides should have been refrigerated but power to the refrigeration system was lost as a consequence of Hurricane Harvey. The fire resulted in the destruction of 350,000 pounds of organic peroxide. More than 200 residents living within 1.5 miles of the site were evacuated and could not return home for a week.

The Arkema Crosby facility is located within both the 100-year and 500-year flood plains. Extensive flooding following heavy rainfall caused the power supply and back-up power to fail. Consequently, the refrigeration system shut down, and the temperature in the peroxide storage containers rose to the point where the self-accelerating decomposition temperature (SADT) was reached and the products ignited.

4.7 WHAT CAN GO WRONG?

There will always be unusual and bizarre incidents. Not all will be predictable, and some may be a "once in a lifetime event". Many will be the result of carelessness and the lack of application of existing knowledge.

In some parts of the world venomous snakes and insects can be a problem and at least one refinery played host to alligators in the fire water pond. There have been several incidents of jellyfish blocking water inlets to cooling water systems and shellfish blocking suction pipes on fire water pumps. Vegetation grows abundantly in many climates, blocking drain systems and, when dry, creating a fire hazard. Many incidents are caused by simple, easily preventable mistakes. Forgetting to fit a plug or blind on a drain or vent line, leaving a drain valve open or failing to identify or replace a corroded section of pipe are typical examples. More extreme examples are often the failure of systems of work. Occasionally, something occurs which, although explicable, would, arguably, be difficult to identify at the time.

In December 1997,[42] an air separation unit (ASU) in Bintulu, Malaysia, exploded. The ASU had a capacity of 2,500 tonnes of oxygen per day and formed part of a process converting natural gas to hydrocarbons using the Fischer Tropsch process. Prior to the explosion, the ASU was working normally.

Low concentrations of hydrocarbons in the atmosphere, which then accumulate in the ASU, are a known problem but were not, initially, considered to be the likely cause of the Bintulu incident, as hydrocarbons discharged from nearby process plants were flared. An exhaustive investigation revealed that particulates and hydrocarbons from forest fires had concentrated in the ASU distillation system and resulted in the explosion.

As with other aspects of safety, a questioning approach and the identification and correction issues at an early stage are important skills, as well as contingency plans which can be adapted as required.

REFERENCES

1 Environment & Society, *England Passes First Alkali Acts*, www.environmentandsociety.org/tools/keywords/england-passes-first-alkali-acts
2 Frankel M, *The Alkali Inspectorate the Control of Industrial Air Pollution*, Social Audit Ltd, 1974
3 US Environmental Protection Agency, Beck E C, The Love Island Tragedy, *EPA Journal*, January 1979, www.epa.gov/archive/epa/aboutepa/love-canal-tragedy.html
4 Barcelona Field Studies Centre, *The Rhine Floods 1995: Causes of the Floods*, https://geographyfieldwork.com/RhineCauses.htm
5 Krausmann, Natural Hazard Triggered Technological (Natech) Accidents—An Overlooked Type of Risk?, *Loss Prevention Bulletin* (250, IChemE, 2016), www.icheme.org/media/2087/lpb250pg11.pdf

6 Carson R, *Silent Spring*, Houghton Mifflin Harcourt, 1962
7 www.epa.gov/acidrain/what-acid-rain
8 US Environmental Protection Agency, *What Is Acid Rain*, www.eea.europa.eu/themes/climate/ozone-depleting-substances-and-climate-change-1
9 United Nations, Ozone Layer Recovery Is on Track, Due to Success of Montreal Protocol, *UN News*, January 2023, https://news.un.org/en/story/2023/01/1132277
10 Baxter P J, Kapila M & Mfondu D, Lake Nyos Disaster, Cameroon, 1986: The Medical Effects of Large Scale Emission of Carbon Dioxide? *British Medical Journal*, May 1989, www.ncbi.nlm.nih.gov/pmc/articles/PMC1836556/
11 Hardin G, The Tragedy of the Commons, *Science*, 162, 13 December 1968, https://faculty.washington.edu/stevehar/HardinTragedy.pdf
12 Fahey D W, Twenty Questions and Answers about the Ozone Layer, *Department of Environment Food 7 Rural Affairs*, June 2002, https://uk-air.defra.gov.uk/assets/documents/ozone-uv/Unep_11_qa.pdf
13 Harris Robert Paxman J, *A Higher Form of Killing,* Harper Collins, 1983
14 Aljazeera, *OPCW Blames Syria Gov't for 2018 Chlorine Gas Attack in Douma*, 27 January 2023, https://www.aljazeera.com/news/2023/1/27/opcw-blames-syria-government-forces-for-2018-douma-chlorine-gas-attack
15 Kletz, Green Intent Red Result, *Symposium Series 147,* ICHEME, 2000, www.icheme.org/media/10216/xv-paper-01.pdf
16 Crawley F K, Ashton D & Morrell R, Achieving the Balance between Safety and the Impacts on the Environment, *Symposium Series 147,* ICHEME, 2000, www.icheme.org/media/10217/xv-paper-02.pdf
17 Crawley F K & Ashton D, Safety Health or the Environment—Which Comes First, *Journal of Hazardous Materials*, 93, 2002
18 Greenhouse Gas Protocol, *Global Warming Potential Values*, www.ghgprotocol.org/sites/default/files/ghgp/Global-Warming-Potential-Values%20%28Feb%2016%202016%29_1.pdf
19 European Chemicals Agency, *Understanding REACH*, https://echa.europa.eu/regulations/reach/understanding-reach
20 Zhu Z & Xu B, Purification Technologies for NO_x Removal from Flue Gas: A Review, *MPDI*, October 2022, www.mdpi.com/2297-8739/9/10/307
21 UK Health and Safety Executive, Buncefield Explosion Mechanism Phase 1 (RR718) 1, *HSE Books*, 2009, www.hse.gov.uk/research/rrpdf/rr718.pdf
22 Forr F O, *The Effect of Bird Droppings on the Corrosion of Structural Materials Used in Offshore Applications* (Master's Thesis), NTNU, June 2022, https://ntnuopen.ntnu.no/ntnu-xmlui/handle/11250/3026010
23 Behayeddin P & Martin G, The Coode Island Fire Incident and It Consequences, *Australian Civil Defence Symposium & Expo*, 2014, https://docslib.org/doc/12567128/the-coode-island-fire-incident-and-its-consequences
24 ORAU Museum of Radiation and Radio Activity, *Contaminated Rebar from Juarez*, www.orau.org/health-physics-museum/collection/accidents/contaminated-rebar-from-juarez.html
25 UK Health and Safety Executive, *Carbon Capture and Storage*, www.hse.gov.uk/carboncapture/index.htm
26 Tuser C, What Is Biological Oxygen Demand (BOD), *Wastewater Digest*, October 2021, www.wwdmag.com/instrumentation-bod/what-biological-oxygen-demand-bod
27 Hites R A, Dioxins an Overview and History, *Environmental Science & Technology*, 45, 2011, https://pubs.acs.org/doi/pdf/10.1021/es1013664
28 Jasi A, Dam Collapse Tragedy Could Have Been Prevented, *The Chemical Engineer*, 20 November 2019, www.thechemicalengineer.com/news/dam-collapse-tragedy-could-have-been-prevented/
29 U S Environmental Protection Agency, *Learning about Biogas Recovery*, www.epa.gov/agstar/learning-about-biogas-recovery#adwork
30 Bhopal Medical Appeal, www.bhopal.org/continuing-disaster/rethinking-bhopal-a-process-safety-analysis-of-the-union-carbide-bhopal-plant/
31 US Chemical Safety and Hazard Investigation Board, *Explosion and Fire at the Macondo Well Investigation Report Volume 1*, www.csb.gov/assets/1/7/vol_1_final.pdf
32 SINTEF, *Offshore Blowout Database*, www.sintef.no/en/projects/2021/sintef-offshore-blowout-database/
33 World Mine Tailings Failures—From 1915, https://worldminetailingsfailures.org/

34 International Tanker Owners Pollution Federation Limited (ITPOF), www.itopf.org/knowledge-resources/documents-guides/disposal/

35 US Environmental Protection Agency, *The Changing Nature of Fire-Fighting Foams*, www.epa.ie/publications/monitoring—assessment/waste/AFFF-Booklet-Web.pdf

36 Vince I, The Sandoz Warehouse Fire—30 Years On, *Loss prevention Bulletin* (251), IChemE, October 2016, www.icheme.org/media/2080/lpb251_pg24.pdf

37 UK Government, www.gov.uk/guidance/environmental-impact-assessment

38 Fehling D, How Hurricanes that Hit the Texas Coast Can Float Giant Tanks, *State Impact Texas*, 7 October 2013, https://stateimpact.npr.org/texas/2013/10/07/how-hurricanes-that-hit-the-texas-coast-can-float-giant-tanks/

39 Organization for Economic Co-operation and Development (OECD), *Risks from Natural Hazards at Hazardous Installations (Natech)*, www.oecd.org/chemicalsafety/chemical-accidents/risks-from-natural-hazards-at-hazardous-installations.htm#:~:text=Natural%20hazards%20such%20as%20earthquakes,Natural%20Hazards%20Triggering%20Technological%20Accidents

40 International Atomic Energy Authority (IAEA), *The Fukushima Daiichi Accident*, IAEA, www-pub.iaea.org/mtcd/publications/pdf/pub1710-reportbythedg-web.pdf

41 US Chemical Safety and Hazard Investigation Board, *Extreme Weather, Extreme Consequences: CSB Investigation of the Arkema Crosby Facility and Hurricane Harvey*, www.csb.gov/assets/1/20/csb_arkema_exec_summary_08.pdf?16265

42 Anon, Fire and Explosion in an Air Separation Unit—A Chemical Detection Exercise, *Loss Prevention Bulletin* (166), IChemE, 2002, www.icheme.org/media/4899/lpb_issue166s8.pdf

5 Hazard Identification

5.1 INTRODUCTION

A well-designed plant with good process safety features will still be unsafe if operated and maintained in a dangerous fashion. It is wrong to think that a plant can be designed to be completely fail-safe. No single safety system can be guaranteed to never fail, not require maintenance and still operate perfectly in changed conditions. Therefore, *defence in depth* is necessary with a combination of safety systems which are independent of each other and which will not suffer from *common mode failures*. The combination of safety systems will comprise a mixture of engineered solutions together with good operational and maintenance procedures. In a perfect world all barriers will be independent, but there are often factors which defeats complete independence. These frequently include management factors such as style, culture, and motivation—which can affect several barriers. A lax approach to maintenance can affect several barriers simultaneously. The potential lack of independence is not always recognised in risk assessments. The barrier concept is shown graphically as the *Swiss cheese model* in Figure 5.1. Note that not all barriers are equally effective.

Hazard identification involves the identification of hazards using structured procedures with the findings recorded. It is a key element in process design and operation and is fundamental to the safe design and operation of any equipment, be it a process plant, a vehicle, aircraft or other engineered system. The tools used vary but are all rigorous, systematic and depend on the knowledge and expertise of the personnel involved. The following are the most commonly used hazard identification tools:

- Hazard studies
- HAZOP—hazard and operability study
- HAZID—hazard identification study
- Task analysis
- LOPA—layer of protection analysis (see Section 6.6)
- Check lists
- What if?
- FMEA/FMECA—failure mode and effect analysis/failure mode effect and criticality analysis

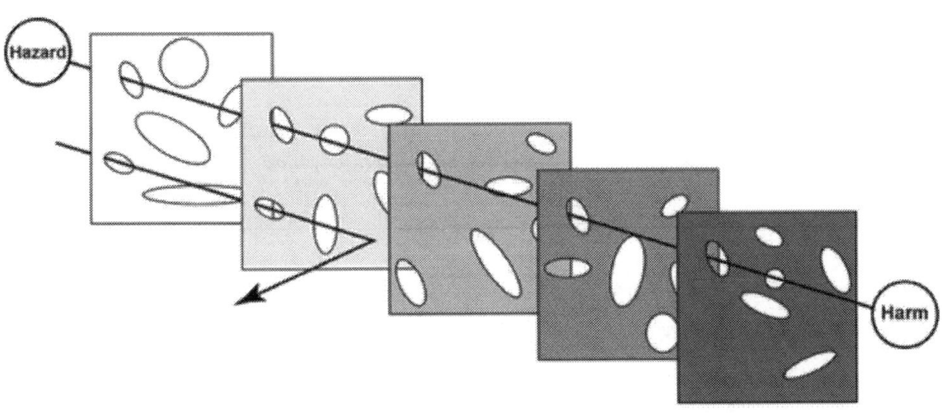

FIGURE 5.1 Swiss Cheese Model

DOI: 10.1201/9781003510680-6

A more comprehensive treatment of risk assessment methodologies is available from Crawley.[1] Note that some techniques such as LOPA (see Section 6.6) can be classified as hazard identification and risk assessment methods.

Audits can be performed at any stage of the plant life cycle. They are probably most useful during the operational phase for examining the drift of management systems from required standards. These are described in Section 21.7, and the more important hazard identification techniques are outlined in the following section.

5.2 HAZARD STUDIES OVER THE PROJECT LIFETIME

The progressive use of hazard identification techniques throughout a project lifetime was formalised into an interlocking series of hazard studies covering most stages of a project by Imperial Chemical Industries (ICI) in the 1970s. The ICI methodology covered phases 1 to 6 of the project lifecycle as defined in Section 1.4 and is used throughout this book with the addition of a phase 0 (inherent safety) and phase 7 (closure, mothballing and demolition safety). Whilst most projects use a classification of this type, there may be changes in the definition and numbering phases, for example, the number of the phases is 1–8 instead of 0–7. This is shown schematically in Figure 5.2 and discussed in detail in Chapters 11 to 17.

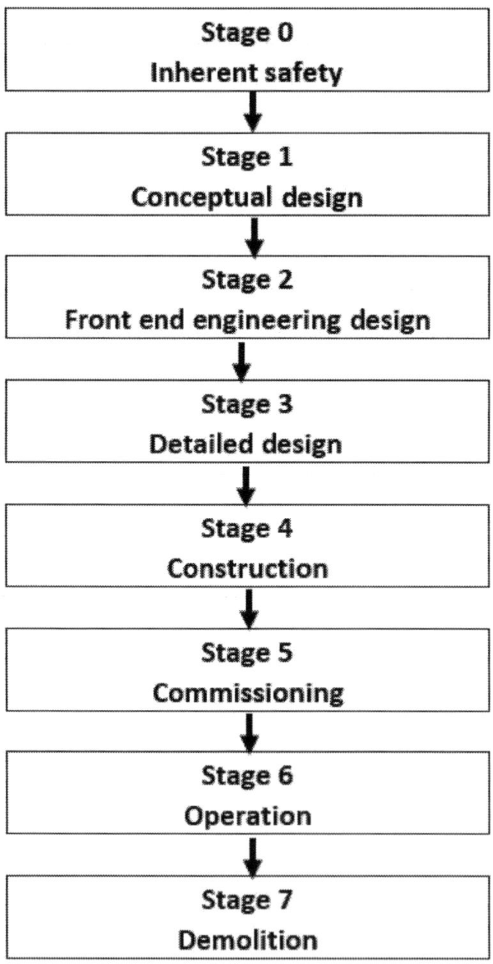

FIGURE 5.2 Relationship of Eight-Stage Process Study System for Project Life Cycle

Whilst the boundaries between the stages might change between organisations, the concept is widely used.[2]

There is often a perception that some of the more detailed studies (see Sections 5.2–5.7) are time-consuming and expensive. However, experience has repeatedly shown that the benefits of identifying problems "on the drawing board" are better than identifying them at a later stage.

The timing and objectives of studies at each stage are shown in Table 5.1.

TABLE 5.1
Project Phases

Phase 0 Inherent Safety	
Objective	To determine if there is a process route, chemistry or unit operations that is inherently safer than the alternatives
Start Point	As early as possible, often at the process development stage (sometimes at theoretical/laboratory stage). Sometimes the concepts are developed before a particular project is identified.
End Point	The identification, or not, of inherently safer options
Comment	Inherently (and environmentally) safe is a concept that must be reviewed in some detail, it requires "thinking outside the box" and is not easy without some depth of experience. Due to pressures on design teams, it is not one of the issues that receives a high priority – hence the necessity of reviewing inherent safety options as early in the project as possible. An inherently safer or "greener" process means a process route which has safety and environmental protection built into the design from the start. There are many ways in which an inherently safer process can be developed. Like many hazard identification techniques, a series of guidewords (such as *intensification, attenuation* and *substitution*) is used, designed to trigger ideas in the mind of the design team. The right decisions need to be made at this point or considerable reworking will be necessary later.
See Chapter	11
Phase 1 Conceptual Design	
Objective	To determine the SHE features which must be addressed during the development of the design and to determine if the concept is viable and the risks are ALARP.
Start Point	After stage 0 has been completed
End Point	The identification of the best process solution (which could be that the project is non-viable).
Comment	Hazard studies at this stage include: • Alternative processes • HAZID studies • Toxic data availability • Reactor kinetics particularly the exothermic properties of reactants and reactions • Availability of feedstock, the means by which it might reach the site and the "risk" to the public during the transfer: • Effluent disposal • Outline studies (hazard indices; environmental impact studies; equipment availability studies) • Reliability Studies on "Safety Critical Items" such as Shut down systems and gas detection systems: • Special materials of construction that might have a safety implication, for example, corrosion.
See Chapter	12

(*Continued*)

TABLE 5.1 *(Continued)*
Project Phases

Phase 2 Front End Engineering Design (FEED)	
Objective	To develop a detailed design, identifying solutions to design issues and, if appropriate, to perform initial risk assessments for the Safety Case.
Start	Once the project is identified as viable.
End Point	Practical and economic solutions are in place for safety issues. It could be that the problems cannot be solved and the project is abandoned or another process route chosen which would require a return to phase 1.
Comment	Basic studies undertaken at this stage are: • Reactor startup and an analysis of the process stability plus safety features • Shut down dynamics and possible impact on safety through the violation of the pressure-temperature envelope • Initial layout • Detailed risk assessments, including the integrity of protective systems (HAZID, HAZOP, What If?, and others as required.) • Product/feedstock movement and storage studies • Requirements for firefighting/protection • Requirements for environmental monitoring, locally or at a distance • Resolution of any problems from phase 1 • Safety case preparation (preliminary) • Environmental impact Assessment (preliminary) Much of the design will now be established and, if done correctly, there should be no major changes before commissioning. If anything significant has been missed, it should be identified in the phase 3 design and hazard studies and may involve potentially expensive design changes and project delay. Outline management systems and operating procedures will be considered at this stage. The project may be delayed or even cancelled at the end of this phase depending on costs and, possibly, difficulties in regulatory approval.
See Chapter	13
Phase 3 Detailed Design	
Objective	To develop the design in adequate detail for construction to take place.
Start Point	Project sanction
End Point	The design is complete and all hazard study recommendations implemented. The safety case (if required) will be developed for review by the regulatory authorities. The minimum of construction should take place until any necessary approvals are received.
Comments	Detailed design will include the following: • Detailed design and specification of equipment • HAZOPs and other hazard identification studies • Overpressure protection and relief and blow down Reviews • Hazardous area classification • Layout • Quantified Risk Assessment (QRA) • Detailed design of protective systems (active or passive) including QRA and LOPA • Process safety systems, including ESD • Fire and gas detection, plus active and passive fire protection systems • Outline operating procedures • Recommendations from Hazard studies • Resolution of any issues from phase 2 • Safety case • Environmental Impact Assessment (to cover both construction and operational phases)
See Chapter	14

(Continued)

TABLE 5.1 ***(Continued)***

Project Phases

Phase 4 Construction	
Objective	To ensure the plant is built as designed and that any changes are subject to appropriate hazard studies and documented.
Start Point	Mobilisation of the construction management team
End Point	The plant is handed over to the commissioning team
Comment	This is a team effort involving personnel from the design, construction, commissioning and operation/maintenance functions. At this stage, there will be frequent quality and safety inspections and audits to ensure that the plant is being built as designed and safe practices are used in construction. Any changes to the design will require review using a MOC procedure.
See Chapter	15
Phase 5 Commissioning	
Objective	To ensure all equipment/systems are correctly installed, procedures are in place and staff are trained to an adequate level of competence.
Start Point	Towards the end of construction.
End Point	On completion of the performance test, however production cannot start until local regulatory requirements have been satisfied.
Comment	Assessments and reviews are often in the form of checklists
See Section	15.7
Phase 6 Operation	
Objective	To identify *lessons learned* (both good and bad) during commissioning (and operation) and incorporate them into day to day activities at the plant
Start Point	Approximately a year after startup and then at regular intervals during the life of the plant.
End Point	Final plant shut down
Comment	There is often a requirement for a major reassessment of the plant and management at intervals (generally every five years). In some countries, this is a legal requirement. During the operational lifetime of the plant much of the equipment and systems will be affected by changes to plant, equipment and operations. These will require routine HAZOPs and other elements of the Hazard Studies
See Chapter	16
Phase 7 Closure, Mothballing and Demolition	
Objective	To identify the hazards that might occur during final shut down, mothballing and demolition of the plant. A preservation programme will be required for mothballing and a detailed plan for demolition. A *Demolition safety case* may be required.
Start Point	Immediately the decision to mothball or demolish the plant has been made.
End Point	Once the plant has restarted (in the case of mothballing) or the site has been restored to a preconstruction condition (in the case of demolition)
Comment	For mothballing, a detailed *preservation plan* will be necessary, including a continuing, but reduced, maintenance requirement. Ideally, the demolition process should be considered during plant design, but this is seldom the case. As in other cases, hazard studies will be necessary to determine: • Structural integrity • Safe methods for size reduction • Cleanliness • Identification of materials which can be recycled • Identification of hazardous material requiring specialised disposal • Verification for equipment due to the sold reduced (including records from the last shut down) • Sequence of demolition (it may not be the reverse of construction) • Disposal routes for steel, catalysts lagging and residual inventory
See Chapter	17

TABLE 5.2
Hazard Identification Method and Indicative Project Life Cycle Phase

Method Used	Project Life Cycle Stage, 0–7							
	0	1	2	3	4	5	6	7
Hazard Studies	X	X	X	X	X	X	X	X
HAZOP			(X)	X			X	
HAZID	X	X	X					X
Task Analysis	X	X	(X)	X			(X)	X
LOPA			X	X			(X)	
Checklists	X	X	X	X	X	X	X	X
What If?	X	X	X					X
FMECA		(X)	X	X			(X)	

The point during the project cycle where each technique is most useful is shown in Table 5.2. X shows that it might be appropriate to carry out that particular study at that phase, but the value of each study should be reviewed on a case-by-case basis. HAZOP at phase 2 might warrant a "Preliminary HAZOP" with a full HAZOP at phase 3. Task analysis in phase 2 might be used for the assessment of maintenance. In phase 6 task analysis should be routine when carrying out maintenance.

5.3 HAZARD AND OPERABILITY STUDIES

HAZOP was originally developed by ICI in the 1970s.[3] The study is a rigorous, systematic, structured technique for identifying potential safety and operating difficulties affecting process systems or equipment. A study should be carried out during the design phase of a project but is periodically repeated during the plant's lifetime. HAZOPs of plant sections are often undertaken as part of the MOC process (see Section 16.9). A well-conducted HAZOP can reduce the commissioning period and sometimes allow plant handover at an earlier date than anticipated by identifying and correcting issues which otherwise would only be found at a later stage. This will boost the economics of the project by establishing an earlier cash flow and preventing additional expenditure on necessary changes not identified at the design stage.

The study starts by splitting the plant into *nodes* and then defining each node in terms of parameters such as *flow, temperature, pressure* and *composition*. Guidewords such as *no, more, less,* and *other than* are then used to establish a deviation and cause. The cause of deviations and the consequence are then recorded for future action. A typical list of indicative parameters and deviations is given in Table 5.3. These may vary between processes and may require adapting for new technologies.

Each parameter is taken in turn and the cause/consequence of each guideword reviewed. For example, when examining FLOW in a pipe the consequences of NO flow (possibly caused by a blockage or closed valve) would be examined and the consequences examined. For NO FLOW in a cooling water line this could cause inadequate cooling in a heat exchanger. This, in turn, could affect the cooling of a reactor in an exothermic process leading to a reactor runaway (see Section 13.5). In addition, the study should look at unusual conditions such as maintenance, natural perils or terrorism. Some protocols have extensive lists of unusual conditions.

An example of specialised keywords for a crystalliser is shown in Table 5.4.

The table can now be put into a matrix which defines credible and incredible combinations of parameters and guidewords. Some keywords such as REVERSE FLOW may be specific to some startup conditions and so should have a "?" and be treated with care. Blind adherence to every parameter and guideword is time-wasting and potentially demoralising for the study team.[4]

TABLE 5.3
HAZOP Parameters and Guidewords

	Parameter			
Guideword	**Flow**	**Temperature**	**Pressure**	**Level**
MORE (M)	X	X	X	X
LESS (Less)	X	X	X	X
NO/NOT (N)	X	X	X	X
PART OF (Part)	X	–	X	X
REVERSE (Rev)	X	–	–	–
OTHER THAN (OT)	X	–	–	–

TABLE 5.4
Specialised Keywords for a Crystalliser

HAZOP Parameters	HAZOP guidewords
Crystal Size	Too big
	Too small
Crystal Density in ML	Too low
	Too high

The study is conducted by a multi-disciplinary team with representatives from the design team, plus operational and maintenance personnel. A facilitator (study leader) is required to spark ideas and control the flow of comments plus a secretary to record the study findings for future action. The HAZOP team must resist the temptation to "design out" problems identified in the study. The timing of the study is critical—too early and the design is still flexible, too late and there is insufficient time to make changes.

HAZOP studies can, sometimes, be degraded because of problems with interpersonal conflicts. Facilitators are sometimes considered as "policemen" intent on criticising the design and so the relationship between the facilitator and the rest of the HAZOP team can break down. An aggressive approach by the facilitator is contrary to the concept of facilitating. The aim of the HAZOP is to identify hazards, not to criticise or find someone to blame. The team must be confident that the facilitator knows how to perform the study. To avoid confusion, the facilitator is fully justified in asking the team to verify that his/her understanding of the issue is correct. Often this can be done by asking the same question in a different way.

Most new projects, and most significant modifications, require a HAZOP study but there is a temptation for these to be rushed or become a "rubber stamping" exercise. If this is the attitude of the project management, the value of the study will be compromised. Performing a HAZOP is mentally tiring, particularly for the study leader and secretary. Attempts to squeeze the study into a very short time frame will cause "HAZOP headache" and result in a poor-quality study. Human factors in hazard studies are discussed in Section 5.9.

Crawley provides a more detailed description,[5] and a simple example is given in Appendix A.

5.4 HAZARD IDENTIFICATION

HAZID[6] examines the impact of one system upon another. It was originally called *3D HAZOP* and is a systematic review of the possible causes and consequences of hazardous events. Whilst HAZOP is

TABLE 5.5
Simple HAZID Example

Effect	To/On	Cause
Collapse	Structure	Use of inappropriate design codes Poor structural design An earthquake measuring 7.8 on the Richter scale damaged parts of Turkey and Syria in February 2023.[7] The official death toll was over 9,000 with over 50,000 casualties. Other sources give higher figures. Thousands of buildings collapsed due to a lack of adherence to construction standards.
Collapse	Structure	Poor construction
Collapse	Structure	Dropped Object
Collapse	Structure	Soil Collapse
Collapse	Structure	Corrosion
Collapse	Structure	Overload—mass of equipment and process fluids, ice and wind
Collapse	Structure	Earth movement (earthquake, poor support or foundation collapse)

cause driven, HAZID is *consequence* driven. HAZID was initially based on a matrix which had three headings:

- Effect
- To/On
- Cause

Under each main head there will be a set of possible items referring to *Effect*, *To/on* and *Cause*. These can be looked at as a set of prompts that must be fully answered. In practical terms it is not unlike a set of building blocks. With the right team it has been possible to identify incidents with a return interval of 10,000 years. A simple example for the collapse of a structure is shown in Table 5.5.

HAZID is a particularly useful form of hazard identification for offshore projects and is often used in the earlier stages of project design, where likely hazards can be identified and removed or mitigated.

5.5 TASK ANALYSIS

Task analysis is a structured assessment of a task. It is done by breaking down the task into detailed steps, with each step being critically examined to identify what might go wrong and what can be done to prevent or mitigate the consequences.

It examines each step of a task and asks the following:

- What are the issues with that step?
- What is the order of each step (hierarchy)?
- What can go wrong during that step?
- What can prevent or mitigate the effect?
- Should the procedure be modified or abandoned?

Task analysis is particularly useful for reviewing operating procedures such as startup, shut down and maintenance activities.

5.6 CHECKLISTS

A checklist should be a carefully compiled, comprehensive list of design features, protective measures, procedural steps, material properties, hazards or good practices have been compiled by experienced personnel for a particular application. Crawley provides detailed guidance on checklists.[8] Checklists are used to systematically check designs, operations, system status, compliance with codes, standards and for other specific requirements.

An easily understood example is the pre-flight checks that take place before every aircraft takes off. Input to the checklist may come from a variety of sources, including experience (often obtained from external sources), task analysis, HAZOP or other sources.

Whilst checklists can be very useful, there is a danger that their completion becomes a "tick-box" exercise. It is highly desirable that there is an option beyond YES/NO answers to ensure that results are not "force-fitted" into a YES/NO box.

5.7 WHAT IF?

Crawley[9] What if? is a brainstorming technique in which a group of experienced engineers and technicians, familiar with the process being studied, review the potential for incidents (cause and magnitude). As with other studies, a study leader/facilitator is necessary to direct discussion, together with a secretary to record the discussion and conclusions. The technique is wide-ranging and can be unstructured, although the study leader may produce guidance based on a preliminary review of the process being studied. More structure can be introduced through the development and use of a checklist. What if? is one of the oldest hazard identification techniques. In some respects, it is a less structured form of HAZOP. Essentially, it involves looking at every system or item of equipment and system and asking, "What if this happens?" It is important to recognise that some equipment (and humans) may fail "safe" or to "danger". Some may "stick" in position. Examples are given in Table 5.6.

Clearly, "what if?" is less rigorous than a full HAZOP but may be justified for a preliminary or "ranging" study, particularly at the early stages of a project.

TABLE 5.6
Examples of What If? Cause/Consequence

Controller	1	What if it fails OPEN? There will be a higher than normal flow—(MORE FLOW in a HAZOP). What is the consequence? Is it acceptable and if not, what is the recommendation of the study?
	2	What if it fails CLOSED? There will be no flow—(NO FLOW in a HAZOP.) What is the consequence? Is it acceptable and, if not, what is the recommendation of the study?
Human Error	1	What if the operator fails to close a valve? The consequence will be case specific, so no action can be recommended in this example. Clearly there is a potential for a loss of containment and an action would be expected.
	2	What if the operator fails to respond to an alarm? The likely consequence is that the condition will drift into an unacceptable regime and, if installed, there will be an ESD (see Section 13.9). If it is not installed, this becomes a recommendation from the study.

5.8 FAILURE MODE AND EFFECT ANALYSIS/FAILURE MODE EFFECT AND CRITICALITY ANALYSIS

A FMEA considers, in a systematic and rigorous manner, all of the possible single failure modes of individual items and elements in a system. The effects of each failure on the item itself and on the system, as a whole, are identified and recorded. The analysis is usually performed by a small team who should have detailed knowledge of the system. In a FMECA the criticality of the consequences and likely frequency are also assessed and recorded. The failure modes associated with each item typically include the following:

- Fail to danger (revealed)
- Fail to danger (unrevealed)
- Fail-safe
- Unrevealed fault (but still functioning)
- Degraded (but still functioning)
- Partial failure

A trivial example would be a nut and bolt on a flanged connection, and the failure modes are shown in Table 5.7.

Typically, the tables of information are more comprehensive and contain more information and, where appropriate, criticality. Studies of this type can assist in the development of operational and maintenance procedures. For example, if corrosion is due to environmental conditions options might be limited, but if corrosion is due to other means, such as dripping from an overhead acid line, protection can be provided by either providing a corrosion-resistant shield or relocating the joint. Corrosion and bolt loosening can be monitored by inspection. Quality assurance checks should indicate if the correct materials are used and a bolt tightening procedure should prevent loosening.

5.9 HUMAN FACTORS IN HAZARD STUDIES

A good variety of hazard identification techniques are available, with different strengths and weaknesses. Unfortunately, many studies are still performed poorly, sometimes due to a lack of management interest or the philosophy that the studies are a contractual requirement and should be completed as quickly and cheaply as possible. Even when this is not the case, studies may be poorly conducted due to the interpersonal dynamics of the people involved or their lack of knowledge and

TABLE 5.7
FMEA for a Nut and Bolt

Scenario	Effect	Failure Mode
Over-tightening	Threads stripped	Fail to danger
Corrosion	Threads fail	Fail to danger (if the corrosion occurs under the nut, this may be unrevealed until failure occurs)
	Corrosion of flats making the nut or bolt impossible to turn.	Possible fail to danger, but the problem will be revealed.
Incorrect Material for Nut or Bolt	Galvanic corrosion	As earlier
Vibration	Gradual loosening of the nut/bolt	Fail to danger, possibly unrevealed
Impact Damage	Sudden failure	Fail to danger

experience. Prior consideration of human factors can make a significant difference to the quality of a study.

Many hazard identification techniques are based on collaborative methods between teams of people. The leadership of the team and interaction between team members play an important part in the success of the study. Generally, the team will consist of a leader (often known as a chairman or facilitator) plus a secretary to record the discussions and recommendations. Other team members will vary depending on the nature of the study and should as a minimum include both design/technical personnel plus operations and maintenance personnel. Failure to include staff from all aspects of the project will affect the integrity of the study. Ideally, the study team should include staff at all management levels. Some of the most useful contributions often come from plant operators and maintenance technicians.

Increasingly HAZOP/HAZID studies are carried out by facilitators (team leaders) who are contracted specifically for the study. Often, team leaders are chosen for their particular skills and may enter the study totally unprepared. The scope, terms of reference and ground rules should be agreed prior to the study and circulated to all participants.

The scope should cover the boundaries of the study, defining what is to be included and what is to be excluded. It should be the result of dialogue between the client and the facilitator and should define roles and responsibilities. The scope should also include reporting procedures and how recommendations should be ranked. It should also address the responsibilities for restudy, particularly following a HAZOP where there may be material changes in the drawings.

The scope, or *terms of reference*, should cover the delivery date, the scope of the study, methodology and resources. Facilitators are sometimes "parachuted in" with no opportunity to discuss the scope of the study. This is obviously undesirable.

Report Contents: Report formats differ widely; most will include an introduction (including a description of the process or equipment being studied), a description of the methodology, dates of the study, participants and recommendations. The worksheets are usually voluminous and form an appendix. It is important that a consistent form of worksheet is used throughout the study, as this will greatly assist understanding and auditing of the completed study.

Study Quality: The quality of a study depends on several factors. Most obviously, the time taken for the study. Many hazard identification methods are mentally exhausting and regular breaks are required. As a rule of thumb, at least an hour should be allowed for a "what if?" study of each process and instrumentation diagram (P&ID). As the study moves forward, the pace will pick up as, often, part of the next P&ID will have been covered by a previous one, and less time will be needed. For a HAZOP, with the exception of very simple drawings, no more than four P&IDs should be attempted during a day.

As in other situations, body language is unimportant. It is a vital clue to team member participation and eliciting information—a key requirement for any kind of collaborative study. These skills take time to develop. An experienced facilitator will have conducted previous studies and should steer the team and, importantly, not supply the answers—that is the function of the team. If answers come from the facilitator the team will become demotivated. The facilitator has probably seen the situation under discussion in a previous study and might want to "steer" the team, particularly if it is not reaching a conclusion. By saying "*Sorry, I did not quite catch the discussion, have you considered this scenario?*" or something similar, time can be saved. It is steering but speeds up the study.

Points to consider include the following:

- Are any of the team getting glassy eyes? This infers either fatigue or lack of interest! Give a direct challenge: "*Jim, what do you think of that?*"
- Is there a "silent member" in the team? The member may feel overawed or not sure of the "pecking order". Include that member, as earlier, with a gentle "*Jim, what do you think of that?*". Once included that member will become more participative.

- Similarly, the facilitator must prevent individuals from dominating the discussion. This is particularly important when a senior manager attempts to take control.
- A team member scratches the head, raises an eyelid or shakes the head. Clearly there is an unvoiced disagreement. Say gently: "*I noted your implied disagreement, would you like to expand on it*?". The facilitator is showing *interpersonal skills* and drawing that member into the discussion. It is possible that that member has a genuine point that has been missed by others. This will be appreciated and the member is more likely to play an active role in future.
- Each facilitator will have developed their own style. For example, sitting when he/she is happy that there is creative discussion, leaning forward to emphasise there has to be a conclusion to the discussion. "*Can someone summarise the issue in a sentence for the secretary*?".
- A good facilitator will be ever alert for subtle messages and for signs of fatigue, indicating that a break is necessary. Give a direct challenge.
  (a) If the former, "*What do you, Jim, think of that!*"
  (b) If the latter, a good facilitator will be ever alert to subtle messages and signs of fatigue, indicating that a break is necessary. *"Time for a break"*.
- Sometimes the facilitator will be aware that a team member is confused and cannot formulate a conclusion. The member may genuinely not understand the issue or may be "tongue-tied" but has a good point to make. There are many ways in which this can be resolved, one being, "*Jim, it looks like you have something to say. Can you put your concerns into a brief statement?*"

Similar interpersonal skills are needed when performing safety audits (see Section 21.7).

REFERENCES

1 Crawley F K, *A Guide to Hazard Identification*, Elsevier, 2020, https://shop.elsevier.com/books/a-guide-to-hazard-identification-methods/crawley/978-0-12-819543–7
2 IChemE Safety Centre, Guidance: Applying Process Safety During Concept Select Phase of a Project, *IChemE Safety Centre*, 2020, www.icheme.org/media/14917/mb-0024_20-applying-process-safety-during-concept-select-phase-of-a-project-guidance.pdf
3 Lawley H G, Size Up Plant Hazards This Way, *Hydrocarbon Processing*, 55, 1976
4 Crawley F K & Tyler B, *HAZOP Guide to Best Practice* (3rd edition), Elsevier, 2015
5 Crawley F K, *A Guide to Hazard Identification* (chapter 4—Hazard & Operability Studies), Elsevier, 2020, https://shop.elsevier.com/books/a-guide-to-hazard-identification-methods/crawley/978-0-12-819543-7
6 Crawley F K, *A Guide to Hazard Identification*, https://shop.elsevier.com/books/a-guide-to-hazard-identification-methods/crawley/978-0-12-819543-7
7 Johnson T, *Turkey Earthquake: Experts Believe Collapse of Buildings Was Preventable 15 February 2023*, www.newcivilengineer.com/latest/turkey-earthquake-experts-believe-collapse-of-buildings-was-preventable-15-02-2023/
8 Crawley F K, *A Guide to Hazard Identification* (chapter 10—Checklists), Elsevier, 2020, https://shop.elsevier.com/books/a-guide-to-hazard-identification-methods/crawley/978-0-12-819543-7
9 Crawley F K, *A Guide to Hazard Identification* (chapter 11—What If?), Elsevier, 2020, https://shop.elsevier.com/books/a-guide-to-hazard-identification-methods/crawley/978-0-12-819543-7

6 Risk Assessment

6.1 INTRODUCTION

Chapter 5 described techniques for the identification of hazards without attempting to estimate the frequency or magnitude of the hazard. In many cases, the extent is obvious—a large storage tank of methanol is obviously a far greater hazard than a 200-litre drum. Calculating the magnitude of a hazard is covered in Chapters 7–10. However, an estimate of frequency is necessary to ensure that the level of risk is acceptable and the resources available to spend on risk reduction are used to best effect. Measurement of risk is covered in this chapter. As in many areas, a standard has been developed to cover risk assessment techniques (IEC 31010).[1]

A "finger in the air" subjective approach to assessing risk was common in the early days of industry but is no longer acceptable. Risks must be assessed using robust tools. Risk assessment can be considered at three levels of detail. These are as follows:

1 Quantitative where the risk is given both magnitude and frequency using tools such as *fault tree* and *event tree* analysis and the assessment of effects such as gas release, fire and explosion. The error bands are often about half an order of magnitude.
2 Semi-quantitative, where the risk both magnitude and frequency are derived from predetermined data sets. Typically, layer of protection analysis (LOPA) and *ranking tools* are used. The error bands can be up to an order of magnitude.
3 Qualitative, where the risk is based on day-to-day experience and judgement. As few people will have experience of more than one major incident, smaller risks are likely to dominate this type of assessment. The error bands are relatively low as the risks are low by definition.

The general hierarchy of risk-reducing measures is as follows:

1 Prevention
2 Detection
3 Control
4 Mitigation
5 Emergency response

Quantitative risk assessment is the evaluation of the likelihood of undesired events and the likelihood of harm or damage being caused, together with value judgements made concerning the significance of the results. It is usually summarised by three questions:

1 How big? Magnitude
2 How often? Frequency
3 So what? Criteria

This chapter addresses the "how often" and "so what" questions. The "how big" (and consequences) are considered in Chapters 7, 8, 9 and 10.

How big? uses calculations and models to determine the magnitude of a hazard and the potential consequences such as fire, explosion and gas (or liquid) dispersion linked to the harm that can be done to plant/equipment, humans and the environment.

DOI: 10.1201/9781003510680-7

How often? uses databases or failure rates of equipment and systems and combines them in *fault trees, event trees* and *bowtie diagrams.*

So what? requires a value judgement and a set of criteria. Typical criteria are multi-faceted and would include, for example:

1 Life and limb—number and catastrophic/chronic
2 Loss of revenue
3 Cost of restoration (and 2 above)
4 Environmental damage
5 Cost of rectification
6 Public reaction (local and national)

Public reaction is the most difficult to judge but could be the worst and can be fatal to a company or even a government's survival! It should not be forgotten that the explosion at the Chernobyl nuclear power plant resulted in three immediate fatalities, 28 within the next fortnight due to high levels of nuclear radiation, and several hundred in subsequent years due to low-level nuclear radiation.[2]

The concept of QRA is often confused, especially by the public. QRA will identify worst-case events (consequences) and an estimate of "likelihood" (usually in the form of a frequency such as once in a million years). Such estimates are based on assumptions which may be inaccurate or commonly used without full justification. The QRA's main value is to compare the risk with criteria of *acceptability of risk* (see Section 1.5). This may be a company's internal criteria or a figure used by the regulatory authorities. Whilst calculations may show an event occurring only once in a million years that day might be tomorrow!

As in other areas there are commonly used terms with precise meanings; some have already been defined in Section 1.9. Others, commonly used in calculations, are defined in Table 6.1.

TABLE 6.1
Common Definitions

Term	Description	Example
As Low as Reasonably Practical (ALARP)	The philosophy used to reduce risk to an acceptable level	See Figure 1.2
Availability	The fraction of time that a piece of equipment or combination of pieces of equipment is available for use.	
Bow-Tie Diagram	A diagrammatic representation of the route to a defined incident and the consequences of the incident occurring	See Section 6.4
Diversity	The use of two or more different types of system to achieve a single purpose. The use of divergent systems will reduce the potential for Common Mode failure.	For example, the use of electrically and diesel-driven pumps to supply fire water
Event Tree	A diagrammatic representation of the individual events following a defined (often called *top*) event.	See Section 6.4
Fault Tree	A diagrammatic representation of the consequences of a defined event occurring.	See Section 6.4
Meantime between Failures (MTBF)	Average time before failure. Note that this is potentially confusing, as equipment is often removed for routine maintenance or due to indications from some monitoring techniques when it had not really failed.	
Meantime to Failure (MTTF)	Average to failure of a non-repairable item.	
Meantime to Repair (MTTR)	Self-explanatory but may include time to isolate, clean the equipment, repair it and test run it after repair.	
Redundancy	The performance of the same function by a number of identical but independent means.	
Reliability	The probability of a particular outcome occurring	An electric motor starting when switched on

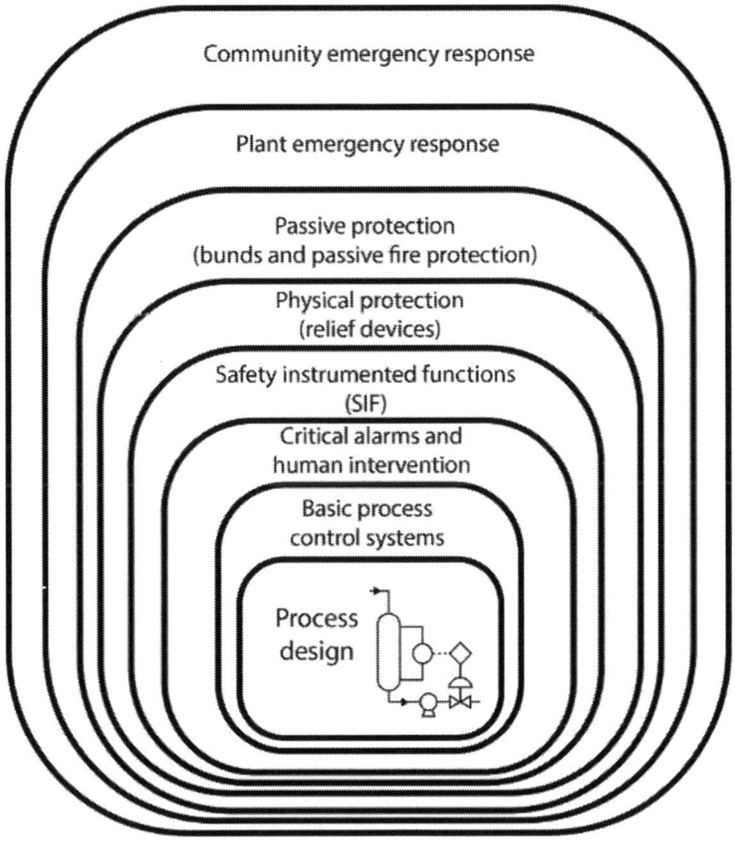

FIGURE 6.1 Standard LOPA "Onion" Model

Reducing frequency and severity of events usually involves a combination of methods. Two commonly used models to demonstrate this are the Swiss cheese model (Figure 5.1) and LOPA demonstrated by the "onion" (Figure 6.1).

Neither model is totally satisfactory. Swiss cheese is soft and malleable so does not immediately appear as a robust form of defence. The LOPA Onion often misses the contribution of *Safety Management Systems* which should fit in as *critical alarms and human intervention* or *site emergency procedures*. Safety management systems include training, *standard operating procedures* (SOPs) such as *permit to work* (PTW) and MOC. Safety management systems are covered in later chapters.

An alternative analogy might be that of *reactive armour*[3] used on modern armoured military vehicles. Provided there are no flaws in the armour, a projectile should be arrested. The armour will become damaged after repeated hits and eventually a projectile will penetrate it and destroy the vehicle. The same is often true with process plants where protective systems often require repair or recalibration after activation. Some components, such as bursting discs, require replacement after activation. Procedures can degrade if not checked and updated when necessary.

6.2 BASIC CONCEPTS

QRA attempts to define the magnitude of the hazard to people, property or the environment. Estimation is in terms of either frequency or probability. Calculation can be extremely complex and involves a great deal of analysis. However, the basics are simple. Probability is a concept which is widely known but often misapplied or misunderstood. An introductory treatment of risk assessment in the process industries is provided by Pitblado and Turney;[4] there are numerous other works which discuss probability in greater detail than required here. For this basic level of understanding, only simple probability will be discussed.

We all make crude but sufficiently accurate estimates of probability every day. If you cross the road what are the chances of getting run over? If you don't buy a sandwich now what are the chances of them being available later when you are hungry? We don't generally process this information in mathematical terms, but a gambler will often look for a mathematical answer: what are the chances of a head or tail in a coin toss or the ball landing on red in a roulette wheel?

Combinations of events require increasingly complicated formula, but for the present purposes, simply considering the addition or multiplication of two probabilities, Pa and Pb, will be adequate.

They can be multiplied together by the formula:

$$P(ab) = Pa \times Pb \tag{6.1}$$

Similarly, probabilities can be added using the formula:

$$P(a + b) = Pa + Pb - Pa \times Pb \tag{6.2}$$

Probabilities have no units but are often expressed as a percentage. A probability cannot exceed unity (when it becomes a certainty). The probability of an accident is often underestimated because there is a misconception that two (or more) events must happen simultaneously. This is not so; one event can be in place before the second occurs. A PSV might be in a failed state for a period of time before an overpressure event occurs.

The following is a simple example. A piece of equipment or system has two states: available (A) and unavailable (U):

$$A + U = 1 \tag{6.3}$$

Instrumented protective systems often have more than one channel, so that spurious failure of the channel does not shut the plant down. The summation of all states must equal unity. For 1 out of 2 systems, the states are given by the binomial theorem:

$$(A + U) \times (A + U) = A^2 + 2AU + U^2 = 1 \tag{6.4}$$

where A^2 = Both channels available
$2AU$ = 1 channel available, 1 channel unavailable
U^2 = Both channels unavailable

Likewise for a 2 out of 3 system = A

$$(A + U) \times (A + U) \times (A + U) = A^3 + 3A^2U + 3AU^2 + U^3 = 1 \tag{6.5}$$

where A^3 All 3 channels available
$3A^2U$ 2 channels available, 1 channel unavailable
$3AU^2$ 1 channel available, 2 channels unavailable
U^3 All 3 channels unavailable

This expression can be expanded further but within the process industries it is unusual to have instrument systems which are more than triplicated. In duplicated systems, only one channel has to operate for the instrument system to activate. For a triplicated system, a voting system with two out of the three systems operating is necessary for the system to activate.

If U = 0.1 (A= 0.9) per system, a 1 out of 2 (1-o-o-2) system has an unavailability of

$$U_{(1\text{-o-o-}2)} = 0.1^2 = 0.01$$

Similarly, availability can be calculated as

$$A_{(1\text{-o-o-}2)} = 2 \times 0.1 \times 0.9 + 0.9^2 = 0.18 + 0.81 = 0.99$$

For a 2 out of 3 (2-o-o-3) unavailability (only one channel available) of the system is

$$U_{(2\text{-o-o-}3)} = 0.1^3 + 3 \times 0.9 \times 0.1^2 = 0.001 + 0.027 = 0.028$$

Availably is calculated by

$$A_{(2\text{-o-o-}3)} = 0.9^3 + 3 \times 0.9^2 \times 0.1 = 0.729 + 0.243 = 0.972$$

The unavailability of a 2 out of 3 system is nearly three times that of a 1 out of 2 system. It is a common fallacy that a 2 out of 3 system is more reliable than a 1 out of 2 system. However, any failure of a 1 out of 2 system would result in a plant shut down and, to allow for the unreliability of instrumentation, the 2-o-o-3 configuration is the process industry standard.

Whilst both reliability and availability are based on probability, they generally have different uses. Reliability issues are most commonly encountered in instrumented safety systems, and availability is most commonly used to determine if equipment will operate for adequate periods of time without failure. For example, will a single pump be adequate to ensure production, or would two pumps (100% capacity each) or three pumps (60% capacity each) be better? Low availability of equipment will negatively impact plant economics, but installing unnecessary spare equipment will increase capital cost.

Most plants consist of several *unit operations* in series. Thus a failure of any unit operation will result in the plant being unavailable. As all unit operations have to be available, the final availability of the plant will be the product of the availabilities of each unit operation. It is easy to see that individual system availabilities may have to be produced with five or six significant figures! As the availability of any single unit may be of the order of 0.995, there is a systematic mathematical error if the values are rounded off at the third or fourth decimal place.

Before any form of risk assessment can be performed, it is necessary to identify the event itself by one or more of the techniques in Chapter 5. Having identified the event (sometimes called the *top event*), it is necessary to apply the three questions posed in Section 6.1. The routes to the top event will usually be well defined. However, sometimes the event is not predicted because there are poor control procedures in place, a lack of critical analysis or a lack of basic knowledge. A simple example is the frequency of significant fires during building refurbishment despite the hazards associated with refurbishment being well known.

In some cases, arbitrary levels of risk are set by, for example, the use of separation distances or maximum storage amounts of hazardous materials. In the case of the risk to life, they are bounded by ALARP and cost-benefit analysis (CBA).

Risk is often measured in frequencies and calculated accident rates are often expressed as 1×10^{-3}/year (once in a thousand years), 1×10^{-5}/year (once in a hundred thousand years), and so on. Calculation of frequency can be complex. It may be a simple matter of determining how many units fail per unit of time, but often *statistical distributions* are used to determine frequency. Failure rates are often expressed in the form of probability distributions, with the normal distribution being the most common. A detailed discussion of the various distributions is outside the scope of an introductory text but is covered in detail by Lees[5] and Green & Bourne.[6]

Probabilities, as shown earlier, can be readily multiplied together but frequencies cannot. The results of a probability or frequency calculation can only be approximate—except for very simple cases, there are likely to be undetected failure modes.

An allied but separate discipline to QRA is reliability engineering, and some of the concepts of QRA may be applicable in reliability analysis—which tends to be highly mathematical.

It is always useful to conduct a reality check, particularly for calculations which are not performed using standard computer models. A frequency cannot be squared or cubed; therefore, any results using these units must be fallacious. Over-detailed or precise analysis is likely to hide the "wood in the trees" and may contain significant errors. The universe is approximately 14 billion (1.4×10^{10}) years old and a frequency estimate of an event occurring once every 10^{-10} years would be suspect. Estimates of this—or greater frequency—are occasionally encountered.

QRA methods have sometimes been used to demonstrate that a risk is safe and no further action is necessary. This is an abuse of the methodology. Any risk must be reduced to ALARP using best available techniques (BAT). The best use of QRA is to determine if the level of safety meets accepted criteria, often used for land-use planning or to help choose between alternative options.

Risk assessments may be required for simple systems, such as instrumented safety systems, through to an assessment of the safety of a complete chemical plant or complex. For simple systems the calculation is relatively straightforward, but to determine the overall level of risk for an entire plant, the first stage is to identify the events which contribute to the overall hazard at the plant. These "top events" (because they are the top events in fault and event trees; see Section 6.4). After identifying the top events, the "how big?", "how often?", "so what?" questions need to be asked. The "how big?" and "so what?" are discussed in Chapters 7–10. The rest of this chapter is devoted to the "how often?".

Whilst manual calculation is possible, and at one time was the only option, most QRA is now performed using one of a number of widely available computer programmes. All but the largest companies will employ a consultant to do this work. For new projects, responsibility for the QRA may rest with the contractor. However, it is essential that the project manager or asset owner checks the accuracy of the calculations—this is a *duty of care*. Checks should include the basic assumptions used which may not be appropriate to a specific system (see Sections 6.3 and 6.8). Poor assumptions and methodology can lead to errors of one or two orders of magnitude.

6.3 DATA

For most QRA studies in the process industries standard scenarios and frequencies (such as leak sizes in pipework) are used. Data is available from a variety of sources, some freely available on the internet, others from commercial sources. A significant problem is ensuring data is accurate and applicable. Data collected, for example, on the reliability of valves in the water industry will not be useful for the calculation of valve failure in high temperature/pressure hydrocarbon duties.

Probability or frequency of failure depends on numerous factors, including environment (temperature, vibration and other factors), plus maintenance regime and age. The latter is of particular importance, with many equipment items having failure rates based on the *bathtub curve*, with failures being related to age and classified as infant mortality (sometimes known as *burn in*), random failures and old age failures (sometimes known as *burn out*). See Figure 6.2.

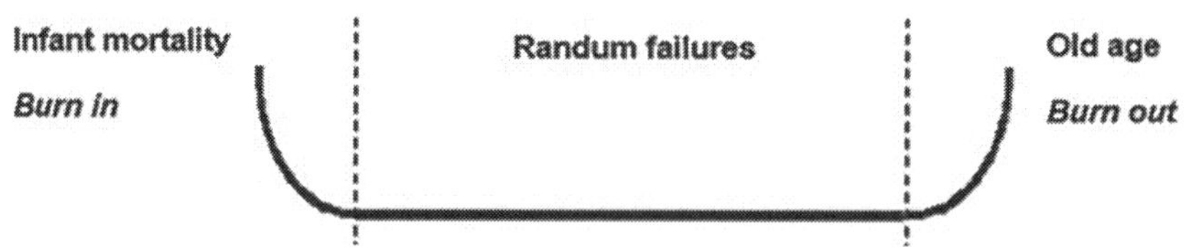

FIGURE 6.2 Bathtub Curve

6.4 TREES AND BOWTIES

With the exception of some incidents related to natural phenomena (now commonly known as NATECH), accidents are generally the result of a number of individual events coming together at one time. This is easily demonstrated by the concept of an *event tree* which shows diagrammatically the combination of events which need to be in place for the top event to occur.

Most trees are based on a combination of "AND" gates where more than one event or condition needs to occur, and "OR" gates where one of several events or conditions will be adequate for the top event to occur. The symbols for AND and OR gates are shown in Figure 6.3. A simple example is shown in Figure 6.4.

Table 6.2 demonstrates how probability is represented by AND and OR gates.

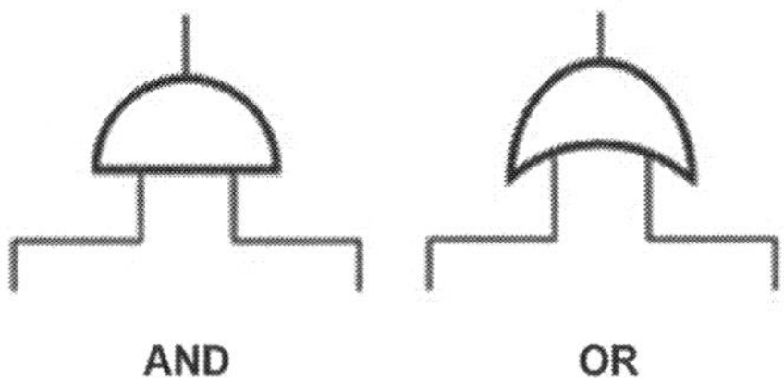

FIGURE 6.3 AND and OR Gate Symbols

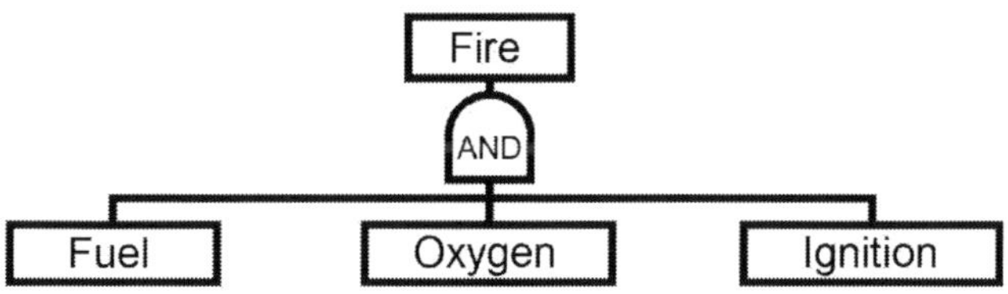

FIGURE 6.4 Simple Fault Tree for a Fire

TABLE 6.2
Use of Probability in AND and OR Gate Symbols.

Gate	Symbol	Meaning	Inputs	Outputs
AND	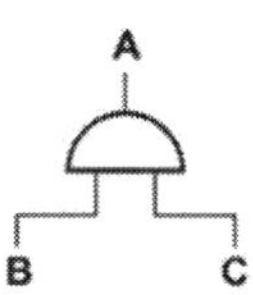	Output exists only if all inputs exist	P_B AND P_C F_B AND P_C F_B AND F_C	$P_A = P_B \times P_C$ $F_A = F_B \times P_C$ Not allowed
OR	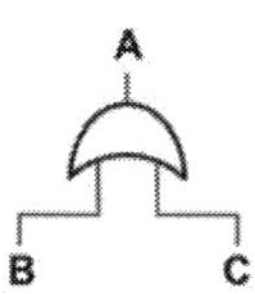	Output exists if any inputs exists	P_B AND P_C F_B AND P_C F_B AND F_C	$P_A = P_B + pc - Pb \times P_C$ Not allowed $F_A = F_B + F_C$

The fire triangle can be used to demonstrate the basis of an event tree. For a fire the following are required:

1 Fuel	Yes
2 Oxygen	Yes, but in the correct proportions with the fluid
3 Ignition	Yes, but at sufficient energy

Remove any element and a fire cannot occur. This can be demonstrated as a simple fault tree (see Figure 6.4).

The top event of a fault tree is the culmination of a number of individual events. An event tree starts with the event and looks at the possible consequences of the event occurring. A simple event tree, based on a minor explosion progressing to a major fire, is shown in Figure 6.5.

Trees can be presented either vertically with the top event at the apex of the tree or, for more complex trees, vertically with the top event to the right-hand side for a fault tree and to the left-hand side for an event tree. Note that very often there may be more than one way of constructing a tree, as shown in Figure 6.5.

Trees can be quantified by assigning probabilities to each event in the fault tree, but problems can arise in evaluating complex fault trees due to the same combination of events occurring in separate parts of the tree.

More recently Bowtie diagrams have been developed which combine a fault and event tree into one diagram, as shown in Figure 6.6, based on a simplified example of work on electrical equipment in a confined area.

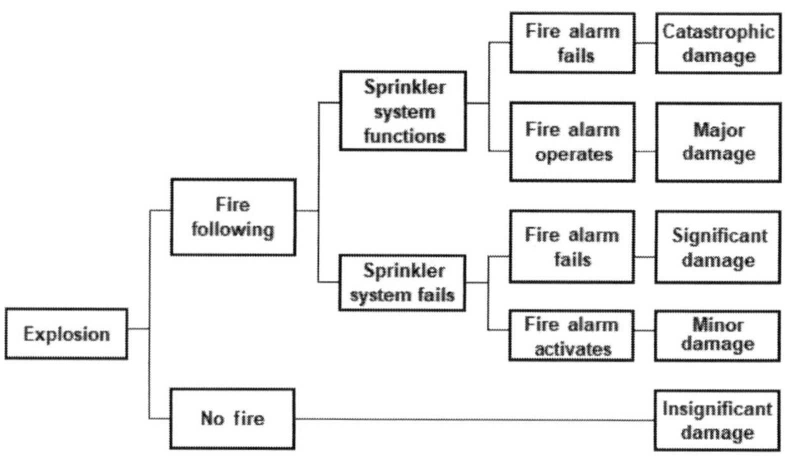

FIGURE 6.5 Simple Event Tree

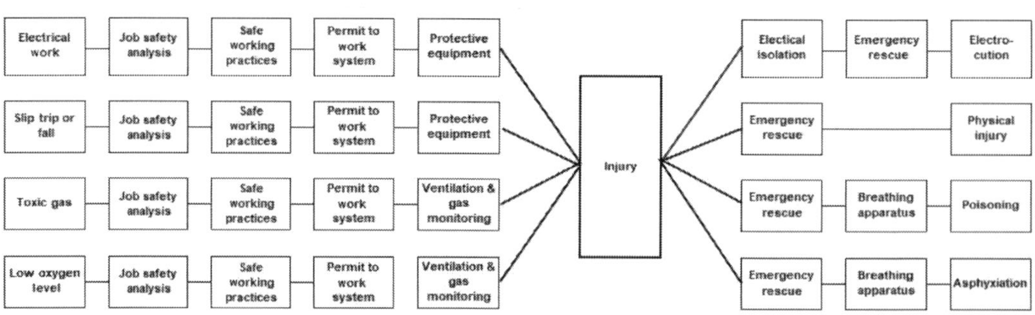

FIGURE 6.6 Bow-Tie Diagram

As with other quantification techniques, the output is only as good as the input. The value of fault trees, event trees and bow-tie diagrams lies in their ability to easily show the relationship between events. They can be a useful tool for accident investigation but are, perhaps, most valuable for reviewing the relative safety of different options. Bow-ties are particularly useful as the various events before and after the top event are shown in the same diagram. The barriers, if effective, will prevent the event occurring or being arrested before the worst outcome is reached.

6.4.1 Quantification of Fault/Event Trees

The quantification of a fault tree is relatively straightforward when dealing with probabilities, using the relationships shown in Section 6.2. However, calculations can go seriously awry if events are double-counted. The use of *cut sets* is a way to avoid double-counting and is illustrated with the simple example of a duplicated set of pumps taking suction from a tank (see Figure 6.7). The pumps have a *Non-Return Valve* (NRV) on the outlet (which is standard practice), isolation valves, normally installed, are omitted for clarity and the potential for maloperation has been ignored. The top event is no flow through to point X as shown in Figure 6.8. This configuration is common in fire water systems.

The tank must be "empty" for the system to fail. If the tank is full, as is credible, the system will work unless there are problems with the pumps and/or NRVs. Flow will stop if pumps A and B do not start and if pumps A and B start but their NRVs fail to open.

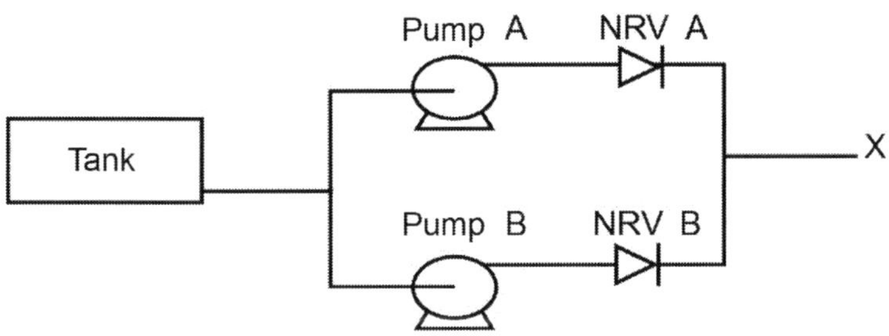

FIGURE 6.7 Duplicated Pumped Fire Water System

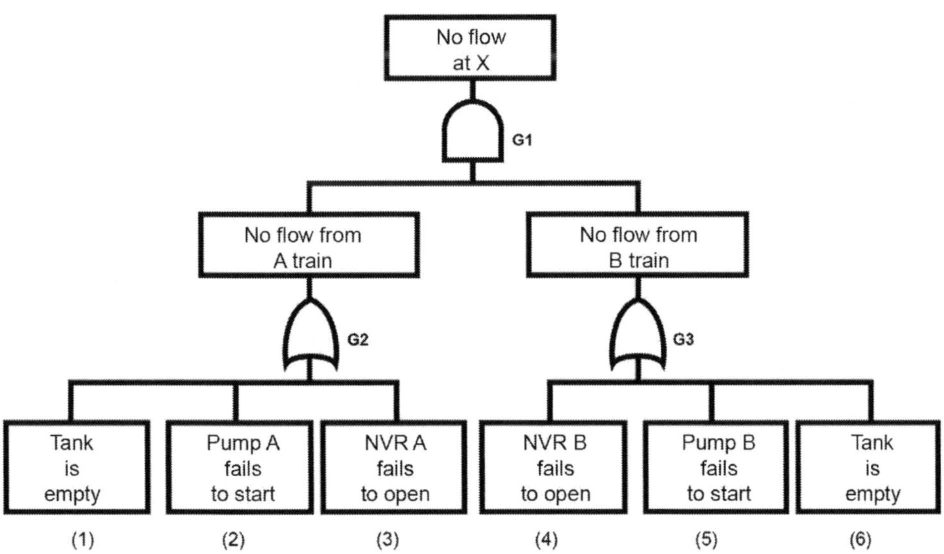

FIGURE 6.8 Fault Tree for Figure 6.7

The simplified fault tree that flows from this system is shown in Figure 6.8.

For the system to fail to work, Gate 2 and Gate 3 must have failed:

Gate 2 = Events 1 + 2 + 3
Gate 3 = Events 4 + 5 + 6

It is now possible to combine the failures of each gate in tabular form as cut sets by taking each entry in turn.

The tree can be evaluated here by expanding the tree in the form of a table. The top event is G1 and can be replaced by the events G2 and G3:

G2	G3

G2 and G3 can be replaced by a vertical row of the events feeding each OR gate, whilst AND gates move horizontally across the table.

1	4
1	5
1	6
2	4
2	5
2	6
3	4
3	5
3	6

However, 1 and 6 are the same event, so the table should be altered to

1	4
1	5
1	1
2	4
2	5
2	1
3	4
3	5
3	1

As the system will fail if the tank is empty the status of the pumps and NRVs is not relevant. As event 1 will, by itself, cause the system to fail, any cut set containing event 1 in combination with any other event(s) is duplication.

The only valid combinations are therefore:

1		Tank is empty
2	4	Pump A fails to start & NRV B does not open
2	5	Both pumps fail to start
3	4	NRV A fails to open & pump B fails to start
3	5	Both NRVs fail to open

Even this simple tree can cause confusion, and the construction of the tree should be considered with care. The fault tree can be redrawn as shown in Figure 6.9.

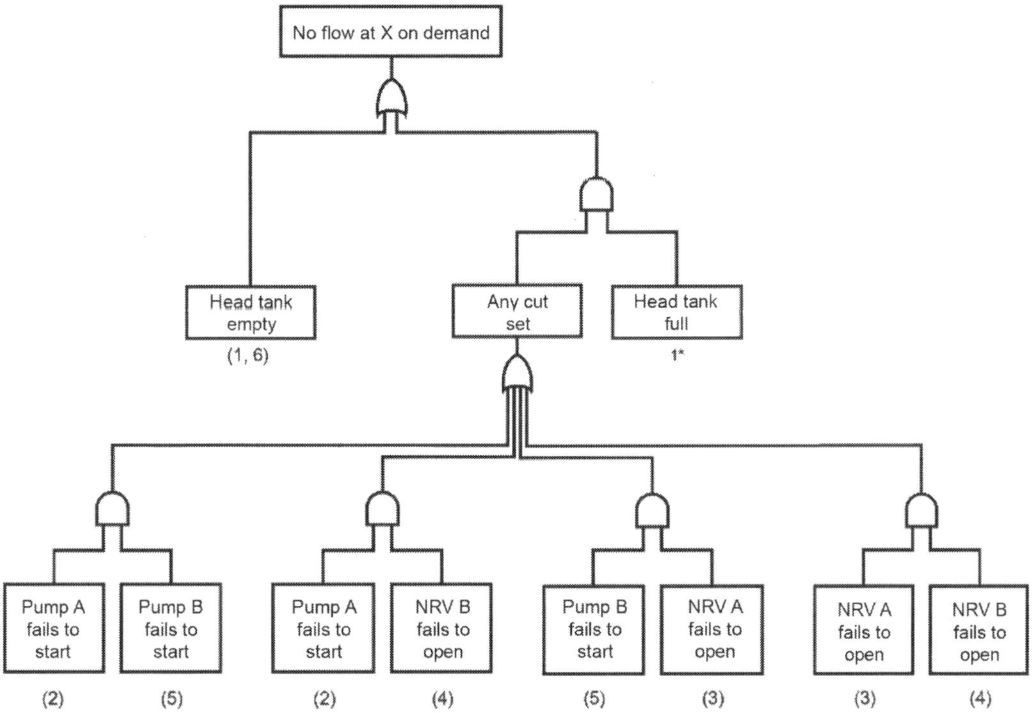

FIGURE 6.9 Simplified Version of Figure 6.8

The tree may vary if a mixture of probability and frequency data is used in the calculation. The tree as shown is correct if the failure of the tank is in terms of probability and the input of the other events in terms of frequency.

A common defect of trees is the exclusion of human failure. As stated in Section 6.3, data must be carefully evaluated. The example given earlier is based on NRVs failing closed. In reality it is much more likely that an NRV will fail open and blind use of failure data without identifying the failure mode can lead to significant errors.

Another common error in the use of trees, and more generally, is mistakes regarding the independence of systems. Failures of these types are defined by the AIChemE and are commonly classified as follows[7]:

Common Cause Failure (CCF) *Concurrent failures of different devices, resulting from a single event, where these failures are not consequences of each other.*

Common Mode Failure (CMF) *concurrent failures of different devices characterised by the same failure mode* (i.e. identical faults).

In the pump example mentioned earlier, if both pumps were supplied from the same power supply and this failed, it would be classified as a common cause failure, and for this reason, fire water systems generally have both electrically and diesel-driven pumps. If the NRV in both pumps failed due to the same design, manufacturing or installation fault, this would be a common mode failure.

The impact of common mode failures is shown in Figure 6.10 which illustrates a simple two-phase separator with level control and a separate low-level shut down. Liquid and gas are fed to the tank, gas discharges from the top and liquid from the bottom. This system is commonly found in oil fields. Level control is maintained by the level control valve (LCV). To prevent "blow through" of gas into the liquid system an independent emergency shut down valve (ESDV) is installed (labelled XCV). Note the arrows on the XCV and CV symbols show the action of the valve on air failure. By convention an arrow pointing towards the valve body shows the valve will fail closed and towards the actuator shows the valve will open.

A simplified fault tree for the system is shown in Figure 6.11.

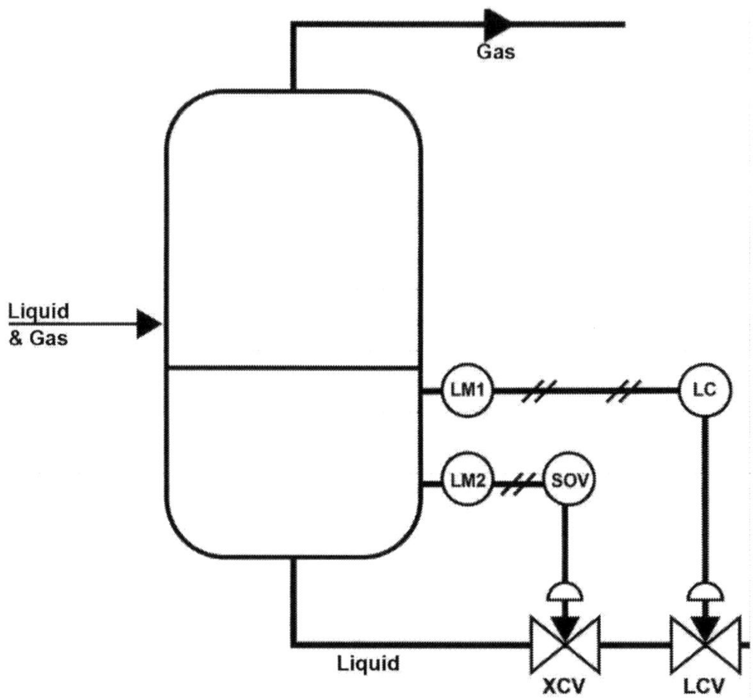

FIGURE 6.10 Simple Two-Phase Separator

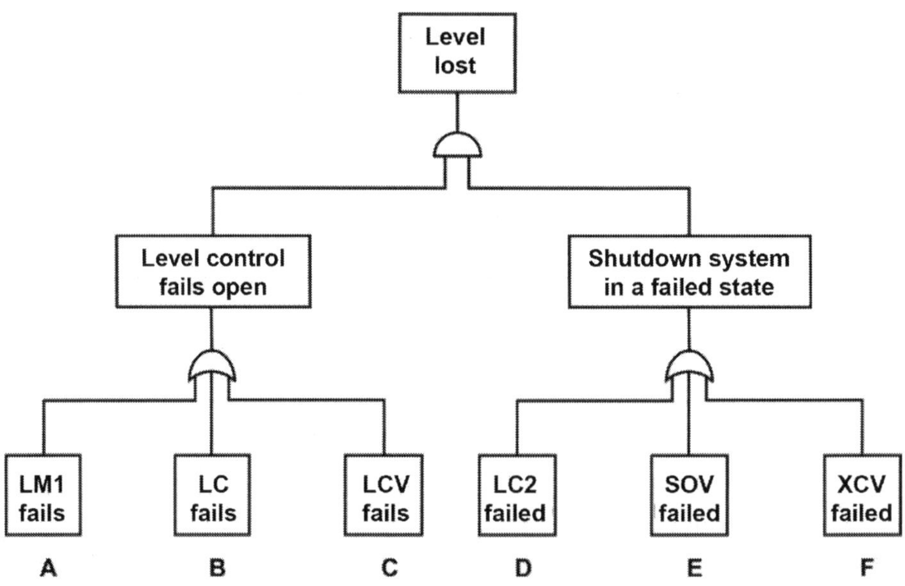

FIGURE 6.11 Simplified Fault Tree from Figure 6.10

However, there are common mode effects that would reflect the reliability of the system. These include blockage in the liquid discharge line and instrument lines which are shown in the revised fault tree, Figure 6.12.

Again, the fault tree can be rearranged as shown in Figure 6.13.

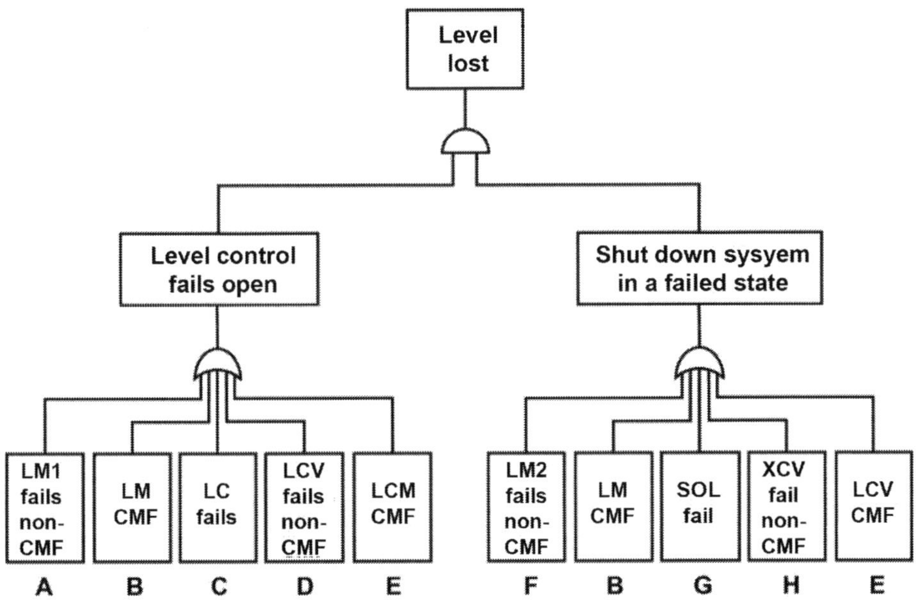

FIGURE 6.12 Expanded Fault Tree from Figure 6.11

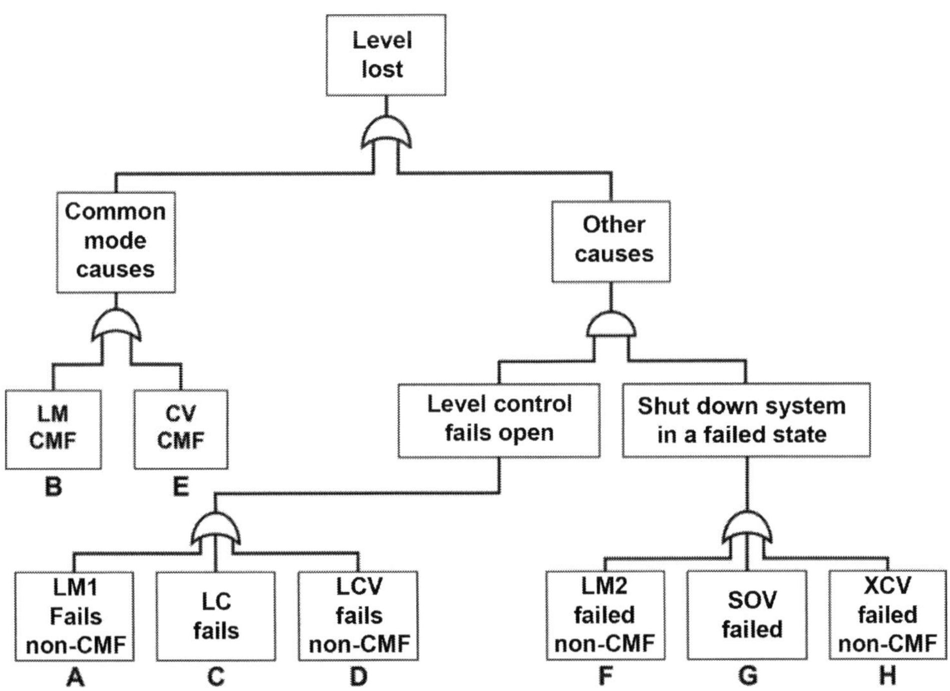

FIGURE 6.13 Rearrangement of Figure 6.12

The fault tree can be evaluated as shown earlier. In reality, many fault trees will be more complex than those shown here. A full assessment would identify the reasons for instrument and valve failures plus consider common cause/mode effects. Developing fault trees requires knowledge of the system, including its failure modes and less obvious effects. For example, solenoid valves in marine environments were historically likely to be less reliable than those in inland locations and would have a higher failure rate.

6.4.2 Converting a Frequency to a Probability

As stated earlier frequencies cannot be multiplied together in the same way as probabilities, and it is sometimes necessary to convert a failure rate into a probability to ensure the integrity of the units (probability or frequency) in the tree. This can be done by converting the frequency into the *probability of failure on demand* (PFD), sometimes called the fractional dead time (FDT).

A system (which will comprise a number of elements) can fail to respond to a demand for three basic reasons:

1 It is in a failed state.
2 It has been taken out of service for maintenance/inspection.
3 Human factors (such as not returning the equipment to service after maintenance).

The sum of all the failure rates of all the elements can usually be obtained from failure databases (information is generally available in the form of failures per year). However, some sources give a value for the total failure rate of the system. Some of the failures are *fail-safe* or *spurious*, which means that the system fails in a safe manner and shuts the process down. *Fail to danger*, F, is the other failure mode which is the parameter of interest when estimating the reliability of a system.

It is standard practice for critical systems to be checked periodically, often between one and four times per year (see Section 19.6). As demonstrated earlier, too frequent testing reduces the reliability of the system. Assuming an exponential distribution, the probability of the system failing on demand can be estimated from

$$\text{Probability of being in a failed state} = 1 - e^{-FT} \tag{6.6}$$

where F = Failure rate of system,/year
IT = Test interval,/year

The expansion of the exponential function is

$$1-e^{Ft}=1-\left[1+\frac{(-FT)}{1!}+\frac{(-FT)^2}{2!}+\frac{(-FT)^3}{3!}+\frac{(-FT)^4}{4!}\cdots\cdots\right] \tag{6.7}$$

Note the divisor is factorial n or n!, so the equation can also be considered as

$$1-e^{Ft}=1-\left[1+\frac{(-FT)}{1}+\frac{(-FT)^2}{2}+\frac{(-FT)^3}{6}+\frac{(-FT)^4}{24}\cdots\cdots\right] \tag{6.8}$$

Clearly, if FT is small (less than 0.1) the second and subsequent terms can be ignored and the failure rate reduces to

$$FT$$

Note that a safety system with low reliability will be of limited value and calculations which show a low value for Ft either have been incorrectly calculated or are in need of redesign.

FT is the value after T years but it is the average value that is of interest. The probability of failure at time t = 0 is obviously zero so a mean value between 0 and t is the *average* or a half of the larger value at time t. The probability that the system fails to danger is

$$\frac{1}{2} FT$$

The derivation of this equation has a number of assumptions open to debate: equipment is not experiencing burn-in/burn-out and the equipment is always correctly returned to service in an "as new" condition.

PFD and FDT have no units and therefore provide a simple and approximate method of converting a frequency into a probability.

As no method of testing is perfect and humans are error-prone, it is prudent to make some allowance for these factors. Assuming human failure as a probability of 0.005 and a test requiring the system to be out of service for 30 minutes (0.5 hours) per test, the FDT becomes

$$FDT = \frac{FT}{2} + 0.005 + \frac{0.5}{(8760T)} \tag{6.9}$$

6.5 RISK MATRICES

Risk matrices are a commonly used method of reviewing hazards and consequences. They range from simple matrices for routine activities to more comprehensive matrices covering an entire site or project. In the latter case, the risk is often considered from a number of viewpoints, such as threat to life, damage to property (and profitability), environmental impact and the reputation of the company. A typical example is shown in Table 6.3.

The increasing darker shading to the top and right shows levels of increasing risk. The level of risk is obtained by multiplying the hazard and frequency scores.

Whilst this is a simple method and has some value, it does have drawbacks, particularly with the assessment of values for the hazard and frequency values. These will be allocated based on the knowledge, experience, and judgement of the team (or individual) producing the matrix. The values chosen by different individuals are likely to vary significantly and are sometimes the subject of controversy. The problem can be reduced by assigning descriptions which are not numerically based. For example, "has never happened in the industry" rather than a frequency of 10^{-6} per year. Matrices are not an alternative to QRA but are used, in a simplified form, to assess low-level risk or as a screening tool to identify high-level risks which may require further assessment irreparable.

6.6 LAYERS OF PROTECTION ANALYSIS

Layers of protection analysis[8] is an analytical procedure that looks at the safeguards present to determine if the protection provided is adequate for every identified hazard. It shows whether additional controls or shut down systems are required by comparing the risk with, and without, these additional elements against a set of predetermined risk criteria. LOPA started life as an order-of-magnitude technique but has developed by using factors such as failure frequencies and probabilities of failure on demand which have previously been recorded or estimated using well-proven methods. LOPA is particularly relevant for human actions (which frequent create hazards) and for the assessment of alarm and shut down systems—which are often a layer of protection. Ideally, LOPA studies should be conducted before a HAZOP takes place.

TABLE 6.3
Typical Risk Matrix

	Impact							Frequency				
Consequence	**Property US$ million**	**Downtime/ Delay Months**	**Product Quality**	**Safety**	**Environment**	**Reputation**		**Very Low (1)**	**Low (2)**	**Medium (3)**	**High (4)**	**Very High (5)**
Catastrophic (5)	>50	>6	Not achieved	Single/multiple fatalities	Massive effect	International media coverage/irreparable. shareholder impact		**5**	**10**	**15**	**20**	**25**
High (4)	20–50	3–6	Substantial effect on objectives	Serious injury resulting in permanent disability	Major effect	National media coverage, substantial stakeholder impact		**4**	**8**	**12**	**16**	**20**
Medium (3)	5–20	1–3	No tolerance for deviations	Serious injury but not resulting in permanent disability	Significant effect	Regional media coverage, moderate stakeholder impact		**3**	**6**	**9**	**12**	**15**
Low (2)	0.5–5	0.5–1	Limited decrease in system performance	Medical treatment of personnel, lost-time incident	Minor effect	Local media attention, minor shareholder impact		**2**	**4**	**6**	**8**	**10**
Minor (1)	< 0.5	< 0.5	Minor degradation of performance	First aid injury	Negligible effect	Negligible media attention or stakeholder impact		**1**	**2**	**3**	**4**	**5**
Key	Unacceptable		Tolerable			Acceptable						
Note 1	Downtime/delay may also be measured in financial loss											

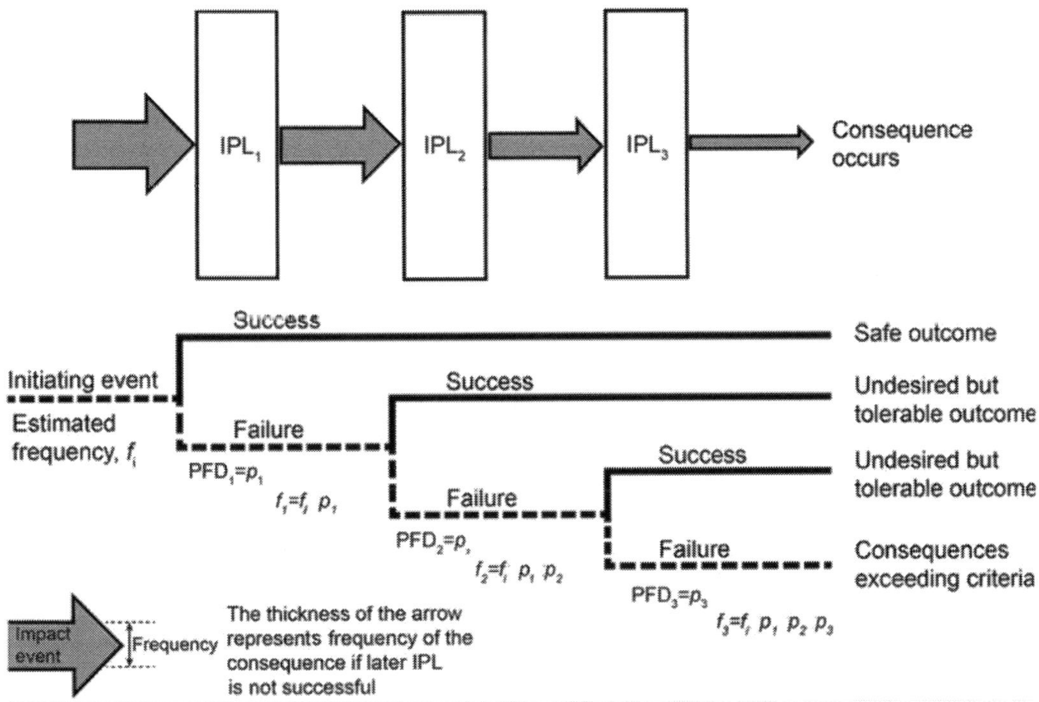

FIGURE 6.14 Example Event Tree

Failure rates are codified in *look-up tables* for equipment items such as control systems, piping, pump seals and other equipment. The look-up tables may be generic and should be used with caution or replaced with more relevant data. The failure rates are then moderated by a series of *independent protective layers* (IPL) in an event tree format using generic probabilities, again from look-up tables. The essential feature of IPLs is that they *must* be totally independent of each other and not subject to common cause or common mode failures (see Section 6.4).

The concept is shown graphically in Figure 6.14 with the inputs from three IPLs in the event outcome tree.

The frequency of f_3 ($f_1 \times p_1 \times p_2 \times p_3$) is then compared to the criteria for acceptability of risk and if the result fails to meet the required criteria, extra layers of protection must be added or a more sophisticated *safety instrumented system* (SIS) installed to meet the required *safety integrity level* (SIL).

SIL 1 is a simplex system with a required *probability of failure on demand* (PFD) of 0.1.
SIL 2 is a one-out-of-two voting system with a required PFD of 0.01.
SIL 3 is a more complex voting system with a required PFD of 0.001.

As the complexity of the systems increases the outcome from the LOPA analysis becomes less accurate and eventually if SIL 2 or SIL 3 are required to meet the risk criteria it may be necessary to resort to a full QRA.

A simple example is shown based on Figure 6.10 for a gas/liquid separator working at 5 barg. Separated water is discharged to a drain system. Whilst loss of interface control is acceptable, discharge of oil into the drain is not. A basic process control system (BPCS) is installed with an additional valve which will close if the interface falls to a critical low level.

The assessment will need to cover

1 human injury
2 loss of production
3 cost of repairs
4 potential environmental impact

For illustrative purposes the following figures have been used for calculation:

1	Initiating event—failure of the interface controller (BPCS)	EV	0.1 per annum
2	Failure of the low-level cut-off	IPL_1	0.05 per annum
3	Human intervention to prevent release and be present when the event occurs	IPL_2	0.2 per annum
4	Ignition frequency (variable and dependent on the outflow and flash potential)	IPL_3	0.1 per annum
5	Failure of fire protection (This will be dependent on how long the fire burns and the efficiency of both the fire protection and the speed of operator intervention)	IPL_4	0.3 per annum

For illustrative purposes, the cost of the event is

6	Repair costs:	£25,000
7	Loss of production:	£10,000
8	Environmental damage:	£1,000

The probability of serious injury (PSI) if an operator is present when ignition takes place and trying to take corrective action is assumed to be 0.5.

The results are shown in Table 6.4.

Based on the calculations given earlier, the consequential costs of the incident occurring are likely to be considered of minor financial significance. However, a serious injury rate of 1.5×10^{-5} may not be acceptable.

TABLE 6.4
LOPA Results

Human Serious Injury (*Caught in a Fire*)	$= EV \times IPL_1 \times IPL_2 \times IPL_3 \times IPL_4 \times PSI$ $= 0.1 \times 0.05 \times 0.2 \times 0.1 \times 0.3 \times 0.5$ $= 1.5 \times 10^{-5}$
Cost of Repairs (*Due to Equipment Damage*)	= EV × IPL1 × IPL2 × IPL3 × IPL4 × £250,000 = 0.1 × 0.05 × 0.2 × 0.1 × 0.3 × £250,000 = £7.5 per annum
Loss of Production (*Due to Equipment Damage*)	= EV × IPL1 × IPL2 × IPL3 × IPL4 × £100,000 = 0.1 × 0.05 × 0.2 × 0.1 × 0.3 × £100,000 = £3 per annum
Environmental Damage	= As above £0.3 per annum on average
Total	£10.8 per annum on average

TABLE 6.5
Suggested Proforma for LOPA

Scenario	Comments
Frequency of Unmitigated Consequences	
Independent Protective Layers (with Descriptions)	
SIL Required	
BPCS Actions with Trip	
BPCS Alarm and Operator Response Independent of Trip (IPL_1)	
Other Safety-Related Protection Systems	
IPL_2	
IPL_3	
IPL_4 and so on	
Pressure Relief Device—Overpressure Events	
Total Probability of Failure on Demand for All IPLs	
Final Frequency of Consequences with All Mitigations in Place	

Actions to lower the incident rate could include the following:

1. Install a remote/distant manual shut-off valve away from the site on the drain line.
2. Install a second low-level shut-off valve.
3. Install a low-low interface level alarm.

Option 1 would be acceptable on a cost-benefit basis but option 3 would be equally acceptable provided that there is a manual isolation on the drain line.

Generally, a company will have a set of risk criteria so the financial risk and frequency can be assessed on a systematic basis. Increasingly reputational risk is a factor in the risk assessment and even where the financial costs are limited, the potential damage to a company's reputation may override financial concerns.

The results would be fitted into a proforma in the format shown in Table 6.5.

6.7 COMPUTERISED METHODS

Many QRAs are now conducted by computer with risk contours calculated based on assumptions of leak rates, probability of ignition and meteorological conditions. Again, it is important to check the accuracy of assumptions and avoid a situation where the computer-generated results are considered accurate without critical review.

6.8 VERIFICATION OF RESULTS

Where a consultant or a contractor is employed to perform a QRA some form of review should be conducted to ensure an adequate level of quality is achieved and the study is fit for purpose rather than being a contractual obligation which must be completed. Any review of a study must be conducted by an independent individual or team who can challenge the results from a position of knowledge and experience. Incorrect assumptions will result in incorrect assessments—particularly if basic assumptions are incorrect—which is not unknown. Table 6.6 suggests a list of topics to be reviewed to determine the comprehensiveness and accuracy of the study. The table is not comprehensive and should be reviewed for the type of facility that the risk assessment is appraising.

TABLE 6.6
Checklist to Assess Credibility Hazard Identification Methods

Can the materials of construction handle the extremes of process temperatures (high and low)?
Are the materials of construction suitable for any potential corrosive or erosive conditions?
Are jointing types and gasket materials suitable for the fluids handled?
Is pipework welded wherever possible, with the number of flanged joints minimised?
Is there equipment which is frequently opened to atmosphere (such as strainers, filters and pig traps)?
Are the isolation standards appropriate to the pressure and nature of the fluids handled?
Has consideration been given to changes in flow-rates, temperatures, pressures and vessel inventories during upset conditions? Are pressure relief systems sized for these conditions?
Is there adequate information available to determine reaction kinetics and size the pressure relief system?
Have assumptions been made for failure rates from pump seals based on seal types (for example, single or double seals and canned pumps)?
What criteria have been used for the design of the ESD systems? What level of testing and repair is possible whilst the plant is running?
Has the potential for common mode and common cause failure been assessed?
What assumptions about "human reliability" have been made (both the likelihood of errors and the ability of humans to survive in extreme conditions of thermal radiation and/or toxic gas concentrations)?
What assumptions about maintenance and inspection procedures have been made?
What assumptions have been made regarding the restart of the plant after maintenance or an ESD?
Are the calculations/units sensible? (probability without units, frequencies not multiplied together)

Chapters 7–10 give guidance on the methodology for estimating the magnitude of events. The theory is deliberately simplified as, in most cases, the modelling is carried out on a computer, where the models are more complex.

REFERENCES

1 International Electrotechnical Commission (IEC), *Risk Management—Risk Assessment Techniques*, IEC31010, ISO, 2019, www.iso.org/standard/72140.html
2 Ritchie H, What Was the Death Toll from Chernobyl and Fukushima, *Our World in Data*, https://ourworldindata.org/what-was-the-death-toll-from-chernobyl-and-fukushima
3 Dop S, Explained; How Explosive Reactive Armour Works, *DefenceXP*, 25 February 2021, www.defencexp.com/explained-how-explosive-reactive-armour-works/
4 Pitblado R & Turney E, *Risk Assessment in the Process Industries*, ICHEME, 1996
5 Lees F, *Loss Prevention in the Process Industries: Hazard Identification, Assessment and Control* (4th edition), ed. S Mannan, Butterworth-Heinemann, 2012
6 Green A E & Bourne A J, *Reliability Technology*, Wiley, 1977 (ISBN 0 471 32480 9)
7 Centre for Chemical Process Safety (CCPS), www.aiche.org/ccps/resources/glossary?title=common#views-exposed-form-glossary-page
8 Crawley F K, *A Guide to Hazard Identification* (chapter 7—Layers of Protection Analysis), Elsevier, 2020, https://shop.elsevier.com/books/a-guide-to-hazard-identification-methods/crawley/978-0-12-819543-7

Part B

Consequence Modelling

7 Source Terms

7.1 INTRODUCTION

The potential hazard is linked to *source terms*—basically the mass of hazardous material which can be released. Whilst a sudden vessel rupture is possible, in most cases a release will be through a hole in a vessel or pipework. The size of the orifice leak for risk assessment purposes can be obtained from various databases, which give size and frequency for leaks. The data is usually generic and not case-specific so must be used with care. For example, the leak rate for a corrosive system is likely to be higher than that for a benign system. Thought and judgement will be required. Essentially the source term answers the question "How much of what and for how long?". Generally, small leaks will last a long time before detection, but large leaks should be detected quickly.

7.2 GAS OUTFLOW

At high pressures over 200 kPa the discharge velocity will be sonic, causing choked flow. The standard gas outflow model is

$$Q = C_d A P_u \sqrt{\frac{M\gamma}{RT_u}\left(\frac{2}{\gamma+1}\right)^{\frac{\gamma+1}{\gamma-1}}} \tag{7.1}$$

where Q = Outflow (kg/s)
C_d = Coefficient of discharge
P_u = Upstream pressure (Pascals)
A = Orifice area (m^2)
M = Molecular weight
T_u = Upstream temperature (degrees Kelvin)
R = Universal gas constant—8,314 (Joules/Mole/°K)
γ = Ratio-specific heat of the gas at constant pressure and volume (C_p/C_v)

At high pressures (greater than 1 MPa) the equation still holds, but adiabatic expansion of the gas reduces the temperature (increasing the density) and hence the exit velocity decreases. As a result, jet mixing velocity is underestimated, in most cases this is not significant. At upstream pressures less than about 190 KPa, Equation 7.1 no longer holds as the gas velocity is subsonic, and the standard equation is now:

$$Q_{LP} = D_d A P_u \sqrt{\frac{M\gamma}{RT_u}\left(\frac{\gamma}{\gamma-1}\right)\left[\left(\frac{P_u}{P_o}\right)^{\frac{2}{\gamma}}\left(\frac{P_u}{P_o}\right)^{\frac{\gamma+1}{\gamma-1}}\right]} \tag{7.2}$$

where P_u = Upstream pressure (Pa)
P_o = Atmospheric pressure (Pa)
Q_{LP} = Outflow (kg/sec) (low pressure)

Other terms as earlier.

DOI: 10.1201/9781003510680-9

7.3 LIQUIDS (NONE FLASHING)

The outflow equation for standard incompressible fluid flow is

$$Q = C_d A\sqrt{2\rho\left(P_u - P_0\right)} \tag{7.3}$$

where Q = Outflow (kg/s)
: C_d = Coefficient of discharge
A = Orifice area (m^2)
ρ = Density (kg/m^3)
P_u = Upstream pressure (Pa)
P_o = Downstream pressure (Pa)

Equation 7.3 assumes negligible liquid head at the point of release. Where there is a significant head of liquid, the pressure equivalent to the head must be added to the upstream pressure. The head (and hence the flowrate) will be reduced as liquid is discharged.

The flow from long pipes/pipelines requires an adjustment for friction losses.[1] This may be important for high-/low-pressure interfaces such as local high-pressure vessels feeding into low-pressure receivers. The friction losses may be very significant and reduce over-pressure in the low-pressure receiver. However, this should not be considered a mitigating measure.

7.4 FLASHING LIQUIDS AT THE ORIFICE (INCLUDING LPG, CHLORINE AND OTHER VOLATILE LIQUIDS)

The outflow equation has to be modified to somewhere between a gas and an incompressible flow equation. This is, again, a generalised equation:

$$Q = C_d A\sqrt{2\rho_c\left(P_u - P_c\right)} \tag{7.4}$$

where ρ_c = Density of fluids at 0.55 P_u (kg/m^3)
P_c = 0.55 P_u

The *flash fraction* (proportion of fluid that vaporises) is discussed in Section 7.4 and can also be determined from thermodynamic data such as enthalpy/entropy diagrams and computer simulations. An approximation of the density, $\rho_{c,}$ can then be determined from physical property data for the liquid and vapour phases multiplied by the proportion of liquid and vapour present at the point of flashing. However, once discharged from the orifice the fluids will continue to flash to atmospheric pressure. Conventionally, when flashing cryogenics into the atmosphere, it has been considered that the mass in the cloud is twice the final flash to make allowances for aerosol formation. Total vapourisation is more likely when the fluids are superheated by more than around 50°C (relative to ambient temperature) at source.

The best method of calculating outflow from flashing flow is by the *homogenous equilibrium method* (HEM) which is complex and requires a detailed knowledge of the physical properties and orifice conditions. The increased accuracy with the uncertainties in any assessment does not justify the use of HEM in an introductory text.

Experimental results suggest that if the effects of flashing are unclear, the following "rules of thumb" are helpful.

Flashing Flow = 0.25 Liquid alone flow
Gas Flow = 0.25 Flashing flow
Flashing Flow = (Gas flow through the orifice × liquid flow through the orifice)$^{1/2}$

There is strong experimental evidence that flashing fluids with a high-pressure difference (over five times atmospheric pressure) and a flash fraction of over 20% do not fall to the ground but become an aerosol due to shearing forces. Visually monitored releases showed little evidence of "rainout" from the spray.

There is also evidence[2] that for gases above their saturation conditions the drop in pressure results in a "moving choke of liquid/gas" such that the pressure at the far end of a long pipeline of over 50 km may not fall for some minutes following a line rupture. This is caused by flashing flows creating a compression wave velocity significantly less than that of the gas; this is complicated by the high pressure drop created by two-phase flow. The compression wave velocity may fall to 100–150 m/s. More significantly the wall temperature could fall below the brittle fracture temperature.

In ruptured piping flashing may take place inside the pipe, upstream of the actual rupture. As a result, there is the complication of a two-phase flow pressure drop which again arrests the flowrate. In simple terms the outflow from a long piece of piping may be significantly less than for the outflow from the equivalent hole in a vessel (see Section 7.3). The flowrate in this situation can only be assessed with sophisticated computer software.

7.5 COEFFICIENT OF DISCHARGE

The highest value for the coefficient of discharge is almost 1 for a well-rounded nozzle entry. It is also the "worst case" discharge for risk assessment. The value for the coefficient of discharge for a sharp-edged orifice can fall to 0.61. If in doubt use the geometric mean for the two values for the coefficient of discharge:

$$(1 \times 0.61)^{1/2} \tag{7.5}$$

This approximates to 0.8, a value used for outflow from a loss of containment. It is recognised that this is also the arithmetic mean, but in the event of uncertainty it is better to use the geometric mean where the error is minimised.

7.6 FLASH FRACTION

If all data is known the physical properties tables should be used; however, the simple formula (Equation 7.6) provides an approximation:

$$\text{Flash fraction} = 1 - e^{\left(\frac{C_p(T_u - T_0)}{\Delta H_v}\right)} \tag{7.6}$$

where C_p = Specific heat (J/kg)
ΔH_v = Latent heat (J/kg)
T_u = Upstream temperatures (degrees Kelvin)
T_o = Orifice or downstream temperature (degrees Kelvin)

Based on experimentation, there are good reasons for believing that superheated, pressure sources with upstream pressures of over 500 kPa produce over 20% flash in the orifice. This would apply to LPG or chlorine stored in spheres or bullets and may result in total evaporation of the two-phase discharge. The liquid droplets are sheared off to form a mist. This results in a significant increase in surface area, resulting in a significant forced evaporation from the fast-moving mist droplets as they move through the air. This, in turn, results in negligible "rain out".

7.7 EVAPORATION FROM THE SURFACE OF PONDS

Pools evaporate[3] at the surface and boil due to heat ingress from the soil/substrate. The evaporation is fairly simple, but the heat ingress from the substrate is complex and involves knowledge of the physical properties of the substrates. This results in a decaying evaporation curve as the substrate chills down.

For a circular pool, a conservative estimation of the "peak" evaporation rate,[4] E, in kg/s can be obtained from

$$E = 3.6 \times 10^{-4} \rho_v \frac{P_v}{P_a}\left(\frac{0.5}{D}\right)^{0.18} \tag{7.7}$$

where E = Evaporation rates (kg/s)
P_v = Vapour pressure of the liquid in the pool (Pa)
P_a = Absolute temperature of the fluid (^{0}K)
D = Diameter of pool (m)

Whilst pools are often circular, many are not, for example, leakage into a bund or rectangular containment. In these circumstances, the equivalent pool diameter, often called the *hydraulic diameter* (D), is calculated using Equation 7.8:

$$D_e = 4\frac{\text{Pool area}}{\text{Wetted circumferance}} \tag{7.8}$$

Boiling due to heat gain from the ground will still occur but at a reducing rate as the ground chills. If the pool can be blanketed with foam or another material which prevents contact between the air and the pool surface, the evaporation rate tends to zero.

7.8 LOSS OF CONTAINMENT

Loss of containment can occur for many reasons; some are obvious, others less so. In QRA, the frequency and size of holes where leakage may occur may not give due consideration to the mechanisms by which leaks occur. Table 7.1 gives the most common causes, but no table can be completely comprehensive. These can be used as an input to the risk assessment methods described in Chapter 6.

The results from leak rate calculations provide the starting point for the assessment of gas dispersion, fire and explosion in subsequent chapters.

7.9 VERIFICATION OF RESULTS

Many QRA assessments provide copies of leak calculations, as always a reality check is useful, and it is prudent to consider the factors shown in Table 7.2.

TABLE 7.1
Common Causes of Leakage

Category	Example	Comments
NATECH	Earthquake	Location specific
	Subsidence	For example, sink holes, landslides, ground heave
	Flooding	Flotation of equipment on water with damage to piping connections
	Temperature	Lack of winterisation of piping and vessels against low ambient temperatures. Excessive snow loading on roofs.
	Wind damage	Atmospheric storage tanks can be subject to damage if almost empty in high wind conditions.
Human Activity	Subsidence	For example, old mine workings or nearby construction
	Dropped loads	These can rupture pipework or storage vessels.
	Structural overload	Possibly as a consequence of corrosion or adding additional equipment without rechecking structural capacity
	Vehicle impact	The installation of barriers will provide protection.
	Marine or aircraft impact	Marine incidents can rupture jetty piping or hoses/loading arms if the vessel suddenly moves off station. There have been limited examples of aircraft impact causing ruptures and fluid releases, but this could occur during refuelling operations.
	Malicious act	Theft from liquid fuel pipelines is relatively common and often causes leakage where the pipe is ruptured to release fuel. Rupture of pipelines by terrorist groups is relatively common in some parts of the world
	Excavation	Procedures are generally in place to prevent uncontrolled excavation close to pipelines, but mistakes still occur.
	Domino affect	Rupture of piping and/or storage vessels by fire or explosion
Piping	Corrosion or erosion	Corrosion of studs and nuts is often overlooked.
	metallurgical failure	For example, stress corrosion cracking and high temperature creep.
	Overpressure	PSVs should prevent damage but design conditions may be exceeded or the PSV failed. Thermal expansion of liquid trapped between valves in direct sunlight (and without thermal relief) must be considered
	Poor construction/ maintenance	For example, misaligned flanges, flanges not adequately tightened or wrong gasket material used. Incorrect design resulting in excessive thermal expansion with flanges "springing"
	Missing plugs or blinds	These are commonly missing on vent and drain lines, particularly after maintenance.
	Releases from surface drains	In principle process fluids should not enter surface drains but spills often occur. Volatile materials which enter the drains can vapourise and—unless the drains are sealed—be discharge from anywhere in the drains system.
Equipment	Incorrect operation	Particularly of pig traps and of filters.
	Pumps and compressors	Seal failures are the most common source of release.

TABLE 7.2
Suggested Checklist for Assessing Loss of Containment

Leak Rate	Is the failure rate in the form of hole size?
	Has full-bore piping rupture been considered?
	Does the discharge rate of leaks make allowance for flashing flow?
Human Error	Have human error scenarios been considered? For example, has opening a pig trap door or a strainer/filter under pressure been considered?
Duration	How has the duration been determined?
	Is allowance made for mitigating factors such as alarms, trips, ESDVs or operator intervention? Are the assumptions realistic?

REFERENCES

1 Levenspiel O, Discharge of Gases from a Reservoir Through a Pipe, *AIChE Journal*, 23(3), May 1977
2 Personal communication.
3 Clancey V J, The Evaporation and Dispersion of Flammable Liquid Spills, *Journal of Process Hazards*, 5, 1974, www.icheme.org/media/12199/v-paper-07.pdf
4 Skelton R, *Process Safety Analysis an Introduction*, IChemE, 1997

8 Gas Dispersion

8.1 INTRODUCTION

Gas dispersion is of fundamental importance to the safe operation of process plant. It is the process which dilutes emissions to tolerable levels by physical processes. Natural air movements and dispersion mechanisms are important in the design process, an obvious example being the elevation and location of vents. Basic designs evolved early in the industrial era and can still be seen in many locations. Perhaps the most obvious examples in everyday life are fossil fuel-fired power stations where the shape of the plume can be clearly seen. Effluent concentrations, even from large plumes, fall to a steady background level giving rise to the belief that "distance lends enchantment". A knowledge of gas dispersion is useful for several reasons, ranging for location and spacing of equipment within the site to determining the path of large-scale accidental releases.

The mechanisms which occur in the dispersion process are complex and continuous. Up to four separate mechanisms may exist simultaneously:

1. **Jet Momentum Mixing**—the highly turbulent mixing process induced by the vortices created by the high rate of shear near to high velocity jets.
2. **Bulk or Cross Flow Mixing**—the injection of one fluid into another due to the relative movement of the two fluid streams as either a rising jet, buoyant plume or a sinking dense plume.
3. **Turbulent Gas Dispersion**—the shearing and mixing of two fluids due to convective currents, loops, and vortices, both large and small. This process dominates when the two fluids reach zero relative velocity.
4. **Molecular Diffusion**—this process is very slow and is usually (and correctly) ignored in normal calculations. However, this mechanism ensures that there is no long-term layering in stagnant atmospheres. Molecular diffusion is the mechanism which can produce the most unexpected effects. For example, lighter-than-air hydrogen may be found in a floor void and heavier-than-air hydrogen sulphide found in a roof space. These possibilities will need to be considered when defining purging and air-sweeping requirements.

The first mechanism is probably the most significant and has been used as a single dispersion mechanism for jets of gases. The second is influential at the start of the plume history; however, it is the third which is the most important after turbulent dispersion for reducing the concentration of odorous, toxic and flammable gases. These mechanisms are shown in Figure 8.1.

Comprehensive descriptions of gas dispersion modelling are available from a variety of sources.[1,2,3] Some simple dispersion calculations can be performed manually or with the use of spreadsheets. However, most modelling is performed using one of the numerous computer models available. Some are specifically designed for real-time use in an emergency. The calculation methods shown here are considered suitable for simple manual calculations to provide "ranging shots" or "sanity checks". These simple calculations—if dispersion calculations can be called simple—will enable plant designers to include features which may be required for specific situations creating emissions, for example odorous releases from vents. These will also assist the software users to input meaningful parameters and avoid "garbage in, garbage out". One essential caveat is that nearly, if not, all software produces results based on a five- or ten-minute TWA. A confidential statistical analysis of dispersion process indicated that the true peak to TWA is often a factor of three

DOI: 10.1201/9781003510680-10

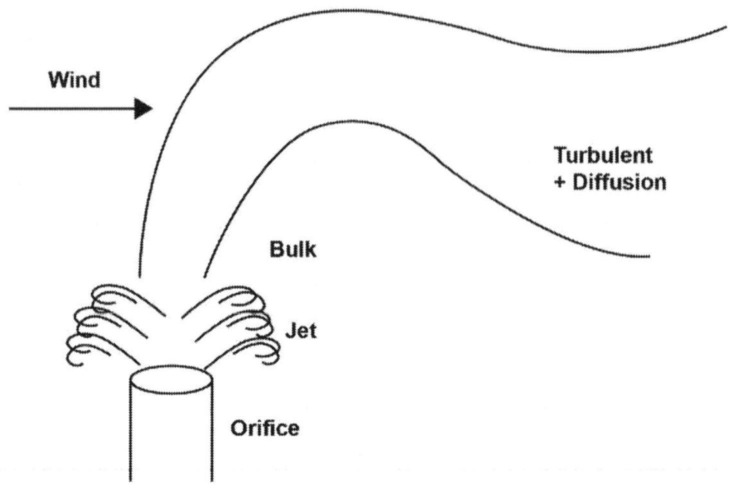

FIGURE 8.1 Graphic Representation of the Dispersion of a Gas Plume

times higher than the TWA[4] (a 1% probability of exceeding the factor of 3). This is critical when assessing flammable and toxic gas dispersion. TWA is discussed in Section 3.6.

8.2 THE BASICS OF METEOROLOGY

Before considering dispersion mechanisms, it is useful to have an appreciation of the meteorological factors which affect dispersion. Air in contact with the surface of the Earth is a *boundary layer* and as a result there is a wind speed gradient rising from the earth's surface upwards, eventually, leading to the jet stream, where wind speeds can reach 200 km/hr. The wind speed gradient is analogous to the boundary layer encountered in piping fluid flow. The shape of the gradient, as would be expected, is influenced by the surface roughness. Cities are very rough, forests less rough and the sea (waves or not) can be classified as smooth. The steepness of the gradient is greatest above the sea. Not only does the roughness affect the gradient; it also affects local turbulence. Above 500 m surface roughness effects are generally negligible. The increase in wind speed with height is noticeable when climbing mountains, as is the reduction in "gustiness". To make meaningful measurements and interpretations, it is normal to quote wind speeds at a standard elevation of 10 m above grade and then use *roughness factors* to develop the gradient.

Clear-air turbulence experienced at a height of about 10 km is the result of the jet stream and turbulence downstream of a mountain range is often experienced. These are noted for completeness and are not essential for the level of detail discussed in this chapter.

The effect of roughness is to increase turbulence at the expense of horizontal velocity. Some examples are given in Table 8.1.

Wind velocity and surface roughness are major factors in gas dispersion together with convective/advective air currents, which are influenced by the atmospheric temperature gradient. The normal *adiabatic temperature gradient* is approximately –1°K/100 m rise. For example, if a balloon of air at sea level was lifted adiabatically to 100 m it would expand and cool by 1°K. If the gradient is –1°K/100 m there would be no tendency for air at any level to rise or fall as the expanded air would have the same density when compressed to the pressure at ground level. The temperature gradient is called *dry adiabatic lapse rate.*

If the lapse rate is greater than –1°K/100 m (for example 0°K/100 m) there will be rising convection and falling air currents which will increase the turbulence in the air. Higher lapse rates, such as 0°K/100 m will cause air to "sit" in pockets and hollows on the ground. The lapse rate is influenced

TABLE 8.1
Values of Roughness (reproduced courtesy of EPSC)

Type	Examples	Plain with a Few Trees
Flat Terrain	Plain with a few trees	0.03
Agricultural Land	Airports, fields, orchards	0.10
Market Garden Area	Area with greenhouses, dispersed habitat and scattered houses	0.30
Residential Area	Built-up area of forests, industrial sites, without significant obstacles	1.0
Urban Area	Town with large buildings, industrial site with large obstacles	3.0

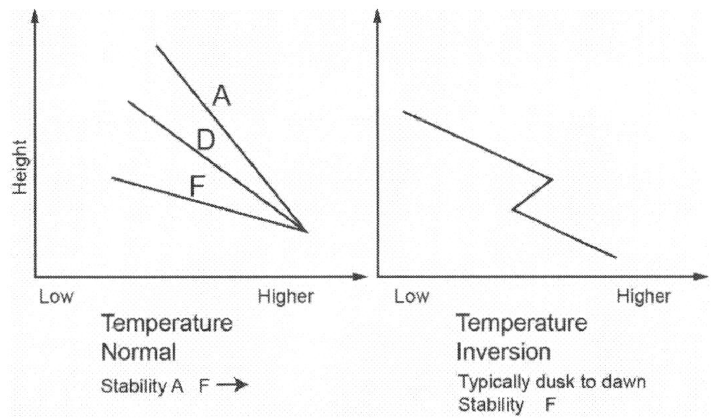

FIGURE 8.2 Atmospheric Temperature Profiles with Height above Ground and Different Stabilities

TABLE 8.2
Weather Stability Categories (reproduced courtesy of EPSC)

Wind Speed (at 10 m elevation)	Day			Night	
	Incident Solar Radiation			Moderately Overcast or Cloud Cover	Cloud
(m/s)	High	Moderate	Low	between 4/8 and 7/8	Cover <3/8
<2	A	A–B	B	F	F
2–3	A–B	B	C	E	F
3–5	B	B–C	C	D	E
5–6	C	C–D	D	D	D
>6	C	D	D	D	D

by the amount of sun falling on the Earth's surface, known as *insolation*. The resulting stability levels have been classified by Pasquill[5,6] and are shown in Figure 8.2. Whilst the Pasquill stability classification is widely used, there are some concerns with its accuracy.[7]

On hot, sunny days with slow wind speeds, there are strong upward *thermal* currents (used by gliders) giving Pasquill A stability. On cold, starlit, still nights the earth is colder than the air with advective or no air currents giving F (or even G) stability. For stability D the lapse rate is equal to the adiabatic lapse rate. For A, B and C the lapse rate is greater than the adiabatic lapse rate, and for E and F, it is less than the adiabatic lapse rate (sometimes even positive with ground frost).

The various stability conditions are typified in Table 8.2. The table also includes the patterns of stability with wind speed and insolation. Observation of wind speed can be made from tree movements. Cloud cover can provide an approximation estimation of stability.

TABLE 8.3
Assessment of Pasquill Category Based on Wind Direction Variance

Approximate Wind Variation (degrees)	Weather Stability Category
60	A
45	B
30	C
20	D
15	E
10	F
5	(G)

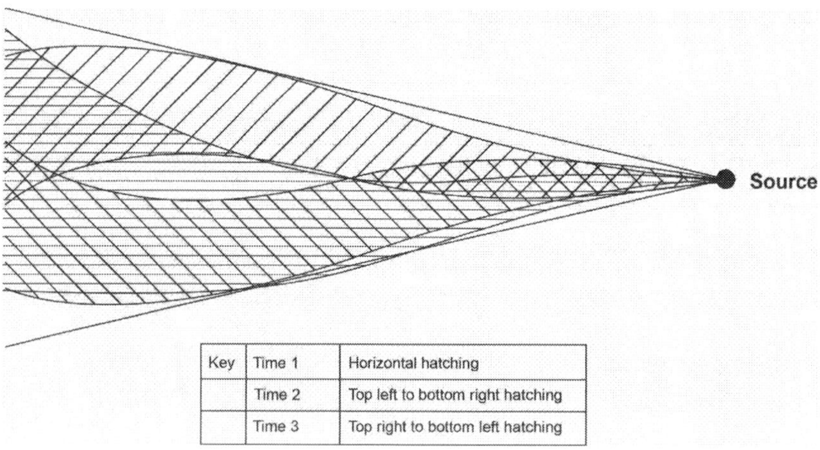

FIGURE 8.3 Hazardous Core of Dispersing Plume Showing Three Snapshots at Different Times

Source: US National Oceanic and Atmospheric Administration Air Resources Laboratory, Pasquill Stability Classes

A flag or weathervane is another, very approximate, method for determining weather stability. The variation of the flag trajectory in degrees is shown in Table 8.3.[8]

The actual path of a gas plume is more complex than the idealised calculations shown in computer plots and, in practical situations allowance must be made for random effects when estimating safe distances from a release. This is illustrated in Figure 8.3.

A real-life example of plume behaviour is shown in Figure 8.4, which shows "loops" in the dispersion process—which are illustrated in an idealised form in Figure 8.3. The loop indicates an "unstable regime". In addition, there is a negative buoyancy element which can only be assessed by complex computation.

In reality, the dispersion process is stochastic and most if not all, computer programmes work on a TWA of all of these plumes. This must be recognised when assessing the toxic dose of gases (see Section 3.6). The integral involves a power which if the TWA is used may result in a significant underestimate in places—potentially putting personnel in a hazardous situation. Figure 8.4 is not applicable to heavier the air gases which are discussed in Section 8.6.

Wind speed and direction are commonly represented in the form of a wind rose, an example is shown in Figure 8.5.

FIGURE 8.4 Dispersion from a Stack Illustrating Looping (reproduced courtesy of DNV)

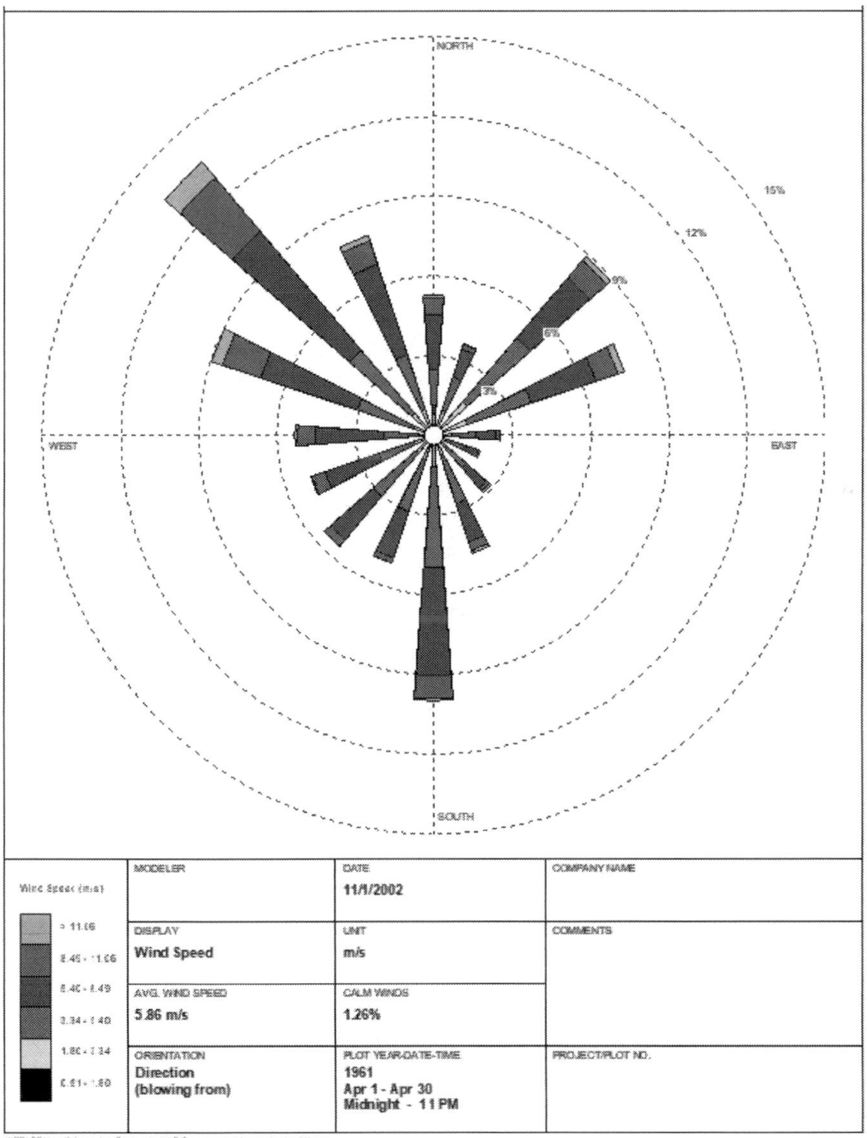

FIGURE 8.5 Typical Wind Rose (Reproduced courtesy United States Department of Agriculture—USDA)

A wind rose is of importance when establishing airport runways, orienting buildings and ensuring that discharges are adequately diluted before reaching habitation in any direction. A visual check on any hedge or tree shows that they are "wind sculptured" at an angle (SW to W in the UK) to the prevailing wind. The rose only includes speed and direction but not atmospheric stability, which is an essential parameter in gas dispersion calculations. Attention must be paid to local perturbations such as hills and valleys which affect dispersion. It is important to note that local features (hills, valleys, tall buildings and so on) will affect wind direction and velocity.

Meteorological data is available from a variety of sources, including some free sources on the internet; however, the level of detail may be limited. The local airport or weather station is probably the most reliable source of data. It must include the spectra of wind directions, wind speeds and atmospheric stabilities. Well over 100 permutations could be developed but, fortunately, many are not credible, such as a highly unstable regime and a high wind speed.

With a knowledge of the wind speed and heat input from the sun, good estimates of weather conditions can be made. Whilst Pasquill D is most common in the UK, *Murphy's law* states that the worst combination of events will occur when least expected, and this combination should be included in any risk assessment. UK risk assessments are usually undertaken at 2 m/s Pasquill D and 5 m/s Pasquill F (known as 2D and 5F). Specific features of relevance are sometimes encountered:

Inversions These occur naturally within the atmosphere. At some elevations the temperature/elevation gradient changes and a "warm" air pocket develops which inhibits natural convection. Under these conditions most gases do not rise buoyantly and are "trapped". This effect can sometimes be seen near industrial stacks at a height of around 1,000 m above sea level (ASL). In this case natural mixing processes are limited, and gases tend to spread horizontally under the inversion layer. If the gases have enough buoyancy (for example methane) or have enough vertical momentum, they may travel through the inversion layer. The effect is also sometimes seen on a cold day where a plume from a chimney appears to spread out at about 300 m altitude.

It is unlikely that inversion conditions will create major problems for normal process plant dispersion but it could be a problem in low-lying areas downwind of a continuous vent such as a power station stack. Due allowance should be made for this fact in design heights and efflux velocities.

Down Wash/Up Wash Buildings or bluff bodies create strong eddies; for example, in rectangular courtyards raindrops may fall on the lee side of the courtyard and rise on the other. Plumes from vents at low elevation, particularly if the efflux velocity is low, can be entrained down the lee side of buildings.

If the vent is higher than approximately 2.5 times the height of the structure or if the structure is very open, the effect is generally negligible.

Wind Channelling Channelling (also known as the *Venturi effect*) occurs when buildings are located close together. Air is channelled through the gaps between buildings, increasing its velocity.

8.3 NEUTRALLY BUOYANT DISPERSION—INSTANTANEOUS RELEASE

Dispersion is usually separated into *heavy* and *neutrally buoyant* gas releases. Heavy gases "slump" on contact with the ground, producing a separate mixing process. The detailed calculation of this behaviour is beyond the scope of an introductory text but is described in readily available sources.[9] Computer software is generally used for these complex calculations.

Instantaneous releases are often given the unfortunate name of *puffs*. These usually result from sudden catastrophic loss of containment, for example the rupture of a pressure vessel resulting in a gas release. The source term for a puff must be defined carefully and may not be as simple as it appears. This can be illustrated by an ammonia release. At ambient temperature ammonia vapour has a relative density of 0.59 compared to air; its relative density at the boiling point of ammonia is 0.74. It may at first be assumed that the release will be buoyant. However, first the fluid will *flash* and mix with air; the air will be cooled by entrained ammonia spray.[10] The Second Canvey Island Report[11] suggests that approximately 60 volumes of air will mix with each volume of flashed vapour- in this case, ammonia. The cold air/ammonia vapour mix is then denser than ambient air. Following the initial flash approximately 20% wt/wt ammonia will enter the vapour phase and mix with 12 volumes of air (20% of 60 volumes). This will result in the vaporisation of most, if not all, of the remaining ammonia mist producing a cloud of approximately 1 volume of ammonia and 12 volumes of air at approximately –30°C. The initial density will be approximately 1.16 relative to air—a heavy gas. However, heat will be gained from the soil and solar radiation. At some point, and the vapour will become buoyant as it warms up. Note that the calculation is complicated by water vapour in the air which will condense to form a white mist. The same phenomena could occur with other gases such as NO_x. Heavier-than-air gases are discussed in Section 8.6.

Calculations based on large-scale dispersion from industrial plant typically have authenticity between 100 m and 10,000 m from the release. It is generally accepted that the theory can be extrapolated, with care, down to 10 m. Thereafter the equations become inaccurate. Calculations are often for 3-, 6- or 12-minute TWA, the average concentration over a set period of time. Unfortunately, this may not accurately predict the release of gases. At any point in space the concentration could vary quite widely from the TWA. As previously stated, it is generally accepted that the ratio of the maximum or peak concentration to the six-minute TWA is in the range of 3.[12]

There are three basic equations used for turbulent gas dispersion. The first two are complex and can be treated together. These were developed by Sutton,[13] Pasquill[14,15] and Gifford.[16] The work of the latter two is defined as the Pasquill–Gifford or *Gaussian* model.

The equations are based on the concentration measured at a point x, y, z at time t on a cloud formed by the instantaneous release of Q^* kg of gas.

8.3.1 The Sutton Equation

The Sutton equation, for an instantaneous release at ground level, has the form:

$$\chi(x,y,z,t) = \frac{2Q^*}{\pi^{2/3} C_x C_y C_z (ut)^{3/2((2-n)}} \exp\left[-(ut)^{n-2}\left(\frac{x^2}{C_x^2}+\frac{y^2}{C_y^2}+\frac{z^2}{C_z^2}\right)\right] \tag{8.1}$$

where X = Concentration (kg/s)
Q^* = Mass of instantaneous release (kg)
Cx, Cy, Cz = Dispersion parameters (dimensionless)
u = Wind speed (m/s)
t = Time (s)
n = Dispersion coefficient (dimensionless)
x, y, z = distances in the plume measured from the moving centre in the downwind, crosswind and vertical directions (m)

Equation 8.1 is often used with the assumption that isentropic conditions exist, in which case, $C_x = C_y = C_z = C$. Values for n and C are available from several sources, but easy-to-work-with parameters have been developed by Clancey[17] and are shown in Table 8.4.

TABLE 8.4
Values for *n* and *C* (reproduced courtesy of IChemE)

Stability Condition	n	C
Lapse	0.17	0.2
Neutral	0.25	0.14
Inversion	0.35	0.09

8.3.2 The Pasquill–Gifford Equation

The Pasquill–Gifford has the form:

$$\chi(x,y,z,t)=\frac{2Q^*}{(2\pi)^{3/2}\sigma_x\sigma_y\sigma_z}exp\left[-\frac{1}{2}\left(\frac{(x-ut)^2}{\sigma_x^2}+\frac{y^2}{\sigma_y^2}+\frac{z^2}{\sigma_z^2}\right)\right] \tag{8.2}$$

where $\chi(x,y,z,t)$ = Concentration at conditions x, y, z and t–m
σ_x, σy, σ_z = Standard deviation dispersion coefficient—m
x, y, z = downwind, cross wind and vertical distances from the point of release—m

Note the difference in the definition of x, y and z between Equations 8.1 and 8.2 (and the subsequent equations).

σ_x can be calculated from

$$\sigma_x^2=\frac{1}{2}C^2(ut)^{2-n} \tag{8.3}$$

Similar expressions exist for σ_y and σz. Values for C and n are shown in Table 8.4.

8.4 NEUTRALLY BUOYANT DISPERSION—CONTINUOUS RELEASE

The Sutton and Pasquill–Gifford equations can be developed for neutrally buoyant gases to cater for a continuous release.

8.4.1 Sutton Equation

For a continuous release at ground level, the Sutton equation changes to

$$\chi(x,y,z)=\frac{2Q}{\pi C_y C_z ux^{(2-n)}}exp\left[-x^{n-2}\left(\frac{y^2}{C_y^2}+\frac{z^2}{C_z^2}\right)\right] \tag{8.4}$$

where Q = mass release rate (kg/s)

For a continuous infinite line crosswind source at ground level, Equation 8.5 should be used:

$$\chi(x,z)=\frac{2Q'}{\pi^{\frac{1}{2}}C_z ux^{\frac{1}{2}(2-n)}}exp\left[-x^{n-2}\frac{z^2}{C_z^2}\right] \tag{8.5}$$

where Q' = Continuous mass release rate kg/m s

Finally, for a continuous elevated source, Equation 8.6 should be used:

$$\chi(x,y,z) = \frac{Q}{\pi C_y C_z u x^{(2-n)}} exp\left(x^{(n-2)}\frac{y^2}{C_y^2}\right) \text{x} \left[exp\left(-x^{(n-2)}\frac{(z-H)^2}{C_z^2}\right) + exp\left(-x^{(n-2)}\frac{(z+H)^2}{C_z^2}\right)\right] \quad (8.6)$$

8.4.2 The Pasquill–Gifford Equation

If the effects of buoyancy are ignored, the Pasquill–Gifford equation can be rearranged for a point release at ground-level downstream of a steady release of Q kg/s. This is shown in Equation 8.7:

$$\chi(x,y,z) = \frac{2Q}{\pi\sigma_y\sigma_z u} exp\left[-\frac{1}{2}\left(\frac{y^2}{\sigma_y^2} + \frac{z^2}{\sigma_z^2}\right)\right] \quad (8.7)$$

Variations on these equations to consider continuous releases from a ground-level source and under cross wind conditions have also been developed:

$$\chi(x,z) = \frac{2Q'}{(2\pi)^{\frac{1}{2}} \sigma_z u} exp\left[-\frac{1}{2}\frac{z^2}{C_z^2}\right] \quad (8.8)$$

A more complex equation is available to describe the release from a continuous elevated point source:

$$\chi(x,y,z) = \frac{Q}{2\pi\sigma_y\sigma_z u} exp\left(-\frac{y^2}{2\sigma_y^2}\right)\left[exp\left(-\frac{(z-H)^2}{2\sigma_z^2}\right) + exp\left(-\frac{(z+H)^2}{2\sigma_z^2}\right)\right] \quad (8.9)$$

Values for σ_y and σ_z for continuous release have been developed by Turner[18] and are shown in Table 8.5.

TABLE 8.5
Pasquill–Gifford Dispersion Parameters

Pasquill Stability	σy,	σ_z	Range (m)
A	$\sigma_y = 0.493x^{0.88}$	$\sigma_z = 0.087x^{1.10}$	$100 < x < 300$
		$\log_{10}\sigma_z = -1.67 + 0.902\log_{10}x + 0.181(\log_{10}x)^2$	$300 < x < 3{,}000$
B	$\sigma_y = 0.337x^{0.88}$	$\sigma_z = 1.35x^{0.95}$	$100 < x < 500$
		$\log_{10}\sigma_z = -1.25 + 1.09\log_{10}x + 0.0018(\log_{10}x)^2$	$500 < x < 2x10^4$
C	$\sigma_y = 0.195x^{0.90}$	$\sigma_z = 0.112x^{0.91}$	$100 < x < 10^5$
D	$\sigma_y = 0.128x^{0.90}$	$\sigma_z = 0.093x^{0.85}$	$100 < x < 500$
		$\log_{10}\sigma_z = -1.22 + 1.08\log_{10}x - 0.061(\log_{10}x)^2$	$500 < x < 10^5$
E	$\sigma_y = 0.091x^{0.91}$	$\sigma_z = 0.082x^{0.82}$	$100 < x < 500$
		$\log_{10}\sigma_z = -1.19 + 1.04\log_{10}x - 0.070(\log_{10}x)^2$	$500 < x < 10^5$
F	$\sigma_y = 0.067x^{0.90}$	$\sigma_z = 0.057x^{0.80}$	$100 < x < 500$
		$\log_{10}\sigma_z = -1.91 + 1.37\log_{10}x - 0.119(\log_{10}x)^2$	$500 < x < 10^5$

Source: Loss Prevention in the Process Industries, Butterworth-Heinmann 1980

TABLE 8.6
Values for θ

Pasquill Stability	θ (d = 1 km)	Θ (d < 100 km)
A	60	(30)
B	45	(20)
C	30	10
D	20	10
E	(15)	(5)
F	(10)	(5)

Source: Loss Prevention in the Process Industries, Butterworth-Heinmann 1980

TABLE 8.7
Values for *h*

Pasquill Stability	*H*	Range (km)
A	$\log_{10}h = 2.95 + 2.19\log_{10}d + 0.723(\log_{10}d)^2$	$0.1 < d < 2$
B	$\log_{10}h = 2.36 + 1.05\log_{10}d + 0.067(\log_{10}d)^2$	$0.1 < d < 10$
C	$\log_{10}h = 2.14 + 0.919\log_{10}d - 0.017(\log_{10}d)^2$	$0.1 < d < 30$
D1	$\log_{10}h = 1.85 + 0.835\log_{10}d - 0.0097(\log_{10}d)^2$	$0.1 < d < 100$
D2	$\log_{10}h = 1.83 + 0.754\log_{10}d - 0.087(\log_{10}d)^2$	$0.1 < d < 100$
E	$\log_{10}h = 1.66 + 0.670\log_{10}d - 0.100(\log_{10}d)^2$	$0.1 < d < 100$
F	$\log_{10}h = 1.48 + 0.656\log_{10}d - 0.122(\log_{10}d)^2$	$0.1 < d < 100$

Source: Loss Prevention in the Process Industries, Butterworth-Heinmann 1980

Detailed QRAs may involve dozens of simulations and generally only A, B, D and E conditions are examined as these represent the most critical conditions.

A simpler alternative to the Sutton and Pasquill–Gifford methods for continuous point source release at ground level is commonly known the Pasquill model:

$$C_0 = \frac{2.810^{-3}\ Q}{udh\theta} \tag{8.10}$$

where C_o = Ground-level concentration on the axis of the plume (kg/m^3)
Q = Mass rate of release (kg/min)
u = Wind speed (m/s)
d = Distance in downwind direction (km)
h = Vertical spread (m)
θ = Lateral spread (degree)

This equation is suitable for simple hand calculations but is not widely used in computer software.

Values for h and θ were developed by Pasquill[19] and are shown in Tables 8.6 and 8.7.

8.5 JET DISPERSION

Jet dispersion is only significant when the efflux velocity of the jet is significantly higher than the wind velocity—this is often the case during emergency venting where discharge velocities are high, sometimes exceeding sonic velocity. The orientation of the jet with respect to the wind has a significant effect on the path of the plume and the dispersion process. A jet which is directed into the wind will produce a "water fountain effect" as the plume is blown back to its source. There will be some recirculation of returning gases into the outflowing jet. However, provided the returning plume has dispersed to a safe concentration by the time it returns to the jet source this effect can be discounted. Calculation of the history of plumes at various angles to the wind is complex and, as vents are usually pointed vertically upwards, it is appropriate to limit the discussion to this orientation.

Outflowing vapours leaving the vent as a jet will generate vortices around the shearing surface in contact with the air stream. The mixing process is radial, and the concentration profile is again represented by a Gaussian distribution with the maximum value on the axis of the jet.

Relatively simple models have been developed by Bordurtha[20] and Cude.[21] The latter's method is based on the assumption that wind turbulence is low at wind speeds of 2 m/s and the jet will die away below this speed. Cude suggests the equation shown in Equation 8.11 to determine the maximum height of the plume at this point.

$$z = \frac{DV_s}{0.32U}\sqrt{\frac{\rho_{gas}}{\rho_{air}}} \tag{8.11}$$

where z = Height above the orifice at which wind speed and jet velocity are equal (m)
D = Diameter of stack(m)
V_s = Initial velocity of the jet (m/s)
U = Wind speed at altitude z (m/s)
ρ_{gas} = Density of discharged gas (measured at discharge conditions) (kg/m^3)
ρ_{air} = Density of ambient air (kg/m^3)

The concentration in the cross-section of the jet is assumed to follow a Gaussian distribution with the maximum concentration on the x-axis to twice the average value. The concentration on the axis at a given height, h, above the orifice is

$$2/\text{Ca} = 0.32\frac{h}{D}\sqrt{\frac{\rho_{air}}{\rho_{gas}}} \tag{8.12}$$

where h = Height (m)
C_a = Concentration (mass fraction) along the axis—kg/kg

The height of the jet is defined as the height at which the velocity of the wind is equal to the velocity of the jet.

A jet, discharged vertically, will be influenced by the mechanisms of jet mixing and turbulent dispersion. For flammable gases, provided the bulk of fluid in the jet is diluted to its lower flammable limit by the point where the diluted plume velocity is over 2 m/s it will not return to ground level as a flammable mixture. For hydrogen the efflux velocity would be supersonic; however, hydrogen is a buoyant gas, so provided the jet is vertical and at a sufficient high velocity dispersion is certain. Note that hydrogen is readily ignited and high-velocity discharges should be avoided. Other efflux velocities to reach the LEL are given in Table 8.8.[22]

Calculated results must be assessed for accuracy. It is uncertain that the upstream pressure or the vent tip design will always be appropriate, and higher molecular weight materials must be heated well above their boiling points at atmospheric pressure before satisfactory dispersion can be satisfied.

Except in the case of vents designed for high-pressure discharge, jets originating from leaks (for example from flange leaks) may hit an obstruction which will reduce both velocity and mixing

TABLE 8.8
Minimum Efflux Velocities (Vertical Jet) to Attain Jet Dispersion Below the Lower Flammable Limit

Fluid	Efflux Velocity m/s
Methane	145
Ethane	138
Propane	125
Butane	115

capability. As a rule of thumb, following the initial mixing process, there is a fivefold dilution; dispersion can then be further analysed as a point source release using the passive release models in Section 8.3.

8.6 HEAVY GAS DISPERSION

Catastrophic rupture (loss of containment) will produce a large instantaneous release of fluids. The behaviour of the release will be critical to the assessment of hazard. The major factors that might affect dispersion are as follows:

1 Mass of fluid released
2 Amount of fluid which escapes
3 Cloud shape and aspect ratio
4 Cloud density (relative to air)
5 Wind speed
6 Initial dilution during the initial expansion
7 Solar radiation

The first stage in calculation is to determine if the release is naturally buoyant or not. The sudden release of a pressurised fluid at its boiling point will produce a two-phase mixture. The initial expanding cloud will mix with the surrounding air, and the ratio of gas to air will be uncertain. Some of the remaining liquid will "rain out" and fall to the ground; the rest will form aerosol droplets which will continue to evaporate. Heat from the ground will vapourise some or all of release, and the rest will "rain out". Overall, somewhere between 0% and 100% of the fluid will be vapourised. Hydrocarbons (such as methane, ethane[23] and propane) almost completely vaporise. As discussed in Section 8.3, the mixing ratios are not exact, and for every volume of flashed vapour, approximately 60 volumes of air are entrained. In the case of propane this results in approximately 15 kg of air per kg of propane released. Other sources[24] have assumed that for cryogenics, mixing will supply sufficient sensible heat to fully vaporise the fluids at their atmospheric boiling point. This shows that vapours which are naturally buoyant (molecular weight less than air) can be the most non-buoyant at their boiling point and that vapours heavier than pentane could behave in a buoyant manner.

After the initial release of a dense gas cloud, heat may be received by solar radiation, conductivity from the ground or some other means. In some cases, this may reduce the density of the gas to the point where it eventually becomes neutrally buoyant, for example LNG or ammonia; in other cases the density of the gas will always remain high, for example chlorine, propane and heavier hydrocarbons.

The release of a heavier-than-air gas is generally considered as four stages (see Table 8.9).

The heavy gas dispersion process is shown in Figure 8.6.

TABLE 8.9
Stages of Development of a Heavy Gas Cloud

1	**Creation of the Cloud**	Vapour is considered to be instantaneously released and modelled as a cylinder.
2	**Collapse**	The cloud collapses to form a larger diameter, lower height cloud. Even at this early stage approximations will need to be made as collapse (sometimes called slumping) of the cloud will not be uniform due to factors such as bumps or hollows in the ground, the slope of the ground and obstructions such as buildings.
3	**Ground Spread**	The cloud will spread along the ground, mixing with air; the direction of travel will be influenced by heat from the ground, air entrainment, plus wind speed and direction.
4	**Passive Dispersion**	Gas concentration can be calculated by Gaussian distribution

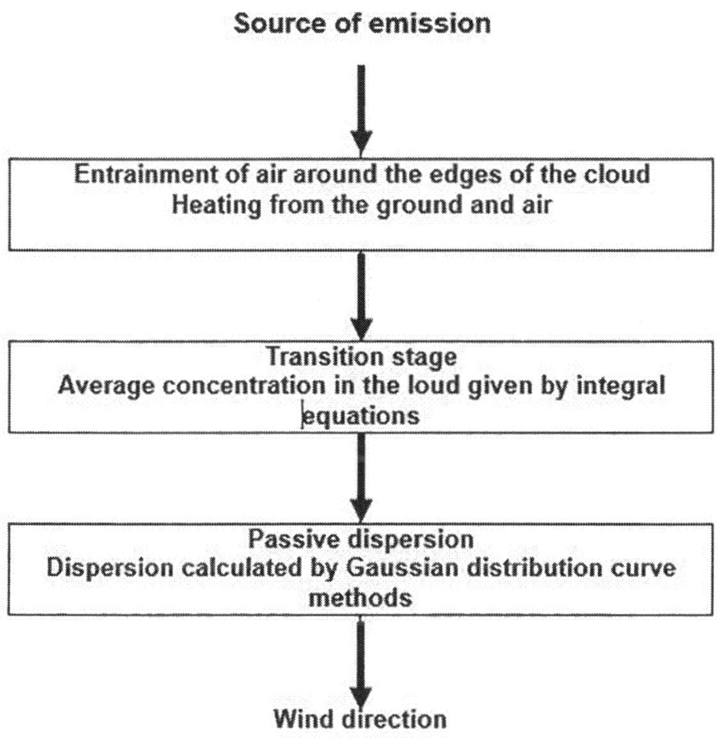

FIGURE 8.6 Development of a Heavy Gas Cloud

The manual calculation of dense gas dispersion is difficult and reliance is generally placed on one of the numerous computer programmes available for dispersion modelling although a complex manual calculation method is available from Britter and McQuaid.[25]

8.7 PRECAUTIONS

Some major gas escapes have occurred due to poor design or operational practices. However, there are a number of measures in common use for mitigation of escapes once they have occurred (see Table 8.10). All these measures require specialist design and training.

TABLE 8.10
Precautions for Gas Dispersion

Method	Details
Detection	There are a range of detection systems available which are commonly installed where leaks may occur. Specialist detectors are available for wide variety of gases and leak detection systems for high-pressure gas systems are also available. The use of oxygen detectors should not be forgotten, although these are generally installed to detect oxygen-deficient atmosphere.
Containment	Piping systems should be designed with the minimum of flanges or screwed joints. Pressures should be as low as possible, and for the large-scale storage of liquefied gases refrigerated, rather than pressurised, storage should be used where practical.
Secondary Containment	Atmospheric storage tanks are generally designed with a secondary containment bund to contain leakages. Most refrigerated storage tanks now have a secondary containment system, although some older single-skin tanks are still in service. Some specialist piping systems transporting high-value or highly toxic materials are double-skinned (pipe in pipe) with detection systems in the annulus.
Ventilation	Systems should be open to atmosphere to allow natural ventilation of leakage. Compressor houses are generally roofed but open sided for this reason. However, in some cases (generally to limit noise or in very cold climates) this is not possible. In these situations, good ventilation systems are necessary.
Barriers	Steam jets and water sprays are very powerful air movers. A well-designed and directed spray system will disperse or contain gases. Water can be used to absorb soluble vapours such as ammonia, chlorine and nitrogen oxides; however, care is necessary as water applied to spills of liquefied gas can provide heat to increase evaporation.
Vapour Suppression	Foam blankets can be used to cover a spillage of LNG, LPG, chlorine and ammonia. This can greatly reduce the evaporation rate. A bed of plastic spheres floating on top of a spill will have the same effect.
Ignition	In the 1970s and 1980s, there were suggestions that flames located around the site boundary could be used to ignite gas clouds before they reached public areas. However, the hazards posed by thermal radiation or a vapour cloud explosion (VCE) from the ignited gas clouds have, thankfully, prevented this from becoming a significant practice. There are however occasional situations where ignition of an escaping gas cloud might reduce overall hazard and this often occurs during oil/gas well blowouts. Automated systems are installed in some cases—particularly where well fluids contain high concentrations of toxic hydrogen sulphide.
Safe Refuges	Where life-threatening gas releases can occur, it is common practice to construct *safe refuges* (sometimes called a safe haven) to act as emergency shelters until gas levels have fallen to a tolerable level. A safe refuge[26] must 1 be a sealed, enclosed environment with no air changes 2 provide a source of breathable air by removing carbon monoxide and carbon dioxide from the internal atmosphere and adding oxygen 3 control the temperature of the refuge 4 have a backup power supply. Most buildings with closed windows and doors will have limited air changes and provide a reasonable level of protection. The physiological effects of increased CO_2 concentrations should not be overlooked. Increased heat creates discomfort, and increased CO_2 increases the heartbeat and breathing rate. There is limited exposure before panic sets in.

8.8 WHAT CAN GO WRONG?

Whilst weather forecasts can give reasonable general information, conditions in a particular location may differ from those in surrounding areas and will also vary from minute to minute. Therefore, any calculation, which will be based on steady state conditions, must be considered approximate and be subject to critical review.

Estimated concentrations will also be affected by changes in source terms. These will often vary over time either because of a liquid spillage completely evaporating or due to reduction in discharge rate from a leak as the pressure drops. As the pressure reduces, the risk may increase as jet mixing will be reduced. Repeat iterations over time will therefore be necessary.

Whilst there are numerous calculation methods, both manual and computerised, available, they all make assumptions which may not reflect reality due to

- wind gusting and changing direction
- land topography
- channelling effects of wind
- changes in ground and air temperature, especially between day and night.

The results of calculations must therefore be considered approximate and not as accurate predictions of concentration. It is prudent to allow a margin for error, especially for situations which may be critical to life, such as planning of emergency escape routes and safe distances between industrial facilities and residential areas.

A "reality check" on the calculations is often helpful, and the following factors should be considered:

- Is the weather data appropriate? Did it have sufficient spread in direction, atmospheric stability and wind speeds?
- Are there any local topological features which might have affected the wind direction spread?
- Has the leak source been specified correctly (velocity, vent size, height, angle of orientation, fluid properties.)?
- Have ambient factors been adequately assessed (ground roughness, solar radiation and so on)?
- Have critical points on the plume been identified?
- Have an adequate number of variables been considered (wind speed, direction, stability)?

In real-life emergency situations dispersion modelling can provide useful guidance, but on-the-ground monitoring should also take place.

REFERENCES

1 Pasquill F, *Atmospheric Diffusion*, Van Norstand, 1962

2 Turner D B, Workbook of Atmospheric Dispersion Estimates, *US Environmental Protection Agency Office of Air Programs*, 1970, https://nepis.epa.gov/Exe/ZyNET.exe/9100JEIO.txt?ZyActionD=ZyDocument&Client=EPA&Index=Prior%20to%201976&Docs=&Query=&Time=&EndTime=&SearchMethod=1&TocRestrict=n&Toc=&TocEntry=&QField=&QfieldYear=&QfieldMonth=&QfieldDay=&UseQField=&IntQFieldOp=0&ExtQFieldOp=0&XmlQuery=&File=D%3A%5CZYFILES%5CINDEX%20DATA%5C70THRU75%5CTXT%5C00000013%5C9100JEIO.txt&User=ANONYMOUS&Password=anonymous&SortMethod=h%7C-&MaximumDocuments=1&FuzzyDegree=0&ImageQuality=r75g8/r75g8/x150y150g16/i425&Display=hpfr&DefSeekPage=x&SearchBack=ZyActionL&Back=ZyActionS&BackDesc=Results%20page&MaximumPages=1&ZyEntry=3

3 Pasquill F, *Meteorological Aspects of the Spread of Windborne Contaminants*, WHO Conference on Protection of the Public in the Event of Radiation Accidents, 1965

4 Personal Communication
5 Pasquill F, The Estimation of the Dispersion of Windborne Materials, *Meteorological Magazine*, 90, February 1961
6 Pasquill F, *Atmospheric Diffusion*, Van Norstand, 1962
7 Kohl J D W & Chapman H L, *Atmospheric Stability Characterization Using the Pasquill Method: A Critical Evaluation*, Vol 187, Atmospheric Environment, August 2018, www.sciencedirect.com/science/article/abs/pii/S1352231018303686
8 US National Oceanic and Atmospheric Administration Air Resources Laboratory, *Pasquill Stability Classes*, www.ready.noaa.gov/READYpgclass.php
9 Britter R E & McQuaid J, Work Book on the Dispersion of Dense Gases, *Research Report No 17/1988*, UK Health and Safety Executive, 1988
10 UK Health and Safety Executive, *2nd Canvey Report—A Review of Potential Hazards from Operations in the Canvey Island/Thurrock Area*, Three Years after Production of the Canvey Report, 1981
11 UK Health and Safety Executive, *2nd Canvey Report—A Review of Potential Hazards from Operations in the Canvey Island/Thurrock Area*, Three Years after Production of the Canvey Report, 1981
12 U S Department of Labour, *Permissible Exposure Limits—Annotate Tables*, www.osha.gov/annotated-pels
13 Sutton O G, *Micrometeorogy*, McGraw Hill, 1953
14 Pasquill F, The Estimation of the Dispersion of Windborne Materials, *Meteorological Magazine*, 90, February 1961
15 Bordurtha F T, *Industrial Explosions Prevention and Protection*, McGraw Hill, 1980
16 Gifford E A, Use of Routine Meteorological Observations for Estimating Atmospheric Dispersion, *Nuclear Safety*, 2, 1961
17 Clancey V J, *The Evaporation and Dispersion of Flammable Liquid Spills*, IChemE Symposium series 39a, 1974, www.icheme.org/media/12199/v-paper-07.pdf
18 Turner D B, *Workbook of Atmospheric Dispersion Estimates*, Environmental Protection Agency Office of Air Programs, 1970, https://nepis.epa.gov/Exe/ZyNET.exe/9101GKEZ.txt?ZyActionD=ZyDocument&Client=EPA&Index=Prior%20to%201976&Docs=&Query=&Time=&EndTime=&SearchMethod=1&TocRestrict=n&Toc=&TocEntry=&QField=&QFieldYear=&QFieldMonth=&QFieldDay=&UseQField=&IntQFieldOp=0&ExtQFieldOp=0&XmlQuery=&File=D%3A%5CZYFILES%5CINDEX%20DATA%5C70THRU75%5CTXT%5C00000022%5C9101GKEZ.txt&User=ANONYMOUS&Password=anonymous&SortMethod=h%7C-&MaximumDocuments=1&FuzzyDegree=0&ImageQuality=r75g8/r75g8/x150y150g16/i425&Display=hpfr&DefSeekPage=x&SearchBack=ZyActionL&Back=ZyActionS&BackDesc=Results%20page&MaximumPages=1&ZyEntry=2
19 Pasquill F, The Estimation of the Dispersion of Windborne Materials, *Meteorological Magazine*, 90, 1961
20 Cude A L, Dispersion of Gases Vented to Atmosphere from Relief Valves, *The Chemical Engineer (No. 290)*, October 1974.
21 Personnel Communication
22 Personnel Communication
23 Personnel Communication
24 Wheatley C J, Crabol B, Carpenter R J, Jagger S F, Massey C, Cleaver R P, Fitzpatrick R D & Prince A J, Comparison and Test of Models for Atmospheric Dispersion of Continuous Releases of Chlorine, *Report No. SRD R438*, Safety and Reliability Directorate UK Atomic Energy Authority, 1988
25 Britter R E & McQuaid J, Work Book on the Dispersion of Dense Gases, *Research Report No 17/1988*, UK Health and Safety Executive, 1988
26 MineARC Systems, *Petrochemical Safe Haven—The What & Why*, https://minearc.com/petrochemical-safe-haven-refuge-chamber-what-are-they/

9 Fires

9.1 INTRODUCTION

Fires are a common occurrence both in process facilities and in everyday life. Fortunately, most are small and easily extinguished without resulting to the emergency services. There are, unfortunately, occasions when serious fires occur with tragic loss of life and damage to property.

Fires in the process industry can be especially destructive due to the large amount of flammable material (note that, unusually, the words flammable and inflammable have the same meaning but the term "flammable" has been used through the text), generally in liquid form, that is available. Whilst fires in process areas can be very destructive, the largest inventories tend to be in storage areas. Both the site and local authority resources for firefighting may be less than those required to extinguish a major fire, and the best option may be to protect the surrounding facility and let the fire burn itself out.

Fires may involve solids, liquids or gases; common examples are shown in Table 9.1.

The obvious hazard from a fire is exposure to high levels of thermal radiation (heat), but there are others, particularly toxic fumes from the smoke plume. A fire can also escalate in intensity or propagate into an explosion. These issues are covered in later sections. A convenient classification for different types of fire is shown in Table 9.2.

9.2 FIRE TRIANGLE

A fire needs three components to exist:

1 Oxygen
2 Fuel
3 Source of ignition

This is commonly expressed in the form of the fire triangle, as shown in Figure 9.1.

If one side of the triangle is removed, the fire will not continue. To initiate a fire, a source of ignition must be present with sufficient energy to ignite the fuel source. The amount of energy required

TABLE 9.1
Typical Fuels

	Examples	Comments
Solids	Any combustible material such as dust, wood, coal, metal powders	In a finely divided form dusts such as flour and metals can explode (see Chapter 10).
	Batteries	The increasing use of batteries as the world moves to a net zero economy has resulted in an increase of battery fires, particularly lithium-ion type batteries.
Liquids	Crude oil, hydrocarbon fuels and many other industrial chemicals	Extended fires in crude tanks can lead to *boil overs* where a layer of water below the crude surface is heated to its boiling point and boils explosively.
Gas including liquefied gases	Natural gas, acetylene, liquefied petroleum gas (LPG)	The major hazard with gases is jet fires, although fireballs may also occur. Explosions which are also a major hazard are discussed in Chapter 10.

DOI: 10.1201/9781003510680-11

TABLE 9.2
Typical Fire Types

Type	Examples	Comments
Solid Fuel Fires	Stacks of timber, buildings	Recycling plants, discarded packaging material and off-spec plastics are a common source of solid fuel fires.
Pool Fires	Storage tanks, liquid spills	Fires in storage tanks occur regularly and may spread to two or more surrounding tanks. In some cases, the tank ruptures with the fire expanding to fill the tank bund.
Jet Fires	Releases from high-pressure pipelines or pressure vessels	Flames from jet fires can be extremely long, and direct flame impingement will destroy unprotected steelwork within a few minutes.
Flash Fires	Drifting release of flammable gas which "finds" an ignition source	Very large amounts of gases may be released before an ignition source is reached. This has happened in a number of oil terminal fires in recent years, with Buncefield[1] in the UK being the best example.
Running Fires	Liquid spills which ignite and then travel either on smooth surfaces or in drains	Examples include ruptures of fuel tanks on aeroplanes and spills when refuelling F1 racing cars
Boiling Liquid Expanding Vapour Explosions (BLEVEs)	Rupture of a pressure vessel with the released gas suddenly expanding and combusting	Whilst a BLEVE is technically an explosion, its major hazard is intense but relatively short-lived thermal radiation. Feyzin[2] and Mexico City[3] are classical examples of a BLEVE.

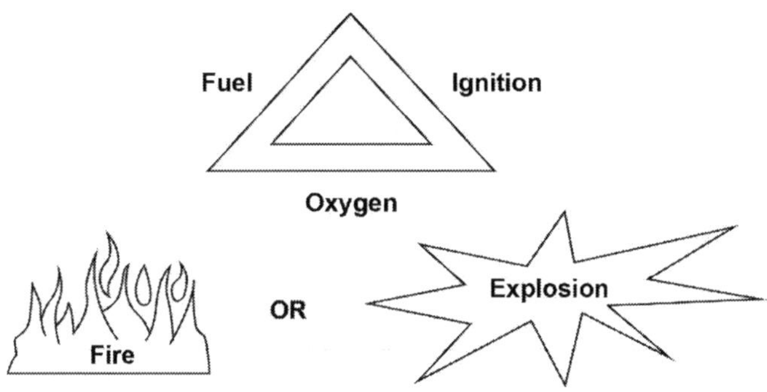

FIGURE 9.1 The Fire Triangle

varies significantly depending on the fuel (or fuel mixture) and the fuel concentration. Ignition energy is discussed in Section 9.3.

9.2.1 Oxygen

Oxygen is naturally present in the atmosphere and therefore is the most difficult component of the triangle to remove, although there are several techniques for separating fuel from oxygen in the air; these are discussed in Section 9.12.

9.2.2 Fuel

Removing the fuel from a fire is a common method of firefighting and is again covered in Section 9.12. The intensity of a fire will depend on the characteristics of the fuel and the quantity. Liquids are

TABLE 9.3
Classification of Flammable and Combustible Liquids[4]

Class	Closed-Cup Flash Point Range	Boiling Point
Class IA	Below 22.8°C (73°F)	Below 37.8°C (100°F)
Class 1B	Below 22.8°C (73°F)	At or above 37.8°C (100°F)
Class 1C	At or above 22.8°C (73°F), below 37.8°C (100°F)	—
Class II	At or above 37.8°C (100°F), below 60°C (140°F),	—
Class IIIA	At or above 60°C (140°F), below 93°C (200°F)	—
Class IIIB	At or above 93°C (200°F)	—

TABLE 9.4
Lower and Upper Flammable Limits for Some Common Gases[5]

Gas	LFL V/V%	UFL V/V%	Minimum Ignition Energy (mJ)
Hydrogen	4	75	0.019
Acetylene	2.5	100	0.02
Benzene	1.4	8	0.22
Methane	5	15	0.29
Ethane	3	12.4	0.24
Propane	2.1	9.5	0.25
Ethylene	2.7	36	0.12

Source: Loss Prevention in the Process Industries, Butterworth-Heinmann 1980

generally classified as *combustible* or *flammable*. The definition of flammable is a liquid which has a *flash point* not exceeding 37.8°C (100°F) using a standard closed cup test method, whilst combustible fluids have a flash point above 37.8°C (100°F). The flash point is defined as the minimum temperature at which sufficient vapour is given off by the liquid to form an ignitable mixture with air. Flammable and combustible liquids are further subdivided as shown in Table 9.3.

Gases and vapours will only ignite and support combustion if the concentration of fuel is between the *lower (LFL)* and *upper flammable limits (UFL)*. If too *lean* or too *rich*, the mixture will not ignite. Examples for some common materials are given in Table 9.4.

The flammability limits for most fuel gases are around half to twice the stoichiometric concentration. A common technique to avoid a flammable mixture is to increase the proportion of nitrogen in the mixture to decrease the size of the flammable envelope. The technique is frequently used to remove a flammable atmosphere when preparing equipment for maintenance. The minimum oxygen content to support ignition is around 10% v/v (about the same as for supporting respiration). A typical flammability envelope is shown in Figure 9.2 where it can be clearly seen in the theoretical plot for the systematic dilution of a mix of 20% v/v methane and 80% v/v nitrogen with air:

1 With one dilution the mix is oxygen deficient
2 With two dilutions the mix is just fuel rich
3 With three dilutions the mix is just flammable
4 With four or five dilutions the mix is fuel lean

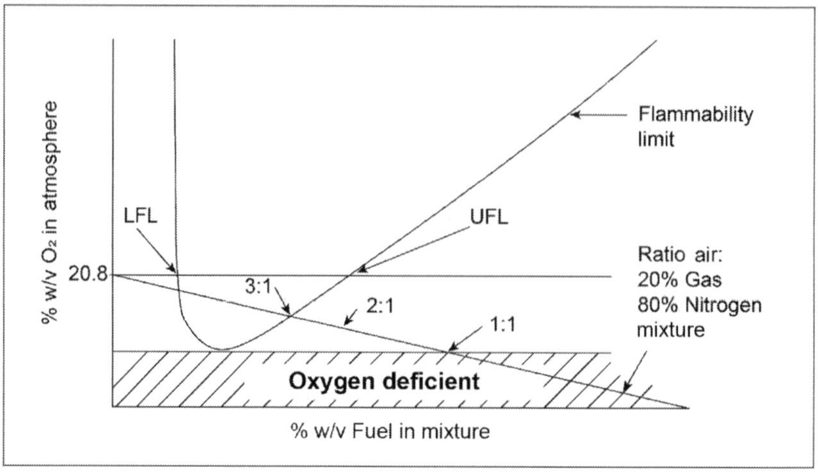

FIGURE 9.2 Typical Flammability Envelope

The fire/explosion hazard of gases is further discussed in Chapter 10.

Purging with air is not considered good practice, as there will come a point where the flammable gas will become diluted to the point where a flammable mixture will exist. This is equivalent to 3 dilutions (as illustrated in Figure 9.2 at point 3.1). It is standard practice to set the maximum concentration of the flammable gas in nitrogen (or other inert gas) at 5% when purging.

The opposite phenomenon is found in gasoline driven cars where an automatic (or in older cars, manual) "choke" is installed to provide an enriched fuel mixture on cold days.

9.2.3 Minimum Ignition Energy

To ignite a flammable mixture the energy source must have sufficient energy. Typical minimal energy values for common gases and liquids are shown in Table 9.4. The ignition energy falls as the oxygen concentration in the atmosphere increases and can reach a condition where spontaneous ignition takes place.

9.3 SOURCES OF IGNITION

There are numerous sources of ignition, and the most common are considered in the subsequent sections. Note that a source of ignition will not necessarily cause ignition. The concentrations of flammable gas within a cloud will vary and, in some cases, may be too lean or rich to ignite. An ignition source under the surface of a flammable or combustible liquid does not have access to oxygen and will not ignite. Additionally, the ignition source must release sufficient energy to cause ignition (see Section 9.2).

It is not always possible to positively identify an ignition source during a fire investigation, and the default finding is generally static electricity. The most common ignition sources, in no particular order, are the following.

9.3.1 Auto-Ignition on Hot Surfaces

The ignition of a fuel by direct contact with hot surfaces such as superheated stream pipes, hot bearings or recently welded metal commonly occurs. Values quoted in the literature are usually between 200°C and 350°C or greater however quoted temperatures may be difficult to achieve in practice. Quoted values are "ideal," and wind cooling or a layer of dust on the surface may result in the actual surface temperature being lower than anticipated. It would be unwise to rely precisely on published figures.

9.3.2 "Spontaneous Combustion" (Chemical Reaction)

Oily rags and some other materials can heat up gradually due to a slow exothermic reaction. Normally the heat will be lost due to conduction or convection but where heat cannot escape it will build up and result in a fire. The reaction rate is initially slow but increases exponentially. Once a high temperature has been reached ignition can take place without warning.

A similar situation can occur with organic materials where a slow chemical reaction can eventually cause ignition A common example is fires involving hay,[6] which are well known to farmers and are likely to be of increasing concern if the use of solid biofuels for power generation increases. A similar phenomenal can occur with coal heaps and is a significant concern with coal stocks at power stations.

There are a number of other materials which can self-ignite on exposure to the atmosphere. Pyrophoric iron sulphide (FeS_2) is a common example in the oil industry but there are a number of similarly reactive substances. Materials such as triethyl aluminium (used as a catalyst in polymer production) spontaneously ignite on exposure to air.

Some materials auto-ignite above a specific temperature. The *auto-ignition temperature* (AIT) is generally in the range of 300°C–400°C but can be as low as 100°C.

9.3.3 Compression

Compression, leading to ignition, is the principle of the diesel engine. Compressors are common items of equipment in both process and utility systems. Compression of mixtures of flammable gases and air in piping systems can generate high temperatures and ignition. This demonstrates again the importance of purging process systems before maintenance and again before the introduction of flammable gases. The temperatures that can be generated during compression are illustrated in Table 9.5.

The potential for fires from this source is not confined to compressors handling flammable gases. Problems can also occur in compressed air systems[8] and the crankcases of lubricated machines or engines if an oil mist is present.

9.3.4 Flame/Spark

Common sources include but are not limited to welding, chipping, grinding and electric faults. Defective equipment, such as bearings, can also generate sparks—these sources are only indicative but give the general idea. Many sources of ignition occur during hot work and a well-designed and implemented *permit to work* (PTW) system is necessary (see Section 16.6). Other sources of

TABLE 9.5
Adiabatic Compression of Air[7]

Pressure (Bar)	Temperature (Max) °C
10	280
20	379
30	461
40	525

flame include boilers and fired heaters within the process and utility areas. Good plant layout and instrumented safety systems will minimise the potential for fixed equipment to act as an ignition source. However, numerous incidents show that the possibility cannot be entirely eliminated even when good practice is used.

9.3.5 Electrical Equipment

Fixed, permanent electrical equipment on plants handling potentially flammable materials has to be *fit for purpose*. The likelihood of flammable materials being present on-site must be assessed area by area and adequately specified electrical equipment installed. The specification is based on *electrical or hazardous area classification* (now more commonly called *ATEX* in Europe). A fuller discussion is given in Section 14.6.

A greater problem may occur with temporary electrical fittings installed for maintenance or construction activities. Whilst these may be of adequate quality when installed cabling can often be damaged by abrasion or bending (particularly close to fittings). Battery-operated electrical equipment such as mobile phones or radios for use in hazardous areas should be *Intrinsically Safe* (IS) and hence unable to provide sufficient energy to ignite a flammable mixture. Note that there have been problems with batteries failing and generating sufficient heat to provide a source of ignition.[9]

In essence, the design should reflect the possibility of flammable materials being present and minimise the sources of ignition. However, there are likely to be unusual conditions where fire hazards increase—particularly during non-routine activities such as maintenance.

9.3.6 Static Electricity

Static electricity is generated when there is charge separation between two phases moving relative to each other. This can be as follows:

1 Solid/solid
2 Solid/liquid
3 Solid/gas
4 Liquid/gas
5 Liquid/liquid

Gas/gas separation appears not to be a source of static generation.

There are many sources of static generation, generally resulting from the shearing of high-resistivity fluids due to flow or pumping. The threshold resistivity for fluids that might result in the formation of static electricity is approximately 10^{12} Ωcm; this includes gasoline, gas oil (diesel) and liquids such as benzene but tends to exclude fuel oils. Two-phase flow—solids, flashing flow or water in oil—enhances the generation of static electricity. Static can accumulate on unearthed structures such as ball valves, filters and even humans. Static is often caused by splash filling, jetting water, carbon dioxide fire extinguishers, steam lances, water settling through hydrocarbons and numerous other scenarios. Static is a very potent source of ignition.

Ignition by static electricity probably occurs more frequently than is commonly realised. When investigating a fire and all other sources of ignition have been eliminated, it is generally assumed that the ignition source "had to be static electricity". In reality, it may have been the main source even when other sources were present.

The two "classic" sources of static ignition are as follows:

Lightning—where liquid (rain or hail) is cycled by strong upward thermals in a cloud. Charges accumulate in the cloud, and static electricity, as lightning, discharges between clouds, within the cloud or from cloud to earth.

Humans—where solids such as clothing rubs on the body or shoes scuff on carpets static is generated. This generally happens on dry days or in air-conditioned buildings where the relative humidity is less than 60–70%. With higher humidity the charge tends to leak away and cannot accumulate.

The human body has a capacitance of about 200 pF and can be charged to around 10,000 volts. The charge energy is about 10 mJ which is more than sufficient to ignite most gas/air mixtures. In one demonstration a Bunsen burner was ignited by an experimenter wriggling in a plastic-coated chair. Another variation is the static generated on a plastic comb following vigorous combing of dry hair (again solid/solid) which can produce small sparks; the charged comb can also be used to "lift" small pieces of paper. The shock when touching a filing cabinet was almost certainly an incentive spark with ignition potential—particularly if a "click" is heard.

When solids are milled, particles move relative to each other and charges can accumulate on the solids; these can result in sparks.

9.3.6.1 Static in Liquid Systems

Static electricity is the result of "charge separation". It is easily shown by the classic physics demonstration of rubbing fur against ebonite as well as the charge on a plastic comb after combing hair.

Liquid/liquid separation is shown graphically in Figure 9.3.

Charge separation will take place in fluids (as in lightning) as well as pipelines and sprays.

Multiphase flow such as slurries, wet steam and sprays has the greatest charge separation potential.

The charge can accumulate inside the fluid or be given up to an unearthed conductor such as a metal filter in the line or a ball valve with soft plastic seats. In these examples the spark is between an insulated conductor and an earthed conductor such as the pipeline itself.

A common example of static generation is the filling of a tank with a non-conducting fluid. The fluid itself is charged and only discharges slowly (charge relaxation) to the tank walls. If the fluid is "dipped" (commonly done to determine level in the tank) with a metal dip tape there could be a local discharge to the tape and a spark causing a fire or, worse, a tank explosion. Fuel road tankers often have a wooden dip rod—not metal—for this reason.

A specific example of static hazards in tankers is "switch filling". If diesel fuel is charged into a tank which previously contained petrol, the diesel carries the charge and ignition takes place. Slurry systems, where solids are transported in a turbulent regime in pipes, are even more potent sources of static electricity due to complex phase movement.

To avoid the build-up of static, it is standard practice to earth all conductors where charges may accumulate. Another rule of thumb is to keep fluid flow velocities below 1 m/s, but this will not work for slurries. Splash filling of tanks should also be avoided.

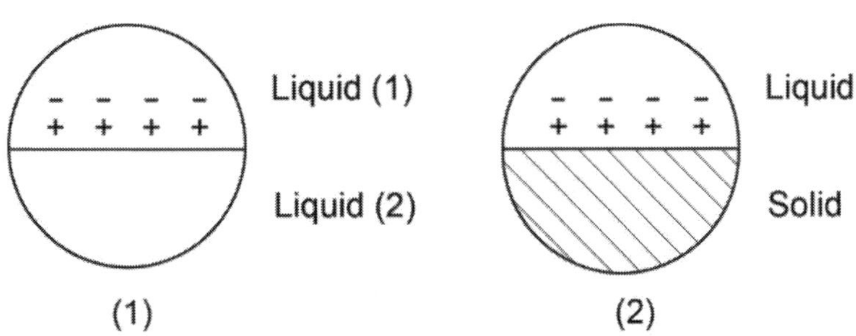

FIGURE 9.3 Charge Separation for Solids and Liquids

9.3.6.2 Droplets

Water droplets draining through non-conducting fluids can also cause charge to accumulate; it is essential that care is taken when water settles out of non-conductive fluids such as "wet" diesel fuel. Flashing fluids can also produce charged clouds. The cloud then induces a charge on an unearthed conductor, which results in a spark where it discharges to an earthed body. Flashing fluids containing solids as slurries are even more powerful sources of static electricity.

Wet steam[10] from steam lances has been known to ignite organic fluids, and under certain circumstances carbon dioxide fire extinguishers have been reported to produce sparks,[11] but a bigger problem is an electric shock to the user of the extinguisher.[12]

9.3.6.3 Splash Filling

Splashing organic fluids can generate static changes due to charge separation (see Figure 9.3). For this reason, it is normal to introduce liquids below the liquid surface. Where this is not always possible (such as top-filling road tankers), filling is conducted at a low rate until the feed pipe into the tanker is fully covered. Splash filling is eliminated in bottom-loading systems for road tankers which are becoming increasingly common.[13] Filler pipes for gasoline/diesel are also electrically earthed through the hose or a separate earthing connection to prevent charge build-up.

Examples of fires started by static are given in the following table:

Steam	An operator was trying to inert a leak of hot benzene with steam using a steam lance to exclude air and prevent ignition. Unfortunately, the steam was wet and generated static which ignited the benzene—the very event the operator was trying to prevent.
Emptying a Polyethylene Sack	An operator was emptying a sack of off-specification polythene into a silo. The sack contained some residual ethylene, and the movement of the polythene (solid) out of the polythene sack (solid) resulted in a spark which resulted in a minor fire. The operator was not injured but was left holding the two corners of the sack—the rest of the sack melted in the short-lived fire.
Sampling	An operator took a sample of organic fluid in a steel bucket—splash filling. The handle was insulated by a plastic coating, so a charge built up on the bucket, and the spark between the pail and sample point caused the fluid to ignite—much to the operator's surprise.
Pump Seal Leak	The seal of a pump handling hot organic fluid containing a solid catalyst failed catastrophically. The hot fluid flashed and created a small charged cloud; this induced a secondary charge on an unearthed sheet of metal insulation cladding. This in turn discharged to an adjacent earthed section across a gap of only 1 millimetre. The spark across this narrow gap ignited the cloud.
Unearthed Valve	A ball valve was fitted with plastic seats. The ball became charged by the organic fluid flowing through the associated piping, and a spark discharged between the ball valve handle and the pipeline. After a period of time the arcing process cut a small hole in the pipe, and the leak was ignited.
Drum Filling	A 45-gallon drum was to be filled with flammable liquid on a trolley fitted with nylon wheels. The drum was therefore insulated. The fill pump was started, but filling was delayed for a few minutes during which time the liquid heated and was subject to shearing in the pump. When the fill nozzle was put into the drum and the fill line isolation was opened, there was an explosion in the drum. It was believed there was a small spark between the pipe and the charged drum. One person was killed.
Steam Leak	An operator walked through a steam cloud released from a leak in steam piping. The operator received a nasty shock when he touched a handrail.
Plastic Coated Handrail	A handrail on a stair was coated with plastic to prevent corrosion. The operators found that they became charged as their hand slid down the handrail and then received a shock when they touched an earthed structure. The operators found a solution—do not use the handrail—but one slid on the stairs and was injured. This is an example of a poorly specified protective system.

9.3.7 Impact Spark

An impact spark can occur when two dissimilar metals (and sometimes other materials) are rapidly forced together. This is the principle of the *flintlock* pistol. Some tools are made from alloys which have low spark potential.

The use of aluminium is restricted on some plants due to the potential for a thermite spark if impact damage with rusty iron (such as a scaffolding pole) occurs. However, wooden scaffold planks are not recommended near boilers or fired heaters for obvious reasons. In specific cases "non-sparking tools" (phosphor bronze) are needed.

9.3.8 Friction

Friction can be generated in several ways such as overheating bearings and seals. A common example is slippage between drive wheels and conveyor belts (this can also generate static electricity). For this reason, it is common practice to install alarms to indicate if the belt is not moving when the drive system is operational.

9.4 EFFECTS OF FIRES

The most obvious output from a fire is high levels of thermal radiation which result in injury to humans and damage to plant, equipment and surrounding areas. However, the toxic hazards from smoke plumes should not be overlooked. Secondary hazards from plumes include reduced visibility.

Increasingly sophisticated methods of calculation are available, many requiring the use of computer modelling, although detailed step-by-step calculations are available in the TNO Yellow Book[14] and other sources. Calculation methods shown in the following sections are simple methods giving approximate answers and are useful for "rule of thumb" calculations and "reality checks" on computer models.

9.4.1 Thermal Radiation

The amount of thermal radiation received by an observer is dependent on the intensity of the flame surface, known as the *Surface Emissive Power* (SEP), and on the distance between the observer and the flame.

The SEP is defined by the Stefan-Boltzmann equation:

$$\varnothing = \sigma \epsilon T^4 \tag{9.1}$$

where $\varnothing$ = Surface emissive power (w/m^2)
σ = A constant (5.67×10^{-8}) (w/m^2 K^4)
ε = Emissivity (or "blackness"). This is 1 for a black body but can be as low as 0.1 for a translucent flame (dimensionless)
T = The absolute temperature (°K)

There is one other modifier which is not essential at a basic level of discussion but may be important in major studies: *atmospheric absorption* of heat by water vapour and carbon dioxide in the atmosphere. In most cases it is very much a second-order effect.

Figure 9.4 shows a graphical representation of the heat balance around a flame boundary for different fuels with different flame characteristics. The top horizontal line is the total amount of heat available in an ideal combustion process. In reality, not all the heat is released as there will be some partial combustion and thermal cracking (pyrolysis) in hydrocarbon fires resulting in the formation of carbon monoxide and soot.

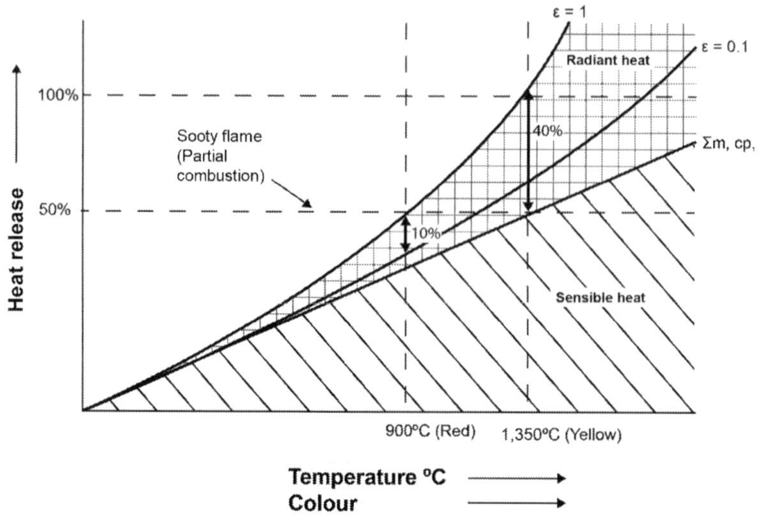

FIGURE 9.4 Heat Balance in a Fire

For illustrative purposes, the 100% and 50% values are the percentage of heat release—sensible and radiant. This will be less than the heat from the combustion of fuel alone due to incomplete combustion which forms carbon—the radiating species and excess air. The 100% intercept at the left-hand side represents a hydrocarbon with a high carbon content. The emissivity is near unity and the rest of the heat is released as sensible heat. The flame is yellow in colour with a temperature of approximately 1,350°C. The right-hand intercept represents methane with little soot. The emissivity is nearer 0.1, the colour is blue and the temperature is nearer to 1,500°C–1,600°C.

50% combustion (taken as an arbitrary value) represents high soot formation and high emissivity. The colour will be red with black sooty patches and a flame temperature of approximately 900°C. Both sensible and radiation heat release are reduced appropriately as shown in the figure. The straight inclined line encloses the sensible heat losses which will also be determined by the excess air present in an atmospheric fire, determined as a first approximation by

$$\Sigma\,(\mathrm{m\ cp}\ \Delta\mathrm{T}) \tag{9.2}$$

where Cp = Specific heat (kJ/kg)
m = Mass of fuel (kg)
ΔT = Temperature difference (degree K)

Specific heat (Cp) changes with temperature, but for this simple model it can be treated as a constant. The curved line represents the radiative heat losses and follows Equation 9.1. Temperature (T) and the emissivity (ε) are not fixed.

In summary, the proportion of heat radiated from a *perfect* black body flame in *perfect* conditions will probably not exceed 40% of the heat available, and the flame temperature will not exceed approximately 1,450°C. In imperfect conditions, such as incomplete combustion, the temperature could be as low as 900°C or even 800°C if the flame is very "sooty" due to incomplete combustion or thermal cracking.

Fuels, which are rich in hydrogen, do not produce soot, and it is the soot which is the black body radiator. The emissivity of gases is significantly less than unity (nearer to 0.1), and so the area between the radiation line and the sensible heat line is reduced. The flame temperature must increase so as to maintain the heat balance. There is a slight converse situation in that *non-smoky* flames are hotter but have a lower level of radiated heat.

TABLE 9.6
Characteristics of Process Industry Fires (reproduced courtesy of HSE)

Type	Size	Duration	Radiated Emissive Power (kW/m²)	Hazard
Pool Fire (Open)	Medium	Long	50–150	Radiation, smoke and toxic gases—CO and CO_2 and by-product reactants such as HCN and HCl
Pool Fire (Severe or Confined)	Medium	Long	100–230	Radiation, smoke and toxic gases—CO and CO_2 and by-product reactants such as HCN and HCl
Jet Fire (Open)	Medium	Medium/Long	50–250	Radiation, smoke Damage to structures and pressure vessels
Jet Fire (Confined)	Medium	Medium/Long	100–300	Radiation, smoke Damage to structures and pressure vessels
Flash Fire	Large	Short	170	Engulfment Damage to vulnerable equipment
Fireball	Large	Short	270	Radiation

A smoky flame may only radiate 10% of the total heat of combustion, but the temperature could be as low as 800°C. At the other extreme oxy-propane and oxy-acetylene cutters do not have (inert) nitrogen from the atmosphere and with the pure oxygen used for combustion, flame temperature will be nearer 2,200°C.

In summary, flame temperatures can be estimated by colours:

Blue = 1,600°C
Straw yellow = 1,350°C
Cherry red = 850°C

This sequence is in the same order as the colours of the rainbow (red, orange, yellow, green, blue, indigo and violet)—information known to blacksmiths for hundreds of years.

The flame surface is not uniform in colour or texture, so, for accuracy, the heat radiated from a surface must be assessed using a radiometer. Reasonably accurate assessments can be made by knowing the amount of flame covered by soot, the visible flame area and the flame colour. SEP for different types of fire is shown in Table 9.6.[15]

A major factor affecting the heat received by an observer is the distance between the flame and an observer. Whilst there are sophisticated calculation techniques, usually relying on computer software, an approximate calculation can be attempted by assuming the flame as a point source halfway along the length of the flame, using Equations 9.3–9.5:

$$Q = 4\pi x^2 \phi_x \tag{9.3}$$

where Q = the total heat radiated (kW)
x = distance from the point to the receiver (m)
ϕ = flux at the receiver (kW/m²)

Equation 9.3 can be rearranged as

$$x = \sqrt{\frac{Q}{4\pi \varnothing_x}} \tag{9.4}$$

and

$$Q = F\,M\,C_v \tag{9.5}$$

where F = Fraction of generated heat which is radiated
M = Combustion rate (kg/s)
C_v = Calorific value (kJ/kg)

F does not exceed 0.4 but can be much lower. Values for a number of common materials are given in Table 9.7. Note that many of these figures are based on small-scale experiments.

This is often called the *point source method* and is shown graphically in Figure 9.5. Until the widespread use of computers, it was typically used for radiation calculation from flares. A similar method can be used for pool fires.

This simple model is useful where the distance between the source and the receiver is at least twice the main dimension (usually the length) of the flame. A more complex approach based on view factors such as that developed by Mudan[17] will give a more accurate estimation of heat received but is impractical as a hand calculation. The view factor approach is most frequently used in the calculation of thermal radiation from pool fires and is discussed further in Section 9.5.

Various publications list the levels of radiation and duration before injury or damage occurs. Typical figures are shown in Table 9.8.

TABLE 9.7
"F" Factor for Some Common Materials (reproduced courtesy TNO)[16]

Material	F Factor
Methanol	0.162–0.177
Methane	0.15–0.34
Butane	0.199–0.269
Gasoline	0.13–0.40
Benzene	0.345–0.36

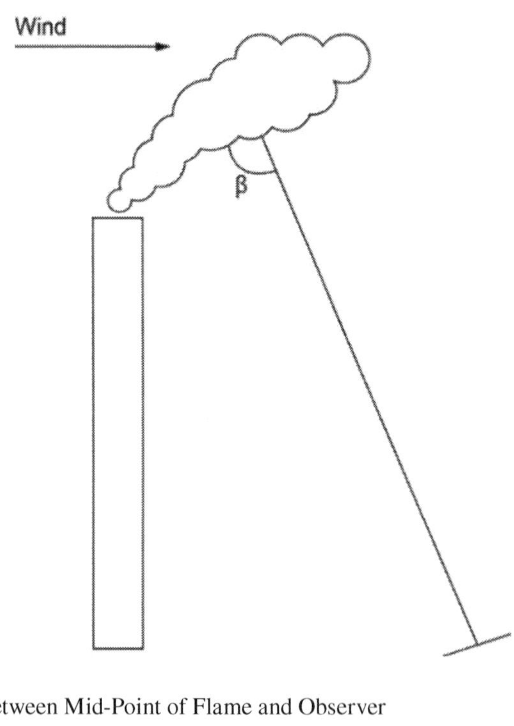

FIGURE 9.5 Distance between Mid-Point of Flame and Observer

TABLE 9.8
Physical Effects of Thermal Radiation Levels

Thermal Radiation Flux kW/m^2	Effect: People	Effect: Other
2	Minimum to cause pain after one minute	
4	Sufficient to cause pain to personnel if unable to reach cover in 20 seconds, blistering of skin likely 0% fatality	
<5	Will cause pain in 15–20 seconds and injury after 30 seconds exposure	
5	Will cause second-degree burns within 60 seconds	
>6	Pain within approximately 10 seconds; rapid escape only is possible	
9.5	Sufficient to cause pain in 8 seconds and second-degree burns in 20 seconds	
10	Pain within 60 seconds, potentially lethal	
12.5	Significant chance of fatality for medium-duration exposure	Thin steel with insulation on side away from fire may reach thermal stress level high enough to cause structural failure
12.5	Fatality	Minimum energy for piloted ignition of wood and melting of plastic tubing
25	Likely fatality for extended exposure and significant chance for instantaneous combustion	Spontaneous ignition of wood after long exposure. Unprotected steel will reach thermal stress temperatures that can cause failure.
25		Minimum energy required to ignite wood at indefinitely long exposure (non-piloted)
35	Significant chance of fatality for people exposed instantaneously	Cellulose material will pilot ignite within one-minute exposure
37.5		Sufficient to cause damage to process equipment

The effect of thermal radiation on humans will vary depending on age, fitness and other factors. Fumes from fires will also have an effect on humans. A comprehensive description of the effects of thermal radiation on humans is given by Hymes, Boydell and Prescott,[18] the Netherlands TNO,[19] the UK HSE[20] and others.

There is a dose relationship between thermal radiation levels and injury based on a $(kW/m^2)^{4/3}s$ relationship. This is illustrated in Tables 9.9 and 9.10.[21]

In addition to the effects of thermal radiation, direct flame impingement or engulfment can also occur. These phenomena can be extremely destructive, and most active firefighting systems will be ineffective. Jet flames are particularly damaging due to the higher temperatures involved by comparison with pool fires. Direct contact with flames or hot surfaces is obviously harmful to humans and other life forms.

9.4.2 Smoke and Fumes

Smoke and the products of combustion (POC) can be harmful in several ways:

1 Soot which chokes the alveoli in the lung leading to asphyxiation
2 Hot combustion products leading to lung damage

TABLE 9.9
Burn versus Physical Dose Relationship (reproduced courtesy of HSE)

Harm Caused	Thermal Dose Unit (TDU) $(kW/m^2)^{4/3}s$ Mean	Range
Pain	92	86–103
Threshold First-Degree Burn	105	80–130
Threshold Second-Degree Burn	290	240–350
Threshold Third-Degree Burn	1,000	870–2,600

TABLE 9.10
Thermal Dose Harm Criteria Guidance (reproduced courtesy of HSE)

Harm Caused	Thermal Dose Unit $(kW/m2)^{4/3}s$
Escape Impeded	290
1–5% Fatality Offshore	1,000
50% Fatality Offshore with Radiation Only to the Front or Back (i.e. Fireball)	1,000
50% Fatality Offshore	2,000
100% Fatality Offshore	3,500

3 Reduced oxygen leading to asphyxiation
4 Combustion products such as CO and CO_2. CO is a toxic product which combines with the blood to form carboxyhaemoglobin, and CO_2 is a potential asphyxiant
5 Combustion products of plastics such as HCN, HCl, and $COCl_2$ leading to toxic effects

Whilst the most obvious danger is from the smoke plumes, particles may be deposited down wind of the fire and decontamination may become necessary. Smoke fumes may also have significant effect on downstream industries. Polluted air from a fire could affect heating, ventilation and air conditioning (HVAC) systems of nearby industries such as food processing and semi-conductor manufacturing plants, which may need to shut down.

9.4.3 Damage to Steel and Other Metals

The reduction in tensile strength of steel is shown in Figure 9.6. The 50% loss of strength is an arbitrary value as there is often a safety factor of two or more. However, clearly at around 450°C structures and vessels may collapse under load. Pressure vessels may rupture.

Forge masters and manufactures of Samurai swords know that at red heat, steels become malleable. The addition of chromium and nickel can increase the strength at higher temperatures but only marginally.

It is important to differentiate the heating of metal by jet flames and pool flames. Jet flames have a heating flux of 200–250 kW/m^2, and pool fires based have a heating flux of between 40 and 45 kW/m^2.

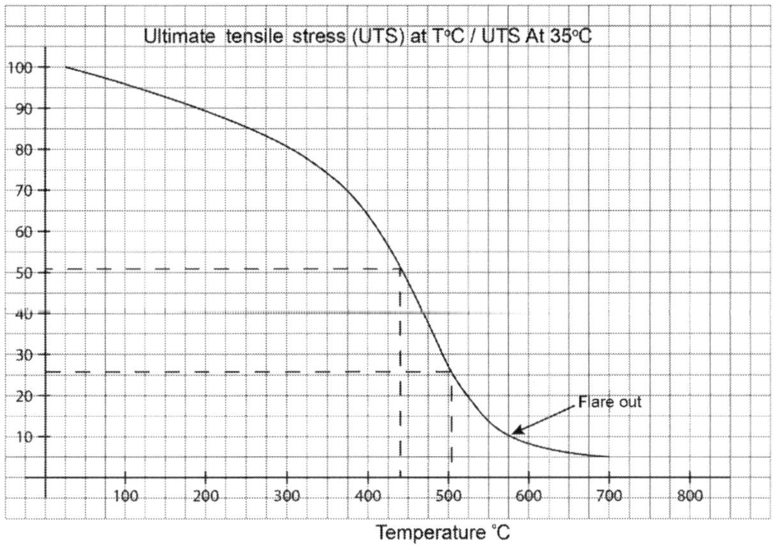

FIGURE 9.6 Reduction in Ultimate Tensile Stress (UTS) for a Typical Steel with Increasing Temperature

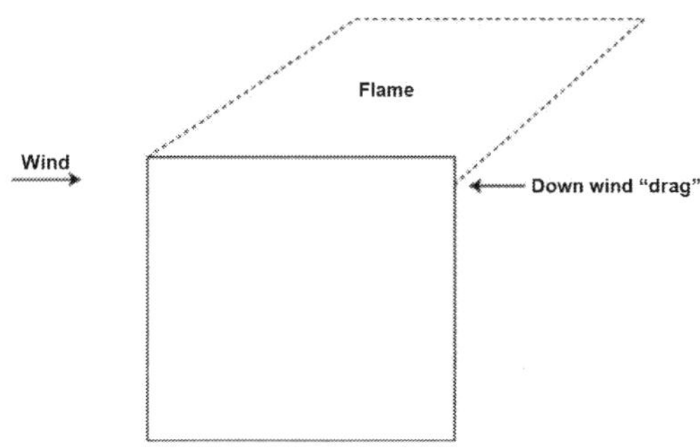

FIGURE 9.7 Pool Fire—Wind Effects

9.5 POOL FIRES

The term *pool fire* is self-explanatory; they range from small fires in depressions in the ground to huge fires in large storage tanks or, following leakage from a tank, in the tank bund. Larger fires from tanks and bunds will have a huge amount of fuel available and may burn out when the fuel is exhausted, rather than be extinguished. Smaller fires, for example from pump seal leaks, are often limited in size with the burn rate matching the leak rate. Flame properties will depend on the nature of the fuel (see Section 9.4). Oil pyrolyses and produces smoke (soot) and does not release the full calorific potential. Liquefied natural gas (LNG) burns relatively smoke free.

The flame from a pool fire is generally modelled as a cylinder. In a no-wind situation the cylinder will be vertical, but the effect of wind results in two effects which increase the thermal radiation load on a downwind observer. One is the tilt of the flame in the down wind direction and the second, more subtle effect is the extension of the flame base downwind, commonly known as *flame drag* or *lift-off*. These effects are demonstrated in Figure 9.7.

The effect of flame tilt is obvious, but the effect of flame "drag", where the base of the fire extends downwind, is often forgotten. The thermal radiation hazard from flame tilt and drag is obvious on the downwind side of a fire. However, the hazards on the upwind side should not be ignored, particularly in a tank farm where thermal radiation on up wind tanks can increase vaporisation of flammable liquids which are blown downwind and are ignited by the flame on the original tank fire. A detailed introduction to pool fires is provided by Stark.[22]

The shape or *aspect ratio* of a vertical pool fire is given by the Thomas equation.[23]

$$H = 42\,D\left(\frac{\dot{m}}{\rho_a\sqrt{gD}}\right)^{0.61} \tag{9.6}$$

where $\dot{m}$ = Burning rate of liquid (kg/m²/sec)
H = Height (m)
D = Pool Diameter (m)
ρ_a = Density of air (approx. 1.22 kg/m³)
g = Acceleration due to gravity (9.81 m/sec²)

In circumstances where the diameter of the pool is based on the liquid leakage feeding the fire, the diameter can be calculated from Equation 9.6.

$$\text{Outflow}(\text{kg}/s) = \frac{\pi D^2}{4}\dot{m} \tag{9.7}$$

The burn rate ($\dot{m}$) can be found from tables but is typically about 0.1 kg/m²/sec. It can be calculated more accurately from Equation 9.7.

$$\text{Burn rate } (\dot{m}) = 0.001\ C_v/L_H \text{ kg/m}^2\text{/sec} \tag{9.8}$$

where C_V = Calorific value (kJ/kg)
L_H = Latent heat of evaporation of the fuel (kJ/kg)

For heavy fuels, in particular, the burn rate may be as low as 0.05 kg/m²/s. This creates a significant issue when extinguishing fires. The surface of the unburnt oil residue can be in the range 400–500°C and as a result fire-fighting reagents may be destroyed on contact with the surface. Further if there is any water under the oil (which is not unusual) a *heat wave* (hot oil layer) descends. Eventually it will come into contact with any water present which then boils semi-explosively producing a *boil-over* which ejects burning fluids out of the tank. The mechanisms of boil-off are complex, but, simply put, the oil is heated by the fire and lighter components boil off and burn. The now hotter oil has a higher density than the underlying unburnt oil and sinks down through the tank contents at approximately 1 m/hr. This is essentially a particular problem with crude oil.[24]

Calculation of the thermal radiation received from a pool fire using the point source method of Equation 9.2 is relatively inaccurate and most calculations are based on a *view factor* (the proportion of the total thermal radiation emitted by the source which is received by the observer) calculated by splitting the cylinder surface into a large number of small segments and calculating the radiation from each of these to an observer. The view factor F_{12} is defined as

$$F_{12} = \frac{1}{A_1}\int_{A_1}\int_{A_2}\frac{\cos\varnothing_1\cos\varnothing_2}{\pi S^2}\,dA_2\,dA_1 \tag{9.9}$$

The terms in Equation 9.9 are shown in Figure 9.8.

The graphical solution of Equation 9.9 is shown in Figure 9.9, where H = height, R = radius and X = distance from the flame centre (all in consistent units).

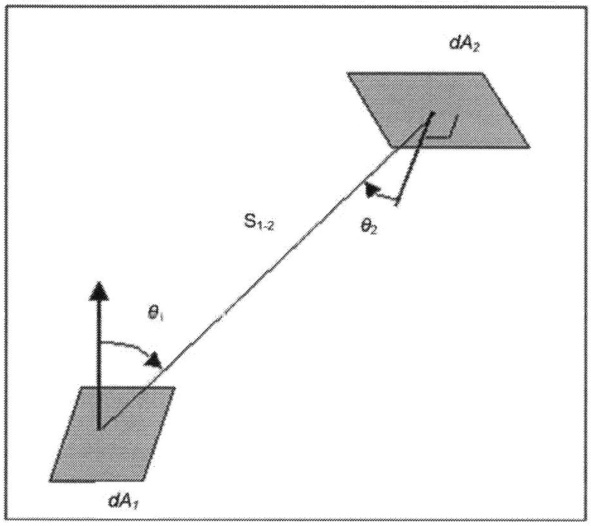

FIGURE 9.8 Graphical Illustration of Terms in Equation 9.9

Calculating the view factor for tilted flames is complex. The tilt must first be assessed, and then a set of tables used to estimate the view factor. This is too complex for an introductory text. Whilst charts for various view factors are available it is no surprise that calculations are generally performed by computer.

Once the view factor has been determined, it is a relatively simple matter to calculate the thermal radiation flux, as shown in Equation 9.10.

$$\varphi_x = F_{12}\, \varepsilon\, \tau \tag{9.10}$$

where φ_x = Flux at point x (kW/m^2)
F_{12} = View factor (dimensionless)
ε = Surface Emissive Power of the source (kW/m^2)
τ = Atmospheric attenuation based on water content in the air (for short distances this tends to be 1)

A simple but real analysis of the emergency plan for fire attack on a storage tank is shown in Figures 9.10 and 9.11. Obviously, the "real event" cannot be staged so a "desk top" approach (in military terms, a war game) has to be adopted. Plots of this type are frequently used for emergency planning (see Section 22.3) and can be used for plant layout.

Figures 9.10 and 9.11 are based on a 10 m diameter, 10 m high storage tank. The fuel is hexane (equivalent to petrol), and a wind speed is 5 m/s. The modelling was done by one of the many available computer programmes. The calculations are based on ground-level heat fluxes—elevated ones will be higher. The surface emissive power of a smoky flame, such as hexane, is 75–100 kW/m^2 (which is often a hidden detail inside the programme). This calculation assumes 75kW/m^2. The flame tilt is 44° to the horizontal.

Figure 9.10 shows the dimensions of the tank and flame, with flame tilt (44 degrees to horizontal) and flame lift off the "down-wind drag" as shown in elevation. The part from –22 m to 0 m at ground level from the tank wall is in the tank shadow.

Figure 9.10 shows the tank wall and a line from the tank wall used to show the limiting maximum heat flux at the base of the tank—it cannot go inside the tank! It also shows that the two

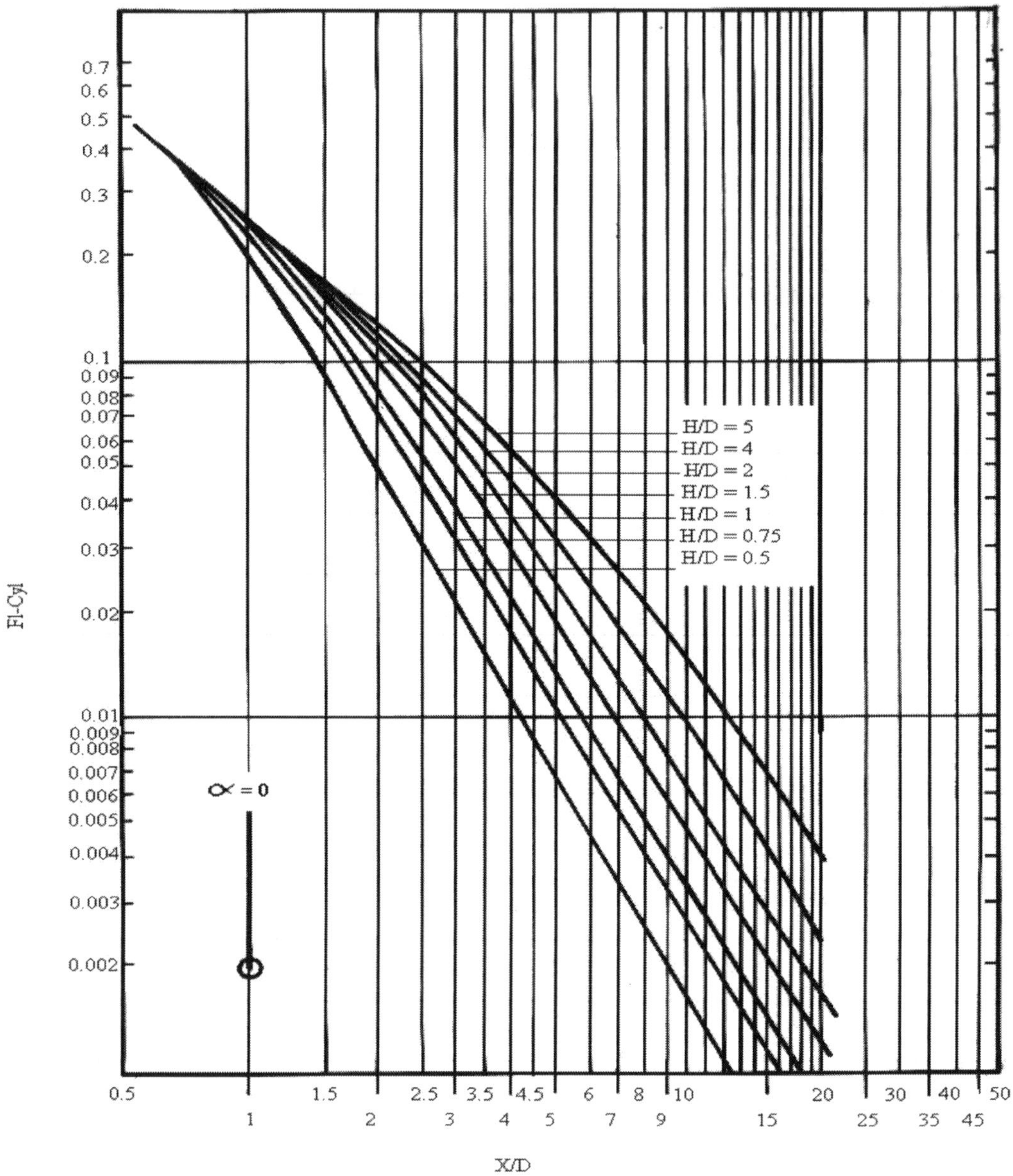

FIGURE 9.9 View Factor for a Vertical Cylinder[25]

elements in the model do not map perfectly. In Figure 9.10, the shadow is –22 m from the tank, but in Figure 9.11, the radiation effects are noticed at –20 m. Once again, the flux is shown at ground level. This simple elevation model ignores the fact that any movement off the wind axis will produce a different view factor which will result in an "*egg shaped*" profile. For example, standing at the tank base at right angles to the wind axis the receiver will still "see" the flame and be subject to thermal radiation. The flux at 8 m from the tank and 20 m from the centre line will be 6.6 kW/m^2.

The faint line starting at $x = -20$ m is the flux on the upwind side, and the bold line is the ground-level flux on the downwind side. The peak flux at ground level is 0.75 kW/m^2 upwind and 19 kW/m^2 downwind side. The peak flux is at +18 m which is the flame centre of mass, as in Figure 9.11.

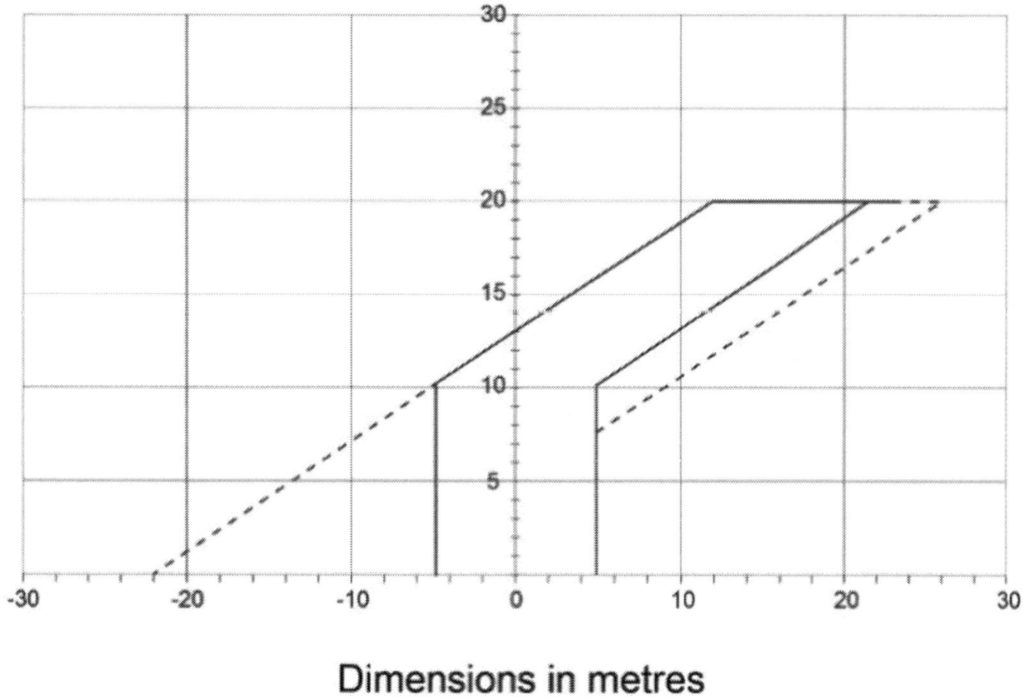

FIGURE 9.10 Graphical Plot of Storage Tank Flame Shape and Size

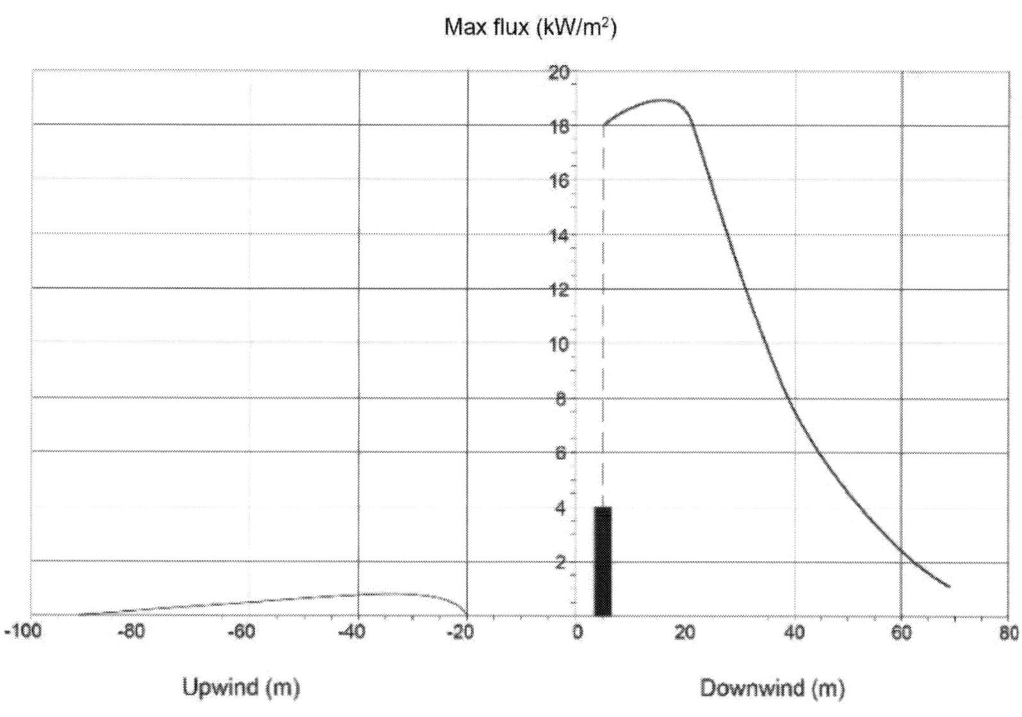

FIGURE 9.11 Heat Flux on the Centre Line of the Wind through the Tank Centre along the Wind Axis

Figures 9.10 and 9.11 demonstrate the advantage of approaching a fire from the upwind direction—where thermal radiation levels will be lower and there will be no products of combustion. For a firefighter the flux at 40 m downwind would be uncomfortable but just tolerable behind a water mist barrier. Upwind the flux based on 2 m above ground will be 1 to 1.5 kW/m^2.

In the example given earlier, hexane has a boiling point of less than 100°C and can be readily extinguished by firefighting foam (see Section 9.12) sprayed into the tank from a safe upwind location.

9.6 JET FIRES

The most common example of a jet fire is a welding torch where good mixing produces a very hot flame with little or no smoke. A simple calculation for flame length from a gaseous release has been provided by Brzustowski.[26]

$$\frac{F_{len}}{D_{jet}} = \frac{1050}{C_d}\sqrt{\frac{M_a}{M_f}} \tag{9.11}$$

where F_{len} = Flame length (m)
D_{jet} = Jet diameter (m)
C_d = Lower explosive limit (vol%)
M_a = Molecular weight of air
M_f = Molecular weight of fuel

Most serious jet fires are gaseous, but small-scale liquid fires may occur, for example, "pinhole" leaks in lube oil systems.

One study[27] of the flame characteristics of Liquefied Petroleum Gas (LPG) gives a correlation of flame length against discharge rate, which can be expressed by the following relationship:

$$L = 12\ W^{0.5} \tag{9.12}$$

where L = Flame length (m)
W = Discharge rate (kg/s)

Alternative estimates by Crawley, based on experiments at Spadeadam in the UK, give the following simple correlations for

$$L = 10\ W^{0.46}\ \text{(methane)} \tag{9.13}$$

$$L = 17\ W^{0.46}\ \text{(LPG)} \tag{9.14}$$

9.7 FIREBALL

A fireball is a sudden large ignition of a flammable gas cloud; it is a short-lived phenomenon but creates very intense levels of thermal radiation for the few tens of seconds during which it exists. A number of formulas[28] have been developed for the size and duration of fireballs; the Martinsen and Marx[29] are commonly used with, the diameter of a fireball can be calculated from

$$D = 5.8M^{1/3} \tag{9.15}$$

where D = Maximum fireball diameter (m)
M = Mass of fuel in fireball (kg)

The duration is calculated from

$$T_d = 0.45\ M^{1/3} \text{ if } M < 37{,}000 \text{ kg} \tag{9.16}$$

$$T_d = 0.26\ M^{1/6} \text{ if } M < 37{,}000 \text{ kg} \tag{9.17}$$

where t_d = Fireball duration (s)

Martinsen and Marx have also provided a simple calculation method for estimating the view factor from a fireball.

$$F = \frac{R^2 \sqrt{((h^2 - R^2)}}{h^3} \tag{9.18}$$

where F = View factor
h = Distance between observer and centre of fireball (m)
R = Radius of fireball (m)

Moorhouse and Pritchard[30] and others have provided a calculation method for the SEP of a fireball, but this can be complex and an arbitrary figure of 235 kW/m^2 is frequently used.

9.8 BOILING LIQUID EXPANDING VAPOUR EXPLOSION

Whilst termed an explosion, the effects of a BLEVE are similar to a fireball although there might be some limited overpressure. A fireball occurs where the metallic surface (usually in the vapour space) of a vessel containing flammable liquid above its boiling point is heated to a temperature where the metal yield strength is exceeded, as shown in Figure 9.6. The vessel will rupture, releasing a large cloud of flammable gas which ignites.

It is often erroneously, and fatally, assumed that the Pressure Safety Valves (PSVs) will operate and protect a pressure vessel exposed to fire from rupture. This is emphatically not the case. The fire sizing of a PSV is based on heat input to the vessel, raising the temperature of the contents and causing boiling with a requirement to relieve the vapours. This assumes that metal temperature of the vessel wall does not increase significantly—which it will if the unwetted area above the liquid level is exposed to very high levels of thermal radiation—generally as a result of direct flame impingement.

Radiation effects of a BLEVE can be calculated in the same way as a fireball.

9.9 RUNNING FIRES

"Running fires" are fires which are running along a surface or structure. Classic examples are an aeroplane on the ground with fuel running from broken fuel tanks, and a cascade down a structure from ruptured containment.

9.10 SOLID FIRES

Fires involving solid materials vary tremendously in size and shape and, unlike most liquid fires, are often located within buildings, which will affect the shape and dimensions of the flames. In most cases, calculation is complex and may involve advanced computer modelling. As well as thermal radiation exposure, the hazards from toxic materials or asphyxiation must be considered.

9.11 FLASH FIRES

Flash fires are the result of a low-level cloud of flammable gases or liquids. The ignition does not create any explosive effects (see Chapter 10) but the flame travels at a few metres per second, and it is unlikely that a person close to the flash fire will be able to escape. Skin and lung damage will result with, potentially, fatal results.

9.12 FIRE PROTECTION

The level of fire protection required at a specific location should be based on an assessment of the potential risk to people, property and the business. Whilst the use of codes and standards can be used as a guide, their application without thought can lead to expensive, non-optimal systems which may be difficult to maintain. It is not unusual to find a high level of protection in some ancillary areas but much less protection in areas where there are high levels of fire hazard or where a fire could result in significant operational difficulties and loss of revenue.

9.12.1 Fire Prevention

Fire prevention is a mixture of common sense, plant design and good operating procedures, including

1 Good housekeeping by preventing the build-up of combustible or flammable materials.
2 Control of ignition sources by
   (a) Smoking controls
   (b) Control of hot work (such as cutting, grinding and welding). This is covered in more detail in Section 16.6.
   (c) Separation of combustible/flammable materials from sources of ignition by separation or the erection of barriers (see Section 13.8).
   (d) Good earthing practices to avoid static discharge (generally with a ground resistance of less than 10 ohms).
3 Removal of ignition sources by the electrical area classification and intrinsically safe systems (see Section 14.6).
4 Good equipment maintenance to prevent leakage of flammable gases and liquids and ignition sources (such as overheating bearings or slipping conveyor belts).

Fires vary tremendously in size, and whilst many fires are relatively minor, they may have significant consequential effects, for example:

Clean Rooms	A minor fire in the clean room (for example, in an electronics manufacturing plant) will likely cause damage to only one or two equipment items, but fumes will have spread throughout the facility via the ventilation system—requiring expensive and time-consuming decontamination.
Control Rooms and Motor Control Centres (MCCs)	A fire in a control room or MCC could require months to repair, and during repair production would be halted. In some cases, alternative control may be possible but is generally less efficient. Even if replacement equipment is easily available, the confined area within these rooms will limit the number of personnel who can be present at one time and thus limit the rate at which repair work can take place.
Electrical Cabling	A fire in a cable tray or underground cable basement will require replacement of the affected cables. Splicing of new cable sections to replace damaged cables is sometimes possible but generally the entire cable length is replaced. For reasons of access, the number of people working on repairs at any one time is limited.

9.12.2 Fire Detection

There are a number of devices used to detect fire, and many also activate an automated firefighting system or a process shut down. In many cases the best fire detector is a human (with sight, sound and smell), but it is impractical for every plant area 24 hours a day. The most common types of detectors are shown in Table 9.11.

TABLE 9.11
Common types of Fire Detection

Parameter	Type	Comments
Smoke[31,32]	Ionic	These are a cheap and common type of detector in widespread use. They contain a small radioactive source which identifies the ionization of smoke particles within a chamber. They are susceptible to false alarms in areas of high humidity.
	Optical	These operate by identifying an increase of light scatter caused by smoke particles obscuring a light beam
	Combined smoke and heat	These generally consist of an optical smoke detector in conjunction with a heat detector, as discussed later.
	Combined Smoke and CO detection	
	Aspirating smoke detectors	These take air samples from an enclosed area for analysis. They give early warning of a fire—generally before smoke is visible. They are often installed in high-value areas with significant amounts of cabling such as control rooms, instrumentation areas and electrical rooms.
	Optical beam detectors	These are a relatively recent development and measure light obscuration by smoke over a distance. These have either a separate transmitter and receiver or a combined unit with the signal returned by a reflector. Earlier systems suffered from a number of faults, but more modern units are reported to be more reliable
Heat	Heat	Instrumented heat detectors are often used where smoke detection is not practical.
	Rate of rise heat detection	There are similar to standard heat detectors but measure the rate of temperature rise. They are typically used in locations such as turbine hoods and kitchens.
	Mechanical types	There are various types of mechanical detectors which fail when a specified temperature is reached. These often activate a fixed protection system. Typical examples are 1 Frangible bulbs (used on sprinkler systems and deluge systems) 2 Heat-sensitive wire (which might fail or raise an alarm by a change in electrical resistance) 3 Melt tubes (generally a pressurised tube which will melt and release internal pressure, activating an alarm)
Flame[6]	Infrared (IR) and ultraviolet (UV) flame detection.	Early flame detectors were ultraviolet (UV) types, but these have largely been superseded by IR or joint UV/IR detectors which are more reliable. Common applications include turbine hoods and compression areas. Flame detectors may spuriously be activated by low-level sun and steam producing a "flicker frequency".

The design and specification of detectors is a specialist activity. Factors to be considered include the following:

1 Environment—heat, humidity and contaminates can affect the operation of detectors.
2 Location—smoke detectors, for example are generally located just below roof level where smoke will be trapped.
3 Signalling—detection systems should alarm to a permanently occupied area where staff are familiar with the alarm panel and can easily identify where an active alarm has been generated as well as being able to identify fault conditions and arrange immediate repairs.
4 Frequency and type of testing.

Most countries will have regulations or standards covering fire alarm systems. Good general information can be found in NFPA 72.[33] As stated earlier, the design, installation and maintenance of fire alarm systems is a specialist activity, and untrained personnel can easily make mistakes. Some examples are given in the following table:

Smoke Detectors	Smoke detectors were located in an open area where smoke would not be trapped, and the detectors would be ineffective. A "hat" to trap smoke close to the detector can be installed.
	Ionizing-type smoke detectors were installed in a paint spray booth and regularly gave false alarms. To prevent this, a plastic bag was placed over the detector to prevent it operating. This problem is often found in illicit "smoking dens" (areas unofficially used by smokers).
Alarm panels	An alarm panel was located in a location which was only occupied during the standard working day. When a fire occurred in an area covered by the panel it was not detected until the fire was well established. The fire occurred in a room containing weld inspection records on a construction site. As a consequence, all the weld inspections needed to be repeated.
	Frequently alarm panels are installed without adequate training of operators and instrument technicians who do not know how to respond when a fault occurs. For smaller organisations a contract with a specialist service company is usually the best option. Larger companies may have suitable resources in-house.

9.12.3 Fire Extinguishing Systems

Whilst the sight of a fire truck travelling to a fire with sirens wailing and lights flashing will be a familiar to most people, a level of fixed protection is often installed to allow a rapid response to a fire or where mobile firefighting resources are unlikely to be adequate.

The design of fire protection systems should start with a review of fire hazards and their consequences, followed by the specification of the protection systems. Protection may be *passive* or *active*. Active systems require manual or automated activation—which may fail. Passive protection requires no action. As with detection, the design of fire extinguishing systems is a specialist activity, and the information provided is for introductory purposes only. Examples of poorly designed fire protection systems are given in Section 9.13. An overview of protections is given in Tables 9.12 and 9.13.

9.12.4 Fire Water

Except for plants in remote, often desert, areas the fire protection will include a fire water system to supply water sprinkler, deluge and foam systems, plus manual firefighting using hose reels, hydrants and monitors. The design of the fire water system has evolved over the years, and although there are differences in detail between different parts of the world, the basics are fairly standard in the following respects:

1 The fire pumps must be able to provide fire water via two independent means—generally electric and diesel pumps, but there are variations such as the use of steam turbine drives or a dedicated generator to back up the main electrical supply.
2 Adequate fire water must be available to fight a fire for a minimum specified period of time. Two hours is a common requirement, but this will be inadequate for many hydrocarbon fires.
3 The fire water main should be looped with sectional isolation valves so that damage to one section of piping will still allow flow via the other side of the loop.
4 The fire main must be protected from accidental damage such as vehicle impact and earthquakes (where applicable). For this reason, fire mains are generally buried, except in earthquake-prone areas. Fire hydrants and monitors are often protected by "corner post" crash barriers.
5 Provision for flushing the fire main and testing of the fire pump head curve on an annual basis is required.

TABLE 9.12
Common Passive Fire Protection Methods

Passive Fire Protection (PFP)		
Type	**Details**	**Common Usage**
Fire-Resistant Coatings	Fire-resistant cement is often applied to structural supports to provide resistance to high levels of thermal radiation. Specialist cements and coatings may have limited impact strength, and it is common practice for the bottom 1–2 m of a vertical support to have a standard concrete covering with PFP above that height. Note that water ingress between steelwork and the covering PFP can occur with results similar to corrosion under insulation (CUI) (see Section 20.5). Where coatings are exposed to a prolonged fire, they will generally require replacement. Whilst large companies will have internal standards, API 2218[34] provides useful guidance.	Protection is often installed on the following: 1 Structural supports where liquid pool fires may occur 2 Equipment supports and saddles where a pool fire may occur 3 Legs of LPG storage spheres
Fire Panels	Standard wall and door components can be used to provide a level of fire resistance. Components are available for internal and external use.	Protection is often installed in industrial areas such as 1 Fire refuges around plant and storage areas 2 Accommodation modules on offshore platforms 3 Protection around critical instrument and electrical cabling. National building codes will often require a level of passive fire protection in buildings, particularly in escape routes.
Intumescent Paint or Coatings	Intumescent paint or coatings are widely applied to the surfaces which may be exposed to fire to prevent temperature rise in vessels and possible rupture. Standard insulation should not be considered as providing fire resistance, although some allowance can be made for insulation when calculating the heat input into a pressure vessel for calculating the size of pressure safety valves (PSVs) on fire-exposed vessels. The original intumescent coatings were not suitable for hydrocarbon fires, but coatings suitable for use in hydrocarbon areas have now been developed and are increasingly being used as a replacement for cement-based coatings. As in all cases, care is necessary in the specification.	Protection is commonly installed on the surfaces of pressure vessels in process areas and can also be used on structural steelwork. Coatings are sometimes fitted retrospectively to protected areas where the fire hazard was not identified at the design stage.
Secondary Containment of Spills	Secondary containment and ground sloping can contain spills of flammable materials or direct them away from sources of ignition.	Bunds around atmospheric storage tanks and sloping ground under LPG storage draining to impounding basins for pressurised storage of liquefied gases. "Pipe-in-pipe" systems where hazardous fluids are transported through an inner pipe with an outer pipe to contain any leakage from the inner pipe. Usually there would be some form of detection system to indicate failure of the inner pipe.

Note: The level of fixed protections will vary with the size and complexity of the site. There is increasing use of fixed systems of all kinds as staffing levels are reduced.

TABLE 9.13
Common Active Fire Protection Methods

Active Fire Protection		
Type	**Details**	**Common Usage**
Sprinklers	These consist of a network of pipework covering the protected area with nozzles located at fixed intervals. The nozzles contain temperature-sensitive bulbs which rupture at a specified temperature, releasing water onto fire area. Systems are either *wet* (full of water) or *dry* (full of compressed air; when the pressure drops following rupture of a heat-sensitive bulb a valve opens allowing water into the sprinkler network). Dry systems are generally installed in areas where freezing may occur. NFPA13[35] is a commonly used guide for the design of sprinkler systems with the spacing of sprinkler nozzles and the water density varying for different levels of hazard. Sprinklers are a very reliable means of firefighting; the major cause of failure is accidental (or deliberate) closure of the main isolation valve. To prevent this the isolation valves are usually locked open. Design of all water-based systems must be considered in conjunction with drainage requirements—undersized drains can result in an expanding pool fire floating on undrained fire water.	Within buildings, especially warehouses and large open structures. Standard sprinklers are not generally used in electrical areas but are sometimes installed in cable basements and cable racks.
Water Deluge	These are similar to sprinkler systems but the nozzles do not contain heat-sensitive bulbs—all the nozzles are open to atmosphere. They are activated either manually or by a heat detection system, most commonly by a compressed air system containing heat-sensitive nozzles or a "melt tube".	These are commonly found in process areas such as: 1 Compressor houses 2 Pump seals 3 Steam turbine bearings 4 Major oil-fired transformers 5 Surface of pressure vessels (especially large storage vessels for liquefied gases such as spheres or "bullets") 6 LPG truck loading areas
Atomised Water	Generation of a highly atomised water spray. These systems operate by generating a mist to isolate fuel from air rather than by cooling.	Lube oil systems and electrical rooms
Foam	These consist of a foam concentrate which is mixed with water in a proportioner and then applied to the surface of a pool fire. Foam systems are defined as fixed or semi fixed. Fixed systems include the entire system from the foam storage and proportioning equipment, through pipework to the point of application. Semi-fixed systems do not contain the foam storage and proportion; a mobile fire truck is connected to the system to supply water and foam. A proportioning device, typically a venturi-type valve, is required to mix the water and foam to the correct ratio. The foam is applied to a fire via pourers, on storage tanks, or a piping network to flat areas such as a liquid fuel tanker loading area.	Fixed foam systems are commonly installed on 1 Storage tanks (on floating roof tanks the foam system will generally only protect the rim seal area). 2 Liquid fuel loading areas.

(Continued)

TABLE 9.13 *(Continued)*
Common Active Fire Protection Methods

Active Fire Protection		
Type	**Details**	**Common Usage**
Carbon Dioxide	Carbon dioxide systems are installed in enclosed areas to "suffocate" the fire by lack of oxygen. The systems are not suitable for locations which will be regularly occupied and provision must be made to isolate the system before entry into any protected area.	Commonly installed in: 1 Gas turbine hoods 2 Electrical areas They are sometimes used in *snuffing systems* to extinguish fires on gas vents which have ignited.
Gaseous Fire Extinguishing Gases	Fixed systems to supply firefighting gases are widely used. There are a number of different types which work either by reducing the oxygen concentration or chemically reacting with the flame.	Generally installed in areas with control, computer or communications equipment but are sometimes used in "snuffing systems" to extinguish fires on gas vents which have ignited.

A comprehensive range of fire protection standards has been developed by the United States National Fire Protection Agency (NFPA);[36] these cover both design and maintenance of firefighting systems.

9.12.5 Firefighting Media

A wide range of firefighting media have been used over the years. Some of the earlier media have been superseded either because improved alternatives are available or because use has been discontinued for safety or environmental reasons. The most common media currently used are shown in Table 9.14.

Whilst a number of other media are available, they are not in widespread use or are used in specialised circumstances.

9.12.6 Firefighting Tactics

Firefighting, particularly of a large fire, is a dangerous and complex business and not something to be attempted by untrained individuals. The long-standing advice is for an individual to report a fire and walk away unless they are confident of being able to tackle it. Most large process sites have pre-plans as part of the overall emergency response plan (ERP) (see Section 22.3). The basics of fighting fires most likely to be found in process plant are shown in Table 9.15. These are broad guidelines only.

In all cases, non-essential personnel should be evacuated to a safe distance which could be several hundred metres away where there is a possibility of a pool fire boilover, fireball or BLEVE. The best action may be to allow the fire to burn itself out whilst protecting adjacent areas and transfer all personnel, including firefighters and other personnel, to a safe distance.

Fires should always be fought from the upwind side as flames and smoke (plus other toxic fumes) will be blown away from the firefighters. Water jets will also travel further and not "feather" in the air.

TABLE 9.14
Common Firefighting Media

Type	Details
Water	This is the most widely used media; it is cheap and widely available in most locations. It is best in carbonaceous fires. In most cases water will not extinguish a fire directly but will remove heat from the fire, often reducing the temperature below that where the fire is self-sustaining (auto-ignition temperature). However, slow exothermic reaction may result in a delayed re-ignition fire. This is why firefighters are often in attendance for some hours after the fire is notionally extinguished to "damp down". Atomised water deluge systems do extinguish the fire by depriving it of oxygen but are only suitable for small-scale fires. The water quality is important and regular flushing of systems is necessary to limit corrosion in pipework and nozzles. Offshore and coastal areas often use seawater in firefighting systems. Where this is the case, comprehensive flushing of areas impacted by firewater will be necessary to prevent salts in the seawater causing corrosion. Alternatively, corrosion-resistant piping materials could be used, but this is generally only economic on offshore locations. There are a few substances which react with water to generate flammable liquids or gases; these should be identified as part of the hazard identification process and alternatives to water used.
Foam[37]	Foam concentrate for use in pool fires is available in several different types with different characteristics. Some types have a good initial "knock down" capability but will break down relatively quickly—possibly allowing the fire to re-establish itself. Standard foam will not work on non-polar solvents such as methanol, and special types have been developed for these materials. Most foam use is based on forming a barrier between the burning liquid fuel and oxygen in the atmosphere, but there are occasional uses of high-expansion foam to totally fill enclosed spaces. Foam can also be used to form a seal over spills of liquefied gases to prevent evaporation. There have been concerns that some foam formulations can contaminate water supplies, and this is an area of current development.
Carbon Dioxide	This is a cheap readily available gas widely used to extinguish fires by depriving them of oxygen. Whilst effective its use is limited to unoccupied areas to avoid asphyxiation hazards.
Firefighting Gases	Gases for firefighting in confined areas have been available for many years. Some of the original gases were implicated in depletion of the ozone layer and are no longer produced, but they may still be found in some older installations. These gases typically interact chemically with the fire to stop combustion by eliminating the "free radicals" which are generated in a fire and are important in the flame propagation process. The availability of some gases may be limited in some parts of the world, and it is important to ensure that systems can easily be refilled after discharge.
Dry Powder	A number of dry-powder-type extinguishing media are available; these are usually delivered from portable equipment rather than fixed systems.

TABLE 9.15
Common Firefighting Tactics

Type	Details
Pool Fires	1 If the pool is being fed from a leak (such as a pump seal), isolate the fuel supply 2 Blanket the pool with foam 3 Cool the surrounding areas to reduce thermal radiation damage
Jet Fires	1 Cool the surrounding areas to reduce thermal radiation damage 2 Isolate the source of fuel supply 3 DO NOT attempt to extinguish the flame—unburnt fuel will continue to be released which is likely to re-ignite, probably at some distance from the original fire.
Fireball	There are no active protective measures possible—prevention is the best option.
BLEVE	If the potential for a BLEVE exists, keep the surface of the affected exposed vessels cool.

9.13 WHAT CAN GO WRONG?

The design of firefighting systems is a specialist activity, and poor designs can result in equipment that is difficult to operate and maintain. The following examples are some of the problems that can occur:

Fire Water storage	A single tank was installed to provide both fire and utility water. Both fire water and utility water offtakes were at the bottom of the vessel which, in times of significant utility water usage, would leave no water available for firefighting. Where a single tank is used for both utility and fire water, the utility offtake should be part way up the tank to ensure a portion of the tank contents is "reserved" for fire water use.
Fire Water Pumps	The design included a turbine flow meter on the discharge of the fire water pump—this would impose a significant pressure drop in the line, possibly affecting the performance of downstream equipment. This is an example of an engineer consulting standards and not applying critical thought.
Foam System	The designer decided to construct a centralised foam system covering several widely spaced tanks. Foam concentrate was stored in a centrally located tank and mixed with water to form foam solution before being delivered to the storage tank on fire. The volume of the distribution piping from the concentrate tank to the storage tank was approximately equivalent to the foam requirement for each tank. By the time the foam in the foam concentrate tank was exhausted, the piping was full of foam, but none had yet reached the tank. Again, this is an example of poor design and was probably more expensive than pumping foam concentrate to a foam/water proportioner close to each tank—the concentrate piping being much smaller in diameter than that required for foam solution.
Carbon Dioxide Fixed System	A new package boiler facility was being designed, and to comply with local fire regulations. A fixed fire protection system was required. The boilers were located within a steel-framed building with metal sheeting roof and walls. Main access to the building was via a number of roller shutter doors which would normally be left open to avoid high temperatures in the building. The building designer decided that the most cost-efficient means of providing fire protection was via a fixed carbon dioxide system. However, it is unlikely that the system would have worked effectively. With the roller shutter doors open, the carbon dioxide concentration would be inadequate to suppress the fire. If the doors were shut and the system was triggered, anyone in the boiler house would probably have been asphyxiated.

The ideal situation is not to have a fire in the first place. However, there will always be the possibility that an unexpected leak or ignition will occur, and firefighting measures must be in place. Many fires start because of ignorance or lack of attention to detail. For example, the burning of rubbish can ignite nearby buildings when the wind changes and the rubbish is more combustible than thought.

The training, competence and knowledge of firefighters might also be suspect, particularly local authority firefighters who have limited knowledge of process hazards. In at least one case, the competence of the local authority firefighters was considered poor and they were kept at the gate of the site whilst shift personnel (successfully) deal with the fire.

REFERENCES

1 UK Health and Safety Executive, *Buncefield Response Reports and Recommendations Arising from the Competent Authority's Response to the Buncefield Incident*, www.hse.gov.uk/comah/buncefield/
2 Refinery Fire at Feyzin, 4 January 1966, www.hse.gov.uk/comah/sragtech/casefeyzin66.htm
3 UK Health and Safety Executive, *PEMEX LPG Terminal*, Mexico City, Mexico, 19 November 1984, www.hse.gov.uk/comah/sragtech/casepemex84.htm
4 www.nfpa.org/news-blogs-and-articles/blogs/2024/04/17/what-is-an-ignitable-liquid-and-how-is-it-classified
5 Lees F P, *Loss Prevention in the Process Industries* (Table 16.4), Butterworth Heinmann, 1991

6 https://www.travelers.com/resources/business-industries/agribusiness/reducing-hay-stack-and-hull-pile-storage-fire-risk
7 Personnel Communication
8 AXA XL Risk Consulting, *Prevention of Explosion and Fire in Compressed Air Vessels and Systems* (PRC.3.1.1), AXA, https://axaxl.com/-/media/axaxl/files/pdfs/prc-guidelines/prc-3/prc311preventionofexplosionandfireincompressedairvesselsandsystemsv1.pdf
9 AIChE, *Process Safety Beacon: Battery-Powered Devices Can Be Ignition Sources*, www.aiche.org/resources/publications/cep/2023/july/process-safety-beacon-battery-powered-devices-can-be-ignition-sources
10 Finke J, Electrostatic Effects of Charged Steam Jets, *Journal of Electrostatics*, 23, 1989, www.econologie.com/dopage-eau/Ionisation_vapeur_detente.pdf
11 Hedlund F H, *Do Not Release Carbon Dioxide If Flammable Vapors Are Present—Static Electricity May Lead to Explosions*, DTU, 2018, https://backend.orbit.dtu.dk/ws/portalfiles/portal/146146696/Do_Not_Release_Carbon_Dioxide_If_Flammable_Vapors_Are_Present_Static_Electricity_May_Lead_To_Explosion_Science_Trends.pdf
12 Butterworth G J, Dowling P D, Electrostatic Effects with Portable CO2 Fire Extinguishers, *Journal of Electrostatics*, 11(1), July 1981, www.sciencedirect.com/science/article/abs/pii/0304388681900759
13 Fishwick T, Loading and Unloading of Road and Rail Tankers—Hazards, Good Practice and Case Studies, *Loss Prevention Bulletin (No. 250)*, IChemE, August 2016, www.icheme.org/media/2088/lpb250pg15.pdf
14 Van Den Bosh C J & Weterings R A P M (eds), *Methods for the Calculation of Physical Efforts (CPR 14E)—The Yellow Book Chapter 6*, TNO, 1996
15 Jagger S & O'Sullivan, Human Vulnerability to Thermal Radiation Offshore (HSL/2004/04), *Health and Safety Laboratory*, https://webarchive.nationalarchives.gov.uk/ukgwa/20230103150842mp_/www.hse.gov.uk/research/hsl_pdf/2004/hsl04-04.pdf
16 Van Den Bosh C J & Weterings R A P M (eds), *Methods for the Calculation of Physical Efforts (CPR 14E) Table 6.6—The Yellow Book*, TNO, 1996
17 Mudan K S, Geometric View Factors for Thermal Radiation Hazard Assessment, *Fire Safety Journal*, 12(2), October 1987, www.sciencedirect.com/science/article/abs/pii/0379711287900245
18 Hyams I, Boydell W & Prescott B, *ICHEME Monograph: Thermal Radiation Physiological and Pathological Effects (Major Hazards Monograph)*, IChemE, 1996
19 Various, *Methods for the Determination of Possible Damage to People and Objects from Release of Hazardous Materials (CPE16E) The Green Book*, TNO, 1989
20 The Steel Construction Institute, *Protection of Piping Systems Subject to Fires and Explosions (Research Report 285)*, HSE, 2004, www.hse.gov.uk/Research/rrpdf/rr285.pdf
21 Jagger S & O'Sullivan S, *Human Vulnerability to Thermal Radiation Offshore (HSL/2004/04)*, Health and Safety Laboratory, https://webarchive.nationalarchives.gov.uk/ukgwa/20230103150842mp_/www.hse.gov.uk/research/hsl_pdf/2004/hsl04-04.pdf
22 Stark G T W Y, Liquid Spillage Fires, *IChemE Symposium Series (no. 33)*, IChemE, 1972, www.icheme.org/media/12132/iv-paper-12.pdf
23 Thomas P H, *The Size of Flames from Natural Fires. 9th Symposium on Combustion*, Academic Press, 1963, https://publications.iafss.org/publications/frn/497/-1/view/frn_497.pdf
24 ENRG Consultants (Project Coordinators), *Lastfire Boilover Research Position Paper and Practical Lessons Learned (Issue 3)*, Lastfire, December 2016, www.lastfire.co.uk/uploads/LASTFIRE%20BOILOVER%20LESSONS%20Issue%203%20December%202016.pdf
25 Personal Communication
26 Brzustonski T A, A New Criterion for the Length of a Gaseous Diffusion Flame, *Combustion Science and Technology*, 6, 1973
27 Kinsman P, *Major Hazard Assessment; A Survey of Current Methodology and Information Sources* (HSE Specialist Inspector Reports Number 29), HSE, February 1991
28 Martinsen W E & Marx J D, An Improved Model for the Prediction of Radiant Heat from Fireballs, *1999 International Conference and Workshop on Modeling Consequences of Accidental Releases of Hazardous Materials*, 1999, www.questconsult.com/pdf/paper55.pdf
29 Martinsen W E & Marx J D, An Improved Model for the Prediction of Radiant Heat from Fireballs, *1999 International Conference and Workshop on Modeling Consequences of Accidental Releases of Hazardous Materials*, 1999, www.questconsult.com/pdf/paper55.pdf
30 Moorhouse J & Pritchard J, *Thermal, Radiation Hazards from Large Pool Fires and Fireballs IChemE Symposium Series 71*, IChemE, April 1982, www.icheme.org/media/25191/the-assessment-of-major-hazards-icheme-symposium-series-71-1982-25-moorhouse.pdf

31 UK Fire Service, *Smoke Alarms*, www.fireservice.co.uk/safety/smoke-alarms/
32 Hind J, *Fire and Gas Detection and Control in the Process Industries*, 2009, www.jonhind.com/fire-and-gas-in-the-process-industry-jon-hind%20paper.pdf
33 National Fire Protection Association, *National Fire Alarm and Signaling Code (NFPA 72)*, 2022, www.nfpa.org/codes-and-standards/all-codes-and-standards/list-of-codes-and-standards/detail?code=72
34 American Petroleum Institute, *Fireproofing Practices in Petroleum and Petrochemical Processing Plants*, RP2218, American Petroleum Institute, 2013, www.api.org/oil-and-natural-gas/health-and-safety/refinery-and-plant-safety/fire-protection/rp-2218
35 National Fire protection Association, *Standard for the Installation of Sprinkler Systems (NFPA 13)*, 2022, www.nfpa.org/codes-and-standards/all-codes-and-standards/list-of-codes-and-standards/detail?code=13
36 National Fire protection Association, www.nfpa.org/
37 Liberty Mutual Insurance, *Firefighting Foams, Aqueous Film-Forming Foam (AFFF), and PFAS*, https://business.libertymutual.com/wp-content/uploads/2023/03/67-5662_AFFFWhitepaper_FINAL.pdf

10 Explosions

10.1 INTRODUCTION

Major industrial explosions occur relatively infrequently, but when they do, damage is usually extensive and can extend well beyond the site boundary. The cost of an explosion, including death, injury, damage to property (including environmental damage) and loss of profit, can be extensive.

The difference between military and civilian explosions must be clearly understood. Military explosions are designed to be destructive; industrial explosives, for example used in quarrying or demolition activities, are less destructive than military "high" explosives. Military explosions are designed to be destructive, and the localised overpressures may be significantly higher than those of industrial explosions. Many civilian explosions are accidental and will have a smaller yield than an equivalent mass of military material.

There are different types of explosions, some relatively minor outside the immediate vicinity of the explosion. Others have overpressure effects which can extend for hundreds of metres, with projectiles travelling even further. The following types of explosions are commonly encountered:

1. Condensed phase explosions (explosion originates from a specific point)
2. Dust explosions
3. Oil mist explosions
4. Pressure vessel rupture
5. Rapid phase transformation
6. Confined explosions
7. Vapour cloud explosion (VCE)
8. Boiling liquid expanding vapour explosion (BLEVE)—whilst technically an explosion, the effects are similar to a fireball and BLEVEs are discussed in Section 9.8.

This chapter deals with accidental explosions and only in passing with intentional explosions such as military explosives and explosives used for mining, quarrying and construction. These nearly always involve condensed phase explosives.

Further details are given in Section 10.2.

10.1.1 Explosions and Fires—A Brief Comparison

Explosions are the result of a rapid release of energy. The main difference between a fire and an explosion is the rate at which the energy is released. This is influenced by diffusion of air into the combustion zone plus the intimate mixing of fuel and air. A pool fire may consume fuel at about 0.1 kg/m^2/s (7.8 kg/s for a 10-metre diameter pool); an explosion may consume many hundreds of kilograms in a fraction of a second. The significance of the rate of release of energy is demonstrated by the following examples:

1. A 10g charge of cordite can propel a 20 g bullet above Mach 1 but will only warm the hands if burnt slowly. The energy release is the same, around 100 J, but the rate of release differs by a factor of at least 10^3.
2. A bicycle tyre pump could be used to inflate a tyre slowly over a few minutes, but if the tyre burst the stored (pressure) energy within the tyre would be released in a fraction of a second, and small fragments from the tyre could become dangerous projectiles.

DOI: 10.1201/9781003510680-12

In a liquid-based fire, the energy can only be released if the fuel is supplied either as an atomised jet (high surface-to-volume ration) or by liquid vaporising from the pool due to heat input from the fire itself. These flames are called *diffusion flames* and are limited by ability of oxygen to diffuse into the combustion process. Many explosions are the result of fuel pre-mixing with air or the presence of an oxidising agent.

10.1.2 Detonation and Deflagration—A Brief Comparison

The blast wave of an explosion is defined as either a *deflagration* or a *detonation.* A deflagration occurs when the flame speed expands at a velocity below the speed of sound. If the flame speed is greater than the speed of sound a detonation results. Most VCEs will be deflagrations, but if there is sufficient turbulence, a transformation from deflagration to detonation is possible. Velocities may reach 2,000 m/s and the overpressures may reach 20–25 bar. Some gases such as hydrogen, ethylene, acetylene and ethylene oxide have an inherent detonation potential under the right conditions. Most condensed phase explosions will be detonations.

Commercial and military explosives are, in general, *condensed phase* explosives and contain both the fuel and oxidant within the explosive. An exception, used for military purposes is the air blast bomb in which a fuel is sprayed over a large area and seeded with detonators. This is used in jungles and against trenches.

Some gases such as hydrogen, acetylene and ethylene have high *fundamental burning velocities* which can lead to a detonation with overpressures reaching approximately 20 bar. These are the results of energetic ignition and/or high turbulence in the flame front. This is further discussed in Section 10.3. Uncontrolled exothermic reactions (often known as runaway reactions) in the process industries can result in detonations—an example being the manufacture of nitroglycerine.

In general, with the exception of ammonium nitrate, a fast deflagration will generate overpressures less than 5 bar.

Probably the largest industrial accidents of non-military explosives are those involving ammonium nitrate in storage. The explosion of a large quantity of ammonium nitrate at Oppau (Germany) in 1921 is well known, but major accidents involving the material still occur with the incidents at Toulouse (France) in 2001 and Beirut (Lebanon) in 2020 also being very destructive.

10.1.3 Blast Wave

The characteristic of an explosion is a very rapid spike in pressure, followed by a longer decline and then a period of negative pressure. Both the positive and negative phases of the explosion contribute to its destructive power. This can produce some apparently odd effects such as glass being "sucked" out of windows. The blast wave for a typical military explosive is shown in Figure 10.1. The peak overpressure depends on the formulation of the explosive.

In Figure 10.1 OP_{max} = Maximum overpressure
I^{pos} = Positive impulse wave
I^{neg} = Negative impulse wave

The pressure versus time curve for a vapour cloud explosion is slightly different from a condensed phase explosion. as shown in Figure 10.2. The shape of the positive wave is more triangular. However, there is still a negative pressure phase.

10.2 TYPES OF EXPLOSIONS

The main types of explosion were listed in Section 10.1, but further details are given in the following sections. Apart from explosive manufacturing, the major explosion hazard in the process industries is generally considered to be VCE which is considered separately in Section 10.3.

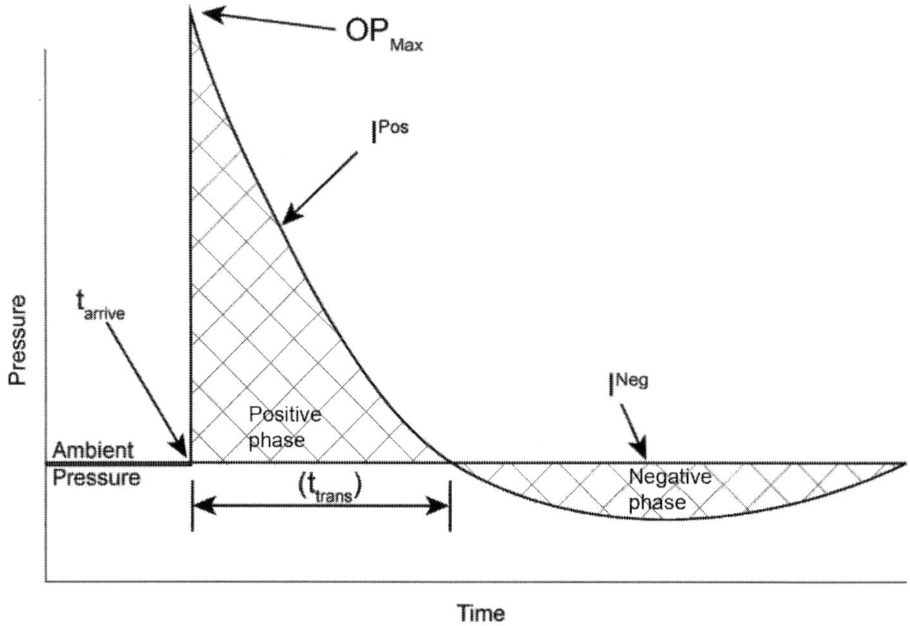

FIGURE 10.1 Peak Side on Pressure Time Curve for a Condensed Phase Explosion

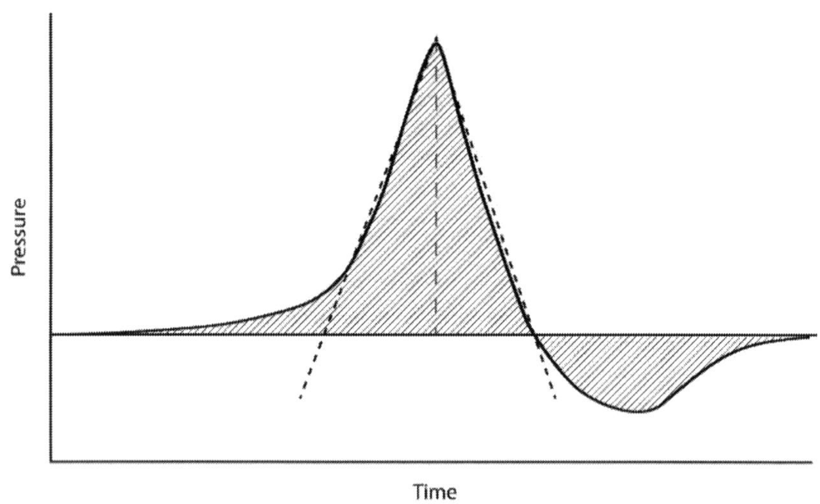

FIGURE 10.2 Peak Side on Pressure versus Time History of a VCE

10.2.1 Condensed Phase Explosions

Commercial and military explosives are, in general, *condensed phase* explosions and contain both the fuel and oxidant within the explosive. These cause *detonations* where the overpressure, at the epicentre, can exceed 1,000 bar. Typical examples include the following:

1. Ammonium nitrate
2. Cordite (a propellant, mix of nitrocellulose and nitroglycerine)
3. Trinitrotoluene (TNT)

4 1,3,5-trinitro-1,3,5-triazine (RDX)
5 Pentaerythritol tetranitrate (PETN)

There are many other explosive materials either developed for specific purposes or obtained as by-products of industrial processes (such as nitrogen trichloride—NCL_3 which can be produced during the electrolysis of brine to produce chlorine or from the reaction of ammonia derivatives and chlorine).

10.2.2 Dust Explosions

Dust explosions frequently occur in two stages. An initial ignition creates turbulence which creates a larger dust cloud which is then ignited from the flame front. The second explosion is far more destructive than the first. Figures for flammability of powders are given in Table 10.1.

Dust explosions occur in a wide variety of industries,[2] where hazards are often perceived as low. Explosions in sugar production, grain handling and woodworking (sawdust) are well known. Dust explosions during the drying of powders are also a well-known issue.

Palmer[3] has provided a detailed explanation of dust explosions.

10.2.3 Oil Mist Explosions

A serious explosion hazard exists in lubricating (lube) oil systems associated with machinery (sometimes called *crankcase explosions*). A fine oil mist can be produced which can explode if ignited. Whilst the risk is generally considered low there have been some serious incidents, particularly involving ship's engines.[4] Protection is commonly provided by inert atmosphere or the provision of "blow out" (rupture) panels. Splash filling should also be avoided (see Section 9.3).

Oil mist explosions can also occur in compressed air systems.[5]

10.2.4 Pressure Vessel Rupture

When a vessel ruptures due to high internal pressure approximately half of the stored energy is released as blast and half propels fragments of the vessel for considerable distances. The hazard exists in gas- or vapour-filled vessels. Liquid-filled vessels are less of a hazard, although a vessel with significant amounts of flammable liquid can result in a BLEVE (see Section 9.8). Modelling a vessel rupture requires knowledge of the vessel size and the burst pressure. Various models attempt to analyse the bursting of a gas-filled vessel but may not be true of a partially liquid-filled vessel. This is a complex topic, outside the scope of an introductory text.

TABLE 10.1
Some Properties of Flammable Solids[1]

Fuel Solid Powder Form	Lower Concentration Kg/m³ (g/m³)	Minimum Ignition Energy mJ
Aluminium	0.045 (45)	50
Magnesium	0.03 (30)	40
Sugar	0.045 (45)	30
Wheat Flour	0.05 (50)	50
Polystyrene	0.02 (20)	15

Source: Dust Explosions and Fires by K N Palmer, Chapman and Hall 1974

10.2.5 Rapid Phase Transformation

This is a rapid, almost instantaneous, boiling of superheated fluids which suddenly nucleate. A common example is the quenching of molten metal with water; the water becomes rapidly superheated (over 100°C) and then nucleates at sites on the molten metal.[6] Similar effects can be obtained by releasing cryogenic fluids, such as LNG,[7] underwater with water being the heat source. The rapid vaporisation can lead to the generation of a blast wave.

10.2.6 Pressure Piling

Pressure piling may occur within a series of connected chambers. The pressure in the first chamber pre-compresses the gas in the second chamber causing an explosion. The explosion in the second (downstream) chamber starts at a higher starting pressure. If there are a series of chambers the piling may become severe.

10.2.7 Explosions in Reactors

A highly exothermic reaction may *run away* if not controlled by temperature, isolation of feedstock or some other means. Many processes with runaway potential are fitted with *kill systems* to stop the reaction. In addition, pressure relief systems will be installed. This is a complex study which is discussed in Section 13.10.

10.2.8 Building (Confined) Explosions

Explosions within enclosed buildings are modelled in a different manner. Typically, walls, windows and roofs will blow out at a fraction of 1 bar overpressure. Windows will, to some extent, act as rupture panels, but glass shards can cause serious injury. Severe building explosions generally only reach about 20–30 kPa as the windows and roof are blown out and act as "explosion relief". Explosions can lead to wall and floor collapse. The design of the structure and the materials of construction will affect the damage profile. Lightweight construction of a steel frame and wall panels (which blow off in the event of explosion) will limit damage.

Whilst a confined explosion can occur in any area where a release of flammable vapours occurs, there are particular concerns in compressors handling flammable gases where a gas escape at high pressure could quickly release tens or hundreds of kilograms of flammable material. To prevent this most compressor houses are partially open to prevent pressure build-up. Where total enclosure is necessary (generally for environmental reasons such as noise limitation in residential areas or very cold environments) explosive rupture panels are installed.

> A major expansion of the process industries took place in the 1950s and 1960s. During this period, some designs included totally enclosed compressor houses. After a number of serious explosions where gas leaks around the compressor ignited, open-sided compressor houses became standard.
>
> Further investigation revealed that a similar serious of explosions had occurred in the 1920s when enclosed compressor houses had previously been a standard feature of plant design (15).

Explosions can also occur in semi-confined areas such as offshore modules where the sides are open but the decks above and below the explosion source provide a level of containment.

For small areas of confinement, the maximum pressure rise is considered to be a factor of eight but can increase to 16 in an oxygen-rich environment.[8] Computational fluid dynamics (CFD) software is frequently used for the complex calculations required for accurate modelling.

10.3 VAPOUR CLOUD EXPLOSIONS

With the exception of incidents based on explosive materials or mechanical failure, there are prerequisites for a VCE. There must be some form of *confinement* such as an enclosed compressor house, or some form of *turbulence* which acts as a *fast feed-back* reaction mechanism. In a confined area, the expanding flame front is constrained and a build-up of pressure occurs. In some cases, structures may be adequate to withstand and divert explosion overpressures (for example in high-pressure reactor bays) but in most cases the blast from the explosion will rupture the walls, ceiling and floor forming the confinement.

In the case of VCEs, the fuel is not concentrated at a single point but is dispersed into a cloud which will burn over a very short period of time. In the middle of a process plant the flame front will be partially blocked and/or confined by process equipment, piper racks and so on. This will result in increased turbulence and fuel-air mixing, which increases the flame speed and overpressure effects. Figure 10.3 shows the turbulence in the form of vortices (as lines emanating from the rear of pipes) resulting from a rapidly advancing flame front on the right. This turbulence mixes the fuel/air mixture and instead of the flame using a diffusion process it utilises turbulence to enhance the combustion process. This results in a high feedback resulting in flame acceleration which eventually results in an explosion.

Much of the early impetus for research into explosions followed the catastrophic explosion at Flixborough in the UK,[9] and the term *unconfined vapour cloud explosion* (UVCE) was then widely used. The term UVCE has fallen out of use as it is considered that a degree of confinement is always necessary and the term vapour cloud explosion or VCE is more appropriate. Since the initial investigation into Flixborough there has been widespread disputes over the conclusions of the report. These refer to the mechanism which caused the release of hydrocarbons and not the effect of the explosion and, therefore, are not considered further.

Explosive modelling has become ever more sophisticated over the years ranging from simple methods to increasingly sophisticated computer methods, and, in some cases, CFD-based computer programmes have been used.

Two simple methods are in common use, and in both cases, accuracy is sacrificed for ease of calculation. The methods are

1 TNT equivalent (TNT_e)[10]
2 Multi-energy (MEM)[11]

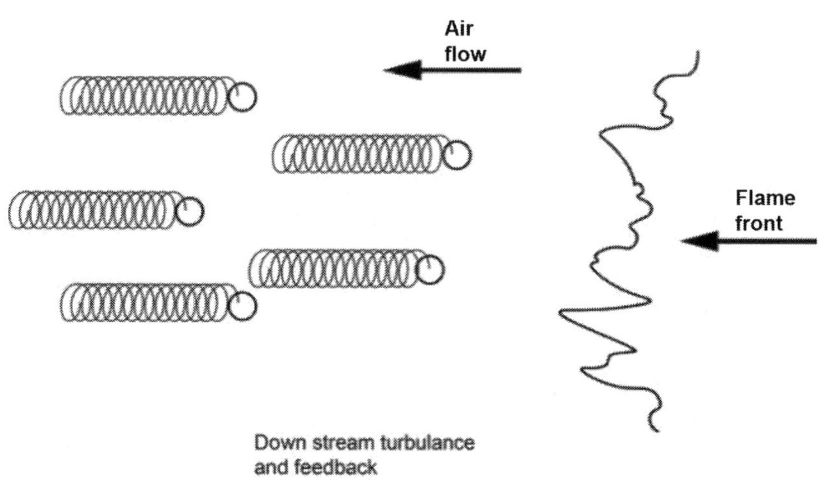

FIGURE 10.3 Generation of Massive Turbulence by the Flame Front

Both these methods assume that the explosion originates from a point source (as in the case of a condensed phase explosive), whereas, in reality, the explosion is cantered within a gas cloud of appreciable volume.

Both TNT and MEM models can be used for manual calculations and as a reality check on more sophisticated methods, often using computer software.

10.3.1 TNT Equivalent Method

The TNT equivalent model is probably the oldest model for the assessment of explosion overpressures. It contains some basic assumptions, which are debatable, but is relatively easy to use and gives approximately correct answers. There are some parallels with the MEM, for example, the energy explosion efficiency is similar to the energy factor in MEM. The ability to alter the conditions for highly reactive fuels and turbulence is, arguably, better handled by MEM.

The method converts the explosive mass in a gas cloud into an equivalent mass of *Trinitrotoluene* (TNT) and models the pressure wave based on information available from studies of TNT explosions. One assumption is that the maximum overpressure in the VCE is limited to 1 bar. However, Figure 10.4 shows some of the damage to distillation columns at Flixborough, which suggests that overpressure exceeded 1 bar. Whilst this is counter to one of the assumptions in the TNT_e model, it does not negate the simplicity of the model.

Probably the biggest source of inaccuracy is the estimation of the mass of fuel in the cloud. Some arbitrary assumptions (possibly based on experience) suggest that the mass released is twice the amount of flashed gas calculated for liquefied gas released from a pressure vessel or pipework (see Chapter 7) to allow for aerosol formation. The release, based on experience,

FIGURE 10.4 Damage to Columns at Flixborough (reproduced courtesy of RSE Consultants)

is commonly considered to be around five minutes after which, it is often assumed, that new material enters the cloud at the same rate as material at the edge of the cloud dilutes to below the LEL. In the event of a vessel rupture, the mass in the cloud is generally considered to be twice the mass of the flashed phase, again to allow for liquid entertainment. Where gas in the cloud is more than 50% of the release, the whole of the release should be assumed to be in the cloud.

The TNT equivalent (TNT_e) of the mass of a gas cloud is given by

$$TNT_e = \alpha\, M_{cloud}\, \Delta_{c(cloud)}/\Delta_{d\,(TNT)} \tag{10.1}$$

where TNT_e = Equivalent mass of TNT (kg)
α = Explosion efficiency (sometimes known as the *yield factor or* energy factor)
M_{cloud} = Mass of flammable material in the cloud (kg)
$\Delta_{c(cloud)}$ = Heat of combustion of the cloud (kJ/kg)
$\Delta_{d\,(TNT)}$ = Heat of detonation of TNT (approx. 4,500 kJ/kg)

The explosion efficiency[12] is often taken as 4% for hydrocarbons and 6% for hydrocarbons with a double bond, such as ethylene, propylene and ethylene oxide. This method can also be used for calculating the effect of condensed phase explosions where the efficiency may be much higher.

In an ideal situation, the pressure in the blast wave will decay in a cube root relationship[13] from the epicentre of the explosion:

$$\text{Distance from epicentre} \approx TNT_e^{1/3} \tag{10.2}$$

This relationship is frequently referred to as *Hopkinson's Law* and is used to define the *scaled distance* from the epicentre:

$$\textit{Scale Distance} = \frac{\textit{Distance}(D)}{\sqrt[3]{TNT}} \tag{10.3}$$

Various charts and curve-fitting equations have been produced, all following the same basic relationship between the mass of TNT in an explosion and the overpressure at a specified distance—known as the *scaled distance*. A typical example is shown in Figure 10.5. The actual overpressure can then be read from the figure based on the actual TNT equivalent of a gas cloud.

The TNT model is still widely used in the insurance industry where ease (and speed) of calculation is important but is largely superseded elsewhere.

10.3.2 Multi-Energy Method

The limitations of the TNT method have long been known, and, in an attempt to produce a more accurate result without resorting to complex calculation methods the multi-energy method (MEM) was developed by TNO in the Netherlands.[14] MEM has some elements of TNT_e but uses a parameter *combustion energy scaled distance* which is plotted as a series of *intensity curves* ranging from detonation to a "soft" explosion. This allows for the modelling of a mix of gases with a range of combustion energies and compositions. The dynamic overpressure and duration are used for the assessment of the "impulse".

The combustion scaled distance is defined as

$$SD = R_o\,(P_d/E)^{1/3} \tag{10.4}$$

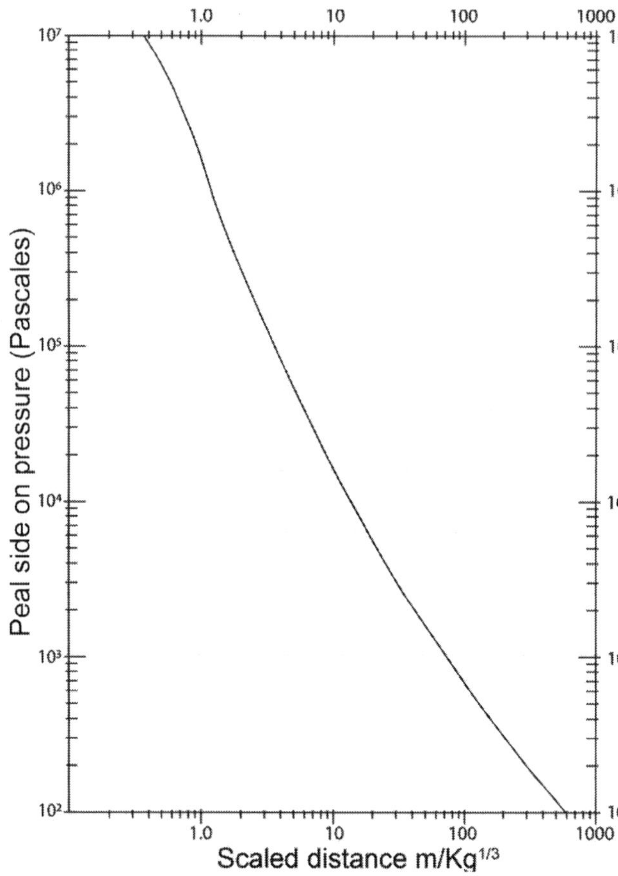

FIGURE 10.5 Scaled Distance versus Peak Side on Pressure Curve[15]. Over the range shown, the curve exhibits an approximately –5/3 slope.

where SD = Scaled distance (m/kg⅓)
R_o = Radius of cloud (m)
P_a = Atmospheric pressure (kPa often taken as 101.325 kpa)
E = Total combustion energy released (kJ)

R_0 is derived from an assessment of the cloud radius:

$$\text{Confined volume} = 2/3\ \pi\ R_0^{\cdot 3} \tag{10.5}$$

As an approximation, the *energy density* of many gases tends to 3.5 kW/m^3.

The overpressure at any distance from the epicentre of the explosion can be estimated on the basis of a plot of 10 lines (see Figure 10.6).

The MEM is considered to provide a better estimate of the overpressure close to the explosion than the TNT equivalent method. Unfortunately, the definition of the curves 1–10 is not well defined and is a function of the reactivity of the gas(es) and the turbulence generated at the flame front. Curve 10 equates to a detonation and curve 1 is a trivial case. Line 7 is likely to be similar to TNT$_c$.

There is a common tendency to use the worst-case line—which will overestimate the explosive effects. This is a topic best dealt with by an expert who can make a judgement for the assessment of turbulence generators.

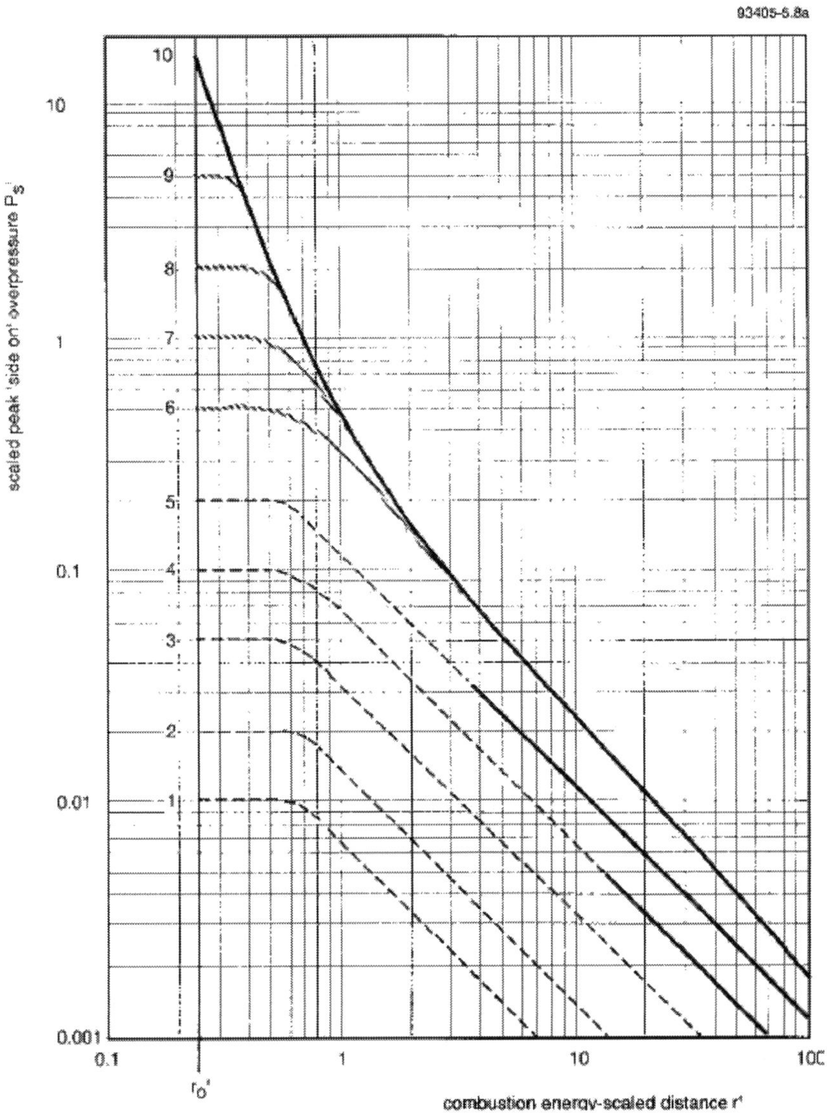

FIGURE 10.6 Multi-Energy Method (reproduced courtesy TNO)

10.3.3 Other Methods

As stated earlier, one drawback of both the TNT equivalent and MEM methods is the assumption that the explosion has an epicentre. With a large cloud of flammable gas the explosion might initiate anywhere within the cloud that is between the lower and upper flammable limits. These point source methods also do not consider the effect of obstacles in the path of the flame front which will affect the speed of burning and hence the overpressure. A number of methods have been developed to increase the accuracy of modelling. As these become increasingly complex it becomes more difficult to perform manual calculations and computer models are increasingly used. Increasing amounts of input data are necessary which requires time for collection and input. The TNO Yellow Book provides a useful guide to the various methods available.[16] Other recently developed methods include the congestion area assessment model[17] and the Baker–Strehlow–Tang model.[18] CFD modelling may also be undertaken.

10.4 PROJECTILES

Projectiles (often called missiles) can be created from drag loading by the blast wave on loose objects or fragments of vessels. A specialised study of projectile travel at Grangemouth (UK)[19] noted that one fragment of an overpressurised vessel travelled over 700 m. From simple ballistics it is likely that it started off as a sheet of steel at an angle to the ground and therefore had some aerodynamic properties. In another incident at Mexico City[20] one end of a pressure vessel travelled over 1 km following a BLEVE. Clearly the damage potential from a pressure vessel rupture can be significant.

Calculation methods for estimating travel distance and direction of projectiles have been developed; much of this relates to fragments from military ordnance, but there have been some studies covering process facilities.[21,22] These calculations are likely to be imprecise due to the potential variation of aerodynamics of the fragments. A simple illustration is the flight of a Frisbee; it can stall and fall short or it can act as an aerofoil and travel further. This is illustrated by the Humber Refinery explosion[23] (UK) where sections of cladding travelled over 500 km and, in some cases, up to 5 km. The shape of the projectile (and hence its aerodynamic properties), which are difficult to predict, may be more critical than any other factor.

A specific example of projectile damage, which is often overlooked, is pressurised gas cylinders where the regulator/valve has become detached. The kinetic energy imparted by the release of the cylinder contents can result in the cylinder achieving significant velocity and travelling considerable distances.

10.5 INJURY

The US Centre for Disease Control[24] (CDC) classifies explosion injuries under four headings but is largely related to military explosives; the classification, with some modifications, is shown in Table 10.2.

TABLE 10.2
Classification of Blast Injuries

Category	Characteristics	Body Part Affected	Types of injury
Primary	Unique to high explosives, results from: The impact of the overpressurisation wave with body surfaces.	Gas-filled organs are most susceptible: 1 Lungs 2 Gastrointestinal tract 3 Middle ear	1 Blast lung (pulmonary barotrauma) 2 Tympanic membrane rupture and middle ear damage 3 Abdominal haemorrhage and perforation 4 Globe (eye) rupture 5 Concussion (traumatic brain injury without physical signs of head injury)
Secondary	Results from debris and fragments	Any body part may be affected	1 Penetrating ballistic (fragmentation) or blunt 2 Injuries 3 Eye penetration (can be occult)
Tertiary	Results from individuals being thrown by the blast wave	Any body part may be affected	4 Bone fracture leading to traumatic amputation 5 Closed and open brain injury
Quaternary	All explosion-related injuries, illnesses, or diseases not due to primary, secondary or tertiary mechanisms. Includes exacerbation or complications of existing conditions.	Any body part may be affected	1 Skin burns (flash, partial and full thickness) 2 Crush injuries 3 Closed and open brain injury 4 Asthma, chronic obstructive pulmonary disease, or other breathing problems from dust, smoke, or toxic fumes 5 Heart failure 6 Hyperglycaemia, hypertension

TABLE 10.3
Human Injury from Pressure Waves[25]

Peak Side on Overpressure (bar)	Probability of Fatality or Injury
0.07–0.21	0.1
0.21–0.34	0.25
0.34–0.48	0.7
Over 0.50	1.0[1]
Note 1	Death may also be due to lung or internal damage or crushing under masonry. Above 3 bar lung damage is certain.
Note 2	The fatality at lower overpressures may be due to physical translation or being buried under debris.

The possibility of death or serious injury is based on a number of factors involving the individual (including health, size and weight) and the level of exposure. Similar variations will apply to plant and equipment damage (construction materials may be more robust than anticipated; conversely, corrosion might have weakened supports). The probability of injury and fatality is shown in Table 10.3.

Overpressure will frequently cause buildings to collapse, and people trapped inside could be crushed. This, unfortunately, is a common cause of injury or death.

10.6 PROPERTY DAMAGE

Damage is a function of the response of the structure and the mass of explosive material. Much will depend on the "impulse versus overpressure response". Lamp posts bend, glass shatters and buildings collapse. Table 10.4[26] is a compilation from several sources and shows some typical damage profiles. High overpressure and low impulse produces damage in "ductile" systems.

The figures earlier are indicative and may vary significantly depending on circumstances. For example, one report[27] cites glass breakage in the range 0.7–6 kPa, but a second report[28] cites the range 0.6–4 kPa (dependent on the size of the explosion and the percentage of glass damage). As glass varies in density, thickness and formulation it is unsurprising that there is variation in the extent of damage. Similarly, there are variations between the two references for displaced telegraph poles. Much will depend on the depth of the post hole, the nature of the soil and any cementing as well as the dimensions of the pole.

Some reports[29,30] suggest that roofs can be lifted off walls and then dropped back in place. This is the result of the positive pressure phase forcing gases into the building and then the following negative pressure phase whilst the gases have not yet exited the building. Detailed assessment of explosion damage generally requires use of computer fluid dynamics (CFD) to consider the turbulence generation, strength of components and venting.

10.7 EXPLOSION MITIGATION

Explosion prevention is the best mitigation and may include the following:

1 Reducing turbulence inducing features during design
2 Maximising open areas and reduce the degree of refinement
3 Designing to allow good gas dispersion
4 Avoiding leaks by using all welded systems (this must be balanced against greater difficulty in dismantling equipment for maintenance)

TABLE 10.4
Suggested Damage Profiles

Damage		Peak Side on Pressure kPa
Domestic Houses		
Glass Failure	5%	0.7
	50%	2.0
	90%	4.5
Building uninhabitable		7
Severe damage to buildings		15
Total Destruction of Buildings		75
Process Buildings		
Serious Damage (major report necessary)		15
Process Equipment		
Instruments displaced (major replacements necessary)		5
Piping Spring		
Storage tank		20
Process piping or Pipe Track		40
Serious Damage		
Storage Tank (90% full)		20 40
System Displaced		
Fired Heater		50
Major piece of Heavy Equipment		40
System Fails		
Fired Heater		60
Major piece of Heavy equipment		50
Missile Flight		
At pressure of 2ka		
Total Destruction		75

It is possible to slow and even arrest a flame front with a suppression system which reacts to the sudden rise in pressure and releases an extinguishant (water, dry powder or gas). Halon was commonly used in suppression systems, but its use has been discontinued due to its effect on the ozone layer.

In areas where a dust explosion potential exists, flat surfaces are sometimes coated with an inert powder (stone dust is commonly used in coal mines). When the first, usually minor explosion occurs, the powder will be mixed into the air by the overpressure wave to prevent or limit the effects of a second more powerful explosion.

Blowout or hinged panels are often installed in confined areas and are designed to rupture or open thus preventing the generation of significant pressures. These have to be properly designed as the release pressure, the panel inertia and size of panel are critical in the assessment. The sizing (volume to area) is also critical.

The US National Fire Protection Association publishes a standard covering various types of explosion prevention.[31]

10.8 WHAT CAN GO WRONG?

No system is perfect, and a continuous questioning attitude is necessary. Changes in operating conditions or poor maintenance can result in protective systems ceasing to be effective. Changes in powder formulation, for example, can result in a change in explosive properties—this has been the cause of more than one explosion in a dryer. Poor maintenance can also make explosion protective systems ineffective. For example, some types of explosion venting are based on swing panels which open if an explosion occurs. These can appear preferable to rupture panels which will need replacing after use. However, if the hinges on swing panels become stiff through lack of lubrication, rust or some other reason they will be ineffective.

A small manufacturing facility was temporarily closed. The facility included an uninterruptible power supply (UPS) utilising lead-acid batteries. The UPS was kept in operation during the shut down but the plant ventilation systems (including the fan to remove fumes from the room housing the UPS) were switched off. Like many UPS systems, this one was located within a small windowless room. During checks prior to restarting the facility, maintenance staff wanted to inspect the UPS system. When the light switch was operated, a spark ignited hydrogen which had been released and accumulated during battery charging but could not escape as the ventilation system was switched off. The resultant explosion caused significant damage.

Either keeping the fan in operation or the provision of correctly specified electrical equipment (see Section 14.6) would have prevented the explosion.

REFERENCES

1 Lees F P, *Loss Prevention in the Process Industries (Table 17.21)*, Butterworth Heinmann, 1991
2 UK Health and Safety Executive, *Prevention of Dust Explosions in the Food Industry*, www.hse.gov.uk/food/dustexplosion.htm
3 Palmer K N, *Dust Explosions and Fires*, Chapman and Hall, 1973
4 Freeston H G, Roberts J D & Thomas A, Crankcase Explosions: An Investigation into Some Factors Governing the Selection of Protective Devices, *Proceedings of the Institution of Mechanical Engineers*, 170(1), June 1956
5 AXA XL Risk Consulting, *Prevention of Explosion and Fire in Compressed Air Vessels and Systems* (PRC.3.1.1), AXA, https://axaxl.com/-/media/axaxl/files/pdfs/prc-guidelines/prc-3/prc311preventionofexplosionandfireincompressedairvesselsandsystemsv1.pdf
6 UK Health and Safety Executive, *The Explosion at the Appleby-Frodingham Steelworks*, Scunthorpe 4 November 1975, HM Factory Inspectorate, www.icheme.org/media/13691/the-explosion-at-appleby-frodingham-steelworks-scunthorpe.pdf
7 Reigstad G A, Lervåg K Y & Aursand L, The Risk for Rapid Phase Transition in LNG Spilled on Water, *SINTEF*, 31 January 2020, www.sintef.no/globalassets/project/predict-rpt/2020-01-sintef-predict-rpt-final-report.pdf
8 Lees F P, Loss Prevention in the Process Industries, Butterworth Heinmann, 1991
9 Parker R J (Chairman), The Flixborough Disaster, *Report of the Court of Inquiry*, HMSO, www.icheme.org/media/17752/the-flixborough-disaster-report-of-the-court-of-inquiry.pdf
10 Phillips H (ed), *Explosions in the Process Industries: A Report of the Major Hazards Assessment Panel Overpressure Working Party (Major Hazards Monograph)*, IChemE, 1994
11 Solberg D M, *Gas Explosions Research Related to Safety of Ships and Offshore Platforms*, International Specialists Meeting on Fuel Air Explosions, November 1981
12 Phillips H (ed), *Explosions in the Process Industries: A Report of the Major Hazards Assessment Panel Overpressure Working Party (Major Hazards Monograph)*, IChemE, 1994

13 Hopkinson R, *British Ordinance Board Minutes 13365*, 1915
14 Van Den Bosh C J & Weterings R A P M (eds), *Methods for the Calculation of Physical Efforts (CPR 14E) Chapter 5—The Yellow Book*, TNO, 1996
15 Phillips H (ed), *Explosions in the Process Industries: A Report of the Major Hazards Assessment Panel Overpressure Working Party (Major Hazards Monograph)*, IChemE, 1994
16 Van Den Bosh C J & Weterings R A P M (eds), *Methods for the Calculation of Physical Efforts (CPR 14E) Chapter 5—The Yellow Book*, TNO, 1996
17 Puttock J S, Fuel Gas Explosion Guidelines—the Congestion Assessment Method, *Second European Conference on Major Hazards Onshore and Offshore*, 1995, www.icheme.org/media/9811/xix-paper-23.pdf
18 Pierorazio A J, Kelly J, Baker Q A & Ketchum D E, An Update to the Baker-Strehlow-Tang Vapor Cloud Explosion Prediction Methodology Flame Speed Table, *Process Safety Progress*, 24(1), March 2005
19 UK Health and Safety Executive, *The Fires and Explosion at BP Oil (Grangemouth) Refinery Ltd*, HMSO, 1989, https://www.icheme.org/media/13700/the-fires-and-explosion-at-bp-oil-grangemouth-refinery-ltd.pdf
20 UK Health and Safety Executive, *PEMEX LPG Terminal*, Mexico City, Mexico, 19 November 1984, www.hse.gov.uk/comah/sragtech/casepemex84.htm#:~:text=At%20approximately%2005%3A35%20hours,killed%20and%20the%20terminal%20destroyed
21 Centre for Chemical Process Safety (CCPS), *Guidelines for Evaluating the Characteristics of Vapor Cloud Explosions*, Flash Fires and BLEVEs, CCPS, 1994
22 Holden P L & Reeves A B, *Fragment Hazards from Failures of Pressurised Liquefied Gas Vessels*, ICHEME Symposium Series 93, 1985, www.icheme.org/media/12184/viii-paper-13.pdf
23 UK Health and Safety Executive, *Public Report of Fire and Explosion at ConocoPhiliips Humber Refinery 16 April 2001*, https://webarchive.nationalarchives.gov.uk/ukgwa/20220701180127mp_/www.hse.gov.uk/comah/conocophillips.pdf
24 Centers for Disease Control and Prevention (CDC), *Explosions and Blast Injuries A Primer for Clinicians (cdc_28957_DS1)*, CDC, latest edition
25 Phillips H (ed), *Explosions in the Process Industries: A Report of the Major Hazards Assessment Panel Overpressure Working Party (Major Hazards Monograph)*, IChemE, 1994
26 Clancy V J, *Diagnostic Features of Explosion Damage*, Sixth International MTG of Forensic Science, 1972
27 Kinsman P, *Major Hazard Assessment; A Survey of Current Methodology and Information Sources* (HSE Specialist Inspector Reports Number 29), HSE, Febr 1991
28 Explosions Monograph, Explosions in the Process Industries, *Overpressure Working Group, Major Hazards Assessment Panel*, IChemE, Rugby, 1989
29 Kinsman P, *Major Hazard Assessment; A Survey of Current Methodology and Information Sources* (HSE Specialist Inspector Reports Number 29), HSE, February 1991
30 Phillips H (ed), *Explosions in the Process Industries: A Report of the Major Hazards Assessment Panel Overpressure Working Party (Major Hazards Monograph)*, IChemE, 1994
31 National Fire Protection Association, *Standard on Explosion Prevention Systems (NFPA 69)*, 2024, www.nfpa.org/product/nfpa-69-standard/p0069code

Part C

Hazard Studies

11 Hazard Study 0—Inherent Safety

11.1 INTRODUCTION

The concept of *inherent safety* originated in the 1970s and can be traced back to the seminar *Major Loss Prevention in the Process Industries*.[1] Whilst not using the phrase *inherent safety* it highlighted many of the principals. Increasingly, the concept of *inherently greener* is also being applied. The application of inherent safety as the first stage of a project is shown in Figure 11.1.

An early example was the development of the Hi-Gee distillation process. Researchers in distillation technology realised that *g* (acceleration due to gravity) was not a constant but could be varied (see Section 11.2). Some common examples of inherent safety predate this, such as the bunding of storage tanks, but the development of the concept broadly falls within this definition.

Inherent (and greener) safety is a concept which requires "thinking outside the box" and, as with many process safety tools, requires some form of checklist or guidewords. These are traditionally the following:

1 Intensification
2 Attenuation
3 Substitution
4 Change
5 Elimination
6 Simplification
7 Getting it right first time
8 Capture and recover

There has been limited progress in the development of inherently safer/greener technology to date but this is likely to change with the new techniques now being developed. Often economic advantages accompany inherently safer/greener concepts.

11.2 GUIDEWORDS AND EXAMPLES

Examples of the use of guidewords are given in the following sections. In some cases, more than one will apply.

11.2.1 INTENSIFICATION

Reducing the working inventory may be counterintuitive. It can be achieved by concentrating the process in a smaller, higher-pressure reactor so reducing the working inventory or total leak potential.

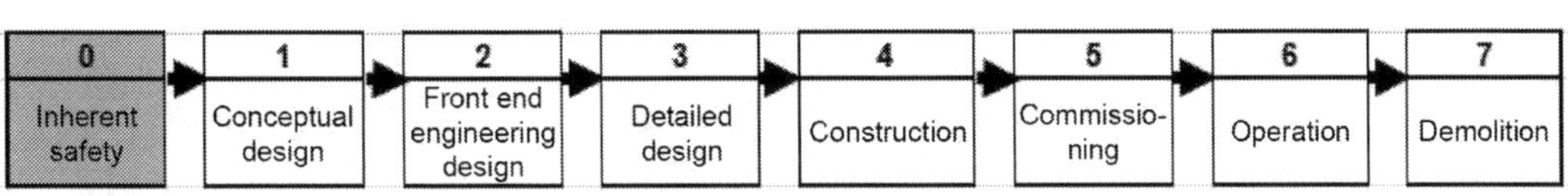

FIGURE 11.1 Project Life Cycle Inherent Safety

DOI: 10.1201/9781003510680-14

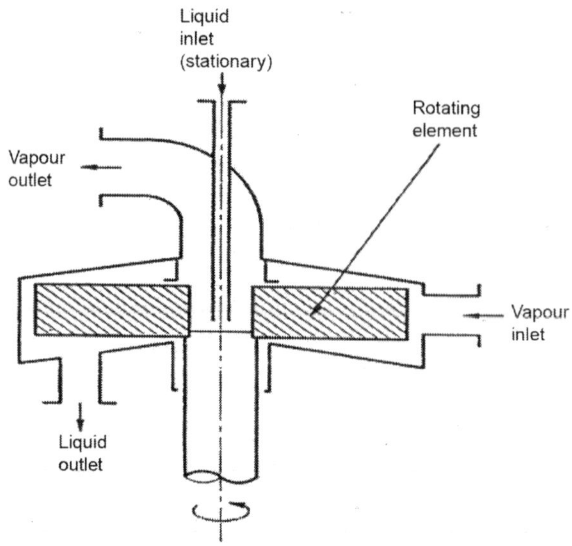

FIGURE 11.2 Hi-Gee Distillation (reproduced courtesy of IChemE)

A simple example might be a high-pressure catalytic reactor which is significantly smaller than a conventional low-pressure reactor. Whilst the potential peak flow-rate through a leaking orifice may be higher, the total release will be significantly lower. In any high-pressure gas reactor, the intensification uses the law of mass action (density of reactants) to produce a smaller, more compact reactor. The lower leak rate, however, must be balanced against the total energy in the reactor—a small high-pressure reactor may contain more energy than a larger, lower pressure reactor. The risk from a leak against a vessel rupture needs to be considered.

A reduction in the volume of process vessels, assuming no change in the temperature and pressure of their contents, will reduce the potential hazard in the event of leakage but can create control problems. For example, if the volumes are too small, level control may be difficult to maintain.

An example of intensification is Hi-Gee distillation[2] where the distillation process can be intensified with a spinning contactor disc and utilises centrifugal forces to intensify mass transfer (see Figure 11.2). The process is essentially a rotating mesh or packed drum with a liquid feed at the centre and a vapour exit at the periphery.

The rotating element is in the form of knit-mesh. The vapour and liquid phases entering are typical of any contactor (absorption or distillation), but the mass transfer is intensified.

Acceleration levels vary across the mesh and are typically 10^4 m/s^2. The effective area is low, but under the high "g" or acceleration the vapour/liquid contact is exceedingly effective with high liquid and vapour loading and low back-mixing. The process has application as a distillation column, a stripping column, an absorption column or a reactor. The intensification is of the order of 10^3, and it is not difficult to imagine a number of processes in series or parallel. In the distillation column it is necessary to have different units for the stripping and rectification section of the column, and if there are side streams each section must be an individual Hi-Gee unit.

The principle is simple and the theory sound but so far it has limited commercial success. There are always concerns over the use of unproven technology and although the inventory of the system might be reduced by comparison with traditional equipment its use is still limited. There are a number of engineering limitations particularly the seal of the drive shaft and overall reliability.

A historical example predating the concept of inherent safety (and the development of basic control systems) is the nitration of glycerine to produce the explosive nitroglycerine. In the original batch process, dating back to the late 19th or early 20th century, the operator watched a thermometer in the nitration reactor.[3] If it reached a critical temperature, he would dump the contents into cold water to stop the reaction. Each reaction batch could contain "hundreds and thousands of gallons" of nitroglycerine. As constantly monitoring a single thermometer was a boring task, the operator sat on a one-legged stool and would fall over and wake up if he fell asleep (see Figure 11.3). Inherent safety could be further increased if the operator had a piece of string attached to his wrist and the dump valve.

The batch process was replaced by Nobel AB (NAB)[4] continuous process, shown in Figure 11.4, with a much lower inventory. In addition, the process could be controlled from a safe location, and function checks could be carried out on the protective system.

The NAB process involves the intimate mixing (dispersion and increased surface area for the reaction) of the reactants in a Venturi. Only one fluid is pumped; the first reactant induces flow in the other reactant and results in intimate mixing and ratio control. The reaction takes place in a

FIGURE 11.3 One Legged Stool for Monitoring Nitroglycerine Production

Source: Nobel Institute

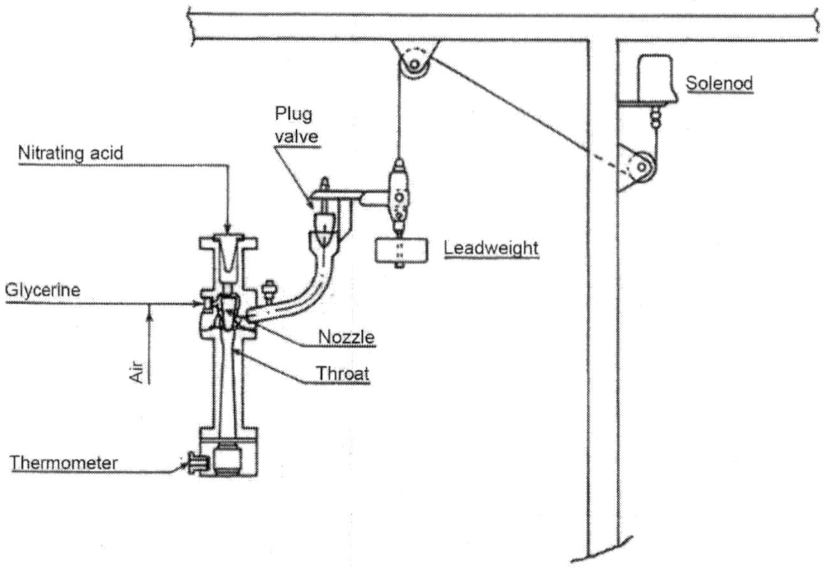

FIGURE 11.4 The NAB Process (reproduced courtesy of IChemE)

"linear reactor" with 100-fold intensification by comparison with the original production process. Another early example of process intensification was the development of the Podbielniak Contactor which combined liquid-liquid extraction and centrifugal separation in a single machine. Whilst originally developed in the mid-1930s it was used in the early production of penicillin during the Second World War.[5]

Improvements in catalyst efficiency, allowing operation at lower temperatures and pressures, can also be considered an example of intensification.

11.2.2 Attenuate

Attenuation involves reducing either the pressure or the temperature such that the outflow following loss of containment is reduced. There are two basic methods:

1 Store materials at a lower pressure and/or temperature. Refrigerated (instead of pressurised) storage of ammonia or LPG are two common examples. The lower temperature and pressure will reduce the flash in the event of rupture. If secondary containment is provided, there will often be sufficient time for a foam solution to be applied over the liquid surface to reduce evaporation.
2 Moderate the thermal mass or the flammability envelope to avoid exothermic reaction run away. A classic example is the production of formaldehyde via air oxidation of methanol vapour over a silver catalyst.[6] In this the flammable envelope and the oxidation temperature are attenuated by live steam.

11.2.3 Substitute

Substitute has some similarities with change (see the following). Examples of substitution include:

1 Substitution of a flammable solvent in an extraction or crystallisation process by water or another benign solvent. The chemistry of the process may limit the options.

An example of substitution is the production of ethylene oxide.[7] One process utilised paraffin as a heat transfer medium—requiring the vaporisation and condensation of large volumes of hydrocarbon to be pumped around a cooling circuit. A safer alternative would be the use of high-pressure water which would remove process heat by boiling to steam.

2 Substitute a catalyst which works at a lower temperature.

11.2.4 Change

It is surprising how often the word *change* can be used beneficially. Electrical heating, heating with a heat transfer medium and heating by a fired heater all have potential risks or hazards. In particular, the control and high-temperature protection has to be designed with care. Steam heating which is easy to control and has a maximum temperature determined by the steam pressure would be a safer alternative.

This is one simple example of how *change* can provide an inherently safer approach.

11.2.5 Eliminate

It is often worth asking "*Why is this done this way*?". This might seem to be a statement of the obvious but the answer is often "*We have always done it this way*!" A review of the process might show steps that can be simplified, substituted or removed. If elimination is not possible, other guidewords might apply. For example, if intermediate storage within the process cannot be eliminated can it be reduced in size?

11.2.6 Simplify

There is a lot to be said for the adage "*Keep it simple stupid!*" (KISS) by reducing complexity as far as possible. There is a tendency for control systems to become overly complex which can lead to operator errors. With the increased use of cheap microprocessors for control, there is a temptation to incorporate an excessive number of alarms which has resulted in the phenomena of alarm flooding[8] (the generation of so many alarms over a very short period of time that operators cannot determine what is going wrong). Alarm flooding has been a factor in a number of incidents.[9,10]

Whilst engineering design is a collaborative exercise, *opinion engineering* can occur, leading to unnecessary complication with little justification.

11.2.7 Getting It Right First Time

Changes in design identified by HAZOP or other means may be identified too late to be incorporate in an inherently safe process route. Equipment orders may already have been placed and time to effect changes could affect the economics of a project. This requires a critical review at the conceptual design stage. Once a project moves into FEED it tends to have its own momentum, making change difficult. This topic is covered in depth in Chapters 13 and 14.

11.2.8 Capture and Recover

This is both an environmental and safety concept. Leaks of process fluids can be corrosive, toxic or flammable. Reducing the size of a leak is covered above, but if one does occur there are ways to minimise the release. One technique used in some industries with high-value materials and products (or where down time is expensive) is the use of an inner pipe for fluid transfer and an outer containment annulus (known as *pipe-in-pipe*). Oil and gas wells are a common example with reservoir fluids flowing through the inner pipe and pressure in the annulus being monitored for leaks. Other examples may be found in industries ranging from semiconductors to effluent treatment.

Modern practice for fuel trucks and tank vents often incorporates vapour recovery or clean-up systems rather than direct discharge to the atmosphere. Although they are environmentally friendly, they do create an additional source of hazard. Containment methods such as bunds around tanks and drip trays under drums have long been standard practice.

11.2.9 Conflicts

There will always be conflicts between process efficiency, safety and the environment. One example was found on an olefins plant. To minimise energy in the refrigeration unit the cooling water leaving the condensers of the propylene refrigeration was reused as coolant in the propylene/propane fractionator. This warmer water required a higher operating pressure in the fractionator, causing a reduction in the relative volatility of the key components. This required the fractionator to have a working inventory of 150 tonnes of hydrocarbons.

REFERENCES

1 Major Loss Prevention in the Process Industries 106th Event of the European Federation of Chemical Engineers, *IChemE*, July 1971
2 Cortes Garcia G E, van der Schaaf J & Kiss A A, A Review on Process Intensification in HiGee Distillation, *Journal of Chemical Technology and Biotechnology*, 92(6), 2017, www.research.manchester.ac.uk/portal/files/65100218/JChemTechBiotechnol_2017_92_1136_1156_A_review_on_process_intensification_in_HiGee_distillation.pdf
3 Lateral Science, *Scottish Nitroglycerin & One Legged Stools*, 13 October 2014, http://lateralscience.blogspot.com/2014/10/scottish-nitroglycerin-one-legged-stools.html
4 Bell N A, *Loss Prevention in the Manufacture of Nitroglycerine*, Major Loss Prevention in the Process Industries IChemE Symposium Series 34, IChemE, 1971
5 Pitt M, Brewing Up a Storm: The Origins of Biotechnology Part 2, *The Chemical Engineer*, May 2023 (No. 983), April 2023, www.thechemicalengineer.com/features/ta/
6 Pickles R G, Hazard Reduction in the Formaldehyde Process, *Major Loss Prevention in the Process Industries*, IChemE Series 34, IChemE, 1971
7 Personal Communication
8 UK Health and Safety Executive, *Better Alarm Handling HSE Information Sheet Chemicals Sheet No. 6*, www.hse.gov.uk/pubns/chis6.pdf
9 UK Health and Safety Executive, The Explosion and Fires at the Texaco Refinery, *Milford Haven*, 24 July 1994, www.hse.gov.uk/comah/sragtech/casetexaco94.htm
10 UK Health and Safety Executive, *Human Factors: Alarm Management*, www.hse.gov.uk/humanfactors/topics/alarm-management.htm

12 Hazard Study 1—Conceptual Design

12.1 INTRODUCTION

Conceptual design is the first major step in the development of the project and its place in the overall project life cycle is shown in Figure 12.1.

Long before any significant engineering activity begins, the project concept will need to be developed and the viability of a manufacturing facility determined. Questions relating to financial viability, stable operation, plus safety and environmental issues will need to be considered. At this stage, there will be relatively little expenditure involved, and studies may indicate that the project is not financially and/or technically feasible. Once FEED starts, project costs will begin to mount up (see Chapter 13). Points to be considered at this stage include the following:

1 Is there a market for the product(s), and is growth and pricing adequate to sustain the plant throughout its working life?
2 Which process is to be used, and is it reliable/proven?
3 Can the product be produced economically?
4 Where can it most economically be produced?
5 Are the final product specifications fully defined, and might they change in the future?
6 Are raw materials available?
7 Are the required utilities available in the quantities required and with adequate reliability?
8 Are there good transport links for import of raw materials and export of products?
9 Are skilled personnel currently available in the area, or will they require any retraining?
10 Is there a local workforce that can be trained as process operators or in other roles at the plant?
11 Are local regulations excessively onerous?
12 Are there tax advantages or restrictions in the area chosen?
13 Is the area politically stable?
14 Are there any terrorism or other security concerns affecting the site or transport routes?
15 Are there any significant features in the disposal of waste or by-products?
16 Are there any environmental/safety considerations which need to be addressed?
17 Is noise and visual impact a potential issue?
18 Are weather conditions suitable (extremes of temperature, precipitation, storm and lightning)?

At this stage the likely plant throughput will probably have been determined. The search for a location may well be for a country or region rather than a specific location—this will come later if the project is deemed viable. The economics of the project will have been through a financial analysis

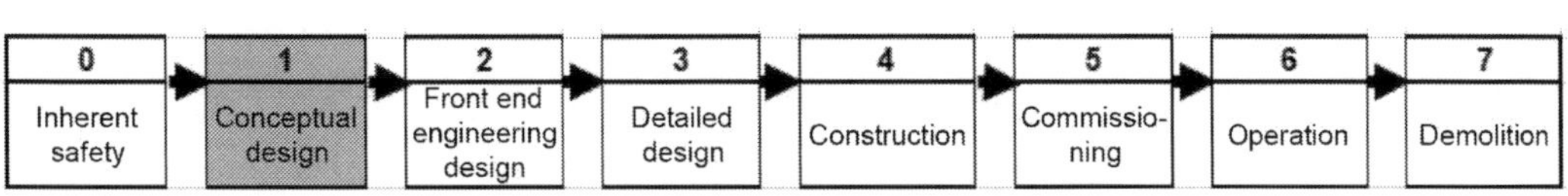

FIGURE 12.1 Project Life Cycle—Conceptual Design

DOI: 10.1201/9781003510680-15

to check for variations in sales, changes in prices of raw materials and utilities, labour costs, debt repayment and other direct and variable costs. Global political factors also have an effect—this is not always realised.

12.2 PROJECT DEFINITION

The analysis earlier will have determined if the project is viable. The next stage provides greater definition of the project's practicality and uncertainties. Whilst the issues identified in Section 12.1 will have acted as a preliminary screening check, a more detailed assessment is now necessary. This will cover some steps previously examined in outline but which must now be examined in detail. Some key features are shown in Table 12.1. Many proposed developments will have industry-specific requirements, and requirements may very through the life cycle of the plant.

TABLE 12.1
Key Factors in Conceptual Project Development

Item		Factor	Comments
Feedstocks	1	Distance from site	Ideally the feedstock and markets should be close to the site, but there are many examples of chemicals being transported significant distances. Long transportation distances are common in the oil and gas industry.
	2	Transportation (pipeline, ship, rail or truck)	Pipelines are generally a safer option, but plant capacity and lifetime will determine if a pipeline is economic. Regardless of economics a pipeline may be impractical for political, geographical or other reasons.
	3	Lifetime of feedstock supply	The project must be adequate to repay the cost of construction and operating costs over its lifetime and still return a profit. Plants often operate beyond their original design life. The capital cost will have been repaid at, or before, the end of design life, and the economics of operation could be transformed.
	4	Quality	The quality (including temperature and pressure) of feedstock must be adequate for the design. Feedstocks taken from hydrocarbon reservoirs will decrease over time. Pressure, temperature and composition may change significantly.
	5	Other materials	As well as raw materials and products, most processes will require storage for catalysts, chemical additives, blending agents and so on as well as spare parts. Loss of supply of these items will result in eventual shut down regardless of the supply of raw materials
Products	1	Transportation	Similar considerations apply to raw materials.
	2	Markets	The viability of the market over the lifetime of the plant needs to be assessed to ensure the economics of the plant. Demand for products may increase or decrease for several reasons. The market may be affected by replacement/substitution and increasingly by regulatory factors. Prior to the *Montreal Protocol* (1987) halogenated hydrocarbons were in widespread use in various applications ranging from refrigeration to firefighting. These materials, when released to atmosphere, were found to be damaging to the ozone layer—which shields the earth from high levels of ultra violet (UV) radiation. The protocol, which has been modified on several occasions, is seen as broadly successful with the damage to the ozone layer now repairing itself.[1] Health and safety effects over the years have led to numerous chemicals being banned or restricted for health reasons. Possibly the best example is the removal of tetraethyllead (TEL), formally an antiknock agent, from gasoline. There are numerous other examples.

(Continued)

TABLE 12.1 *(Continued)*
Key Factors in Conceptual Project Development

Item		Factor	Comments
Storage	1	Basis	The storage requirements will be very variable; in some cases sales volumes are relatively constant. In other cases, for example fertilisers, sales volumes peak at certain time of the year. Storage requirements may also need to consider the ability to supply customers during shut downs for maintenance (commonly called "turnarounds).
	2	Storage	Costs of storage can be a significant part of the plant costs, and decisions will need to be made on the size of storage necessary to ensure that the plant does not suffer repeated shut downs due to a lack of raw materials or product storage areas being full. Storage arrangements must ensure there is no degradation of the product due to high/low temperature, moisture or contamination.
	3	Special storage conditions	Some materials will require specific storage conditions, such as refrigeration, or an inert atmosphere.
Waste Disposal	1	Basis	Generally, some form of treatment facility is built into the plant design. Provision must be made for waste disposal (during both normal operation/ maintenance and emergency conditions). Provisions must include the safely storing of materials in upset conditions if they cannot be disposed of safely.
	2	Solid waste	This may include waste from the process but other materials should not be overlooked, for example: used lube oil, damaged parts, spent catalyst, packaging material and food waste. At one remote Middle East location food waste was stored in flimsy plastic bags for extended periods before being removed for disposal. Inevitably, some of the bags were damaged, and food smells attracted a variety of insects and vermin. The smells also attracted monkeys who were difficult to remove and could become aggressive.
	3	Liquid waste	Much of the liquid discharge is water from cooling and rainfall. Discharges will often require monitoring to ensure they are free from contamination. For example, process fluids from heat exchanger tube leaks entering the cooling water system. Other effluents might require chemical or biological treatment. Fire water will also enter the waste disposal system and may be contaminated. It may be necessary to store fire water runoff in a buffer storage facility and check for contamination before discharging.
	4	Gaseous Waste	Carbon dioxide from combustion processes is likely to be the major source of gaseous waste, but consideration should be given to leakages of process materials. The design should consider emergency situations which require flaring or the inclusion of scrubbers to prevent toxic chemicals from reaching the atmosphere.
Plant Location (see also Section 12.5).	1	Nearby residents	Few people in the developed world are likely to want a large chemical plant located close to their homes—even if there are no safety concerns, other factors such as visual intrusion, noise and increases in traffic will often make the proposed plant unwelcome. This differs from many developing countries where plant construction acts as a magnet for shanty towns providing services for migrant construction workers and, later, permanent jobs.
	2	Areas surrounding the proposed plant site	Where there is a concentration of industrial facilities an incident at one location may affect others. In some instances, this can lead to a *domino effect* with the original event causing fires and explosions at neighbouring locations. More frequently, an incident at one location will only cause limited damage to surrounding industrial facilities. On 21 September 2001 a major explosion of ammonium nitrate occurred at the AZF site at Toulouse, There were 30 fatalities among workers and nearby residents. Around 10,000 people received injuries with 14,000 suffering post-traumatic stress. 27,000 flats and houses were damaged[1,2].

(*Continued*)

TABLE 12.1 *(Continued)*
Key Factors in Conceptual Project Development

Item		Factor	Comments
			There is sometimes a need for a *cordon sanitaire* separating the proposed development from other industrial, commercial or residential sites to mitigate offsite effects of an onsite incident.
	3	Nuisance problems	Obvious examples will include noise, night-time illumination and odours from the plant. There are also secondary considerations such as increased traffic (with noise and pollution) both during construction and the operational phases.
	4	Topography	It is generally desirable for plants to be located on a flat area, although minor terracing is sometimes necessary. In a few cases plants have been built on the side of hill to harness hydropower or to use gravity to transport liquids, thus avoiding the need for pumps.
			Natural perils (sometimes known as NATECH) around the plant site need to be considered (earthquake, storms, hail, lightning, rain and so on) (see Section 4.6). Good design can minimise the potential for problems that can occur but some may be outside the control of the plant designers. For example, flooding might be controlled on-site but might cause effluent containment pits to overflow. The potential for landslides on a flat site is limited but could be a problem on an approach road or where the site is adjacent to hills.
			The site might need *future proofing*, for example, by increasing the site elevation above sea level to allow for an increase in water depth during a flood. Deforestation on land next to the site can also cause problems, as can future developments close to the plant.
	5	Soil surveys	These will need to be produced to determine the load-bearing capability of the soil and the need for compaction, piles or other features.
	6	Previous site use.	On brownfield sites,* records should exist of previous use, but these may be incomplete or inaccurate particularly for older locations. Surveys to check for contamination may be necessary.
			Even on greenfield sites problems may occur if site development reveals archaeological remains or military ordnance in former battlefield areas.
	7	Other surveys	It will be necessary to ensure there are no previous underground workings or utility systems crossing the site. These concerns will also apply to pipelines.
			Planned expansion of a farm included the construction of a new milking parlour close to a crude oil pipeline. Permission to construct the milking parlour was refused due to concerns that effluent might result in corrosion of the pipeline.
Process Technology	1	General	There may be more than one process route to the product and the economics, operability, maintainability, safety and environmental impact of the alternatives will need to be compared.
			For many existing large-scale processes, there is often a dominant technology with minor variations provided by different process licensors. The *Haber* process for ammonia production is an example.
			The degree of novelty in the process is an important factor. In principle, operators prefer to remain with well-proven technology, but there is often a degree of innovation in new designs, such as an updated control system or new design of pump seal. The larger the amount of new technology, the greater the need for detailed assessment to ensure safe and reliable operation.
			See also Section 12.3.
Utilities	2	General	The key utilities are generally power and water. For large facilities, these may be produced on-site; smaller sites may make use of imported power and water. Other utilities are generally produced on-site. The quantities required will need to be assessed to determine the plant economics.

(Continued)

TABLE 12.1 *(Continued)*
Key Factors in Conceptual Project Development

Item		Factor	Comments
	2	Water	Water may be required for process use but will, in most cases, be required for cooling and fire water requirements as well as washdown and "domestic" use. Adequate volumes of water may be available from a reliable source but quality is also important—particularly if there is a high mineral content.
	3	Electrical power	Larger facilities will often include their own power generation facilities partly for economic reasons but also to ensure a high level of reliability. On-site emergency power will also be necessary for most facilities.
Local Infrastructure	1	Transport	The local infrastructure, particularly transportation, is important not only to provide easy access during construction and operation for employees and visitors. Transport of components, spare parts, chemicals and other materials must also be considered. In many cases, raw materials or products will also arrive or depart by road or rail. The adequacy of the existing facilities will need to be assessed to ensure they are suitable for increased traffic flow and for the large/heavy loads that may be transported, particularly during construction. In some cases, long loads, such as large distillation columns or wind turbine blades, may require bends in the road to be realigned. Bridges and the like may need strengthening for heavy loads. Access for emergency services will need to be assessed, possibly at the same time as a site evacuation is taking place. The potential for noise and fumes from traffic must also be considered.
	2	Technical support	Whilst most sites will have a level of in-house technical capability, there is likely to be a need for external technical resources, for example, vendor support for major equipment and specialised inspection activities. The availability of specialist technical services and workshops for maintenance of equipment may be an important consideration.
Staffing	1	General	Ideally the workforce should be available locally, but often senior and technical staff may be required to relocate to the site.
	2	Accommodation	In locations where there is not a pool of existing trained personnel, operators and technicians may need to relocate to the site. In small communities, the influx of new arrivals can cause unease and may overload schools, recreational and medical facilities, and so on. In remote locations standalone accommodation, social and medical facilities will need to be built.
Cultural	1	General	Cultural sensitivities vary around the world and even between different parts of the same country. Many rural communities, for example, are not used to shift working.
Political/ Terrorism	1	Strikes	The potential for strike action can exist anywhere, but the potential for civil unrest, covering strikes, rioting and site blockades, is greater in some areas than others.
	2	Terrorism	The terrorism threat in some parts of the world is extreme and may require such stringent security measures that the project becomes uneconomic. Additionally entire countries may be potentially subject to regional or international sanctions. Even if the raw materials are received and products sold within one country, problems may still occur with the supply of spare parts and money transfers.
Regulation	1	General	Regulations in some countries may be particularly cumbersome and act as a disincentive to invest. There could also be difficult reputational risks for a company which knowingly invests in a country with low safety standards and then suffers a major accident.

* "Brownfield" refers to locations which have previously been developed and are now being redeveloped. Sites which have not previously been developed are referred to as "greenfield" sites.

12.3 PROCESS TECHNOLOGY

Products can often be produced by various routes using different feedstocks and process technology. Each route will have positive and negative features and will have different by-products and effluent issues. The various process routes will need to be assessed to determine potential safety and environmental impacts as well as economic viability. Difficult choices will need to be made if, for example, one route is more hazardous than others but has significant environmental advantages (see Section 4.2). The choice of process route will have been predetermined or some routes rejected without detailed analysis due to economic, technical, safety or environmental risks.

12.4 DESIGN ACTIVITIES

Design activities at this stage will determine the viability of the proposed development and establish the basic parameters. Preliminary process flow diagram will be developed together with heat and mass balances and a plot plan. This will allow basic equipment parameters to be determined (surface area of heat exchangers, number of trays or height of packing in columns, and the power requirements of pumps and compressors). In addition, utility requirements, feedstock and product handling, plus storage requirements need to be determined. This level of detail will be adequate for an estimate of the preliminary project budget cost. This will probably be within the range of +/– 30%.

In addition to the project finances, there will be numerous factors affecting the practicality of the project. Typical questions to be answered include the following:

1. Is the local power grid sufficiently reliable, or should power generation facilities be installed?
2. Will emergency power be required for critical equipment (for example, to prevent a runaway reaction) and control systems?
3. What level of spare process and utility equipment should be installed: single unit, two units (one operating, one spare) or multiple units?
4. What period of storage for raw materials and products is required?
5. Will additional facilities be required for maintenance and shut downs, such as catalyst regeneration (or change-out) and decoking of furnaces?
6. What should be the size and number of storage units (including buffer storage and storage of *off-spec* materials)?
7. How many and what types of facilities are needed for receiving raw materials and dispatching finished products?
8. What type of process cooling will be used (air coolers or a cooling water system)?
9. To what extent will maintenance be conducted on-site or contracted to other companies?

At this point the project could be defined as either viable or non-viable.

12.5 LOCATION

Determining the plant location is probably one of the most difficult issues at the conceptual design stage and there have been cases where work has started at one location but was then abandoned and a new location chosen. The challenges will vary from place to place. Construction close to an affluent residential area is likely to meet with a "Not In My Back Yard" (NIMBY) response, but in an area of high unemployment, a new industrial facility is likely to be welcomed. The same is true of plants proposed in developing countries, but the temptation to reduce safety and environmental standards must be resisted.

Regardless of location, the impact of a new plant can be minimised by landscaping and tree planting, but note trees to become involved in a fire or explosion. The plant's impact on the local environment can be more difficult. During the construction phase, there is likely to be large influx of workers, requiring temporary accommodation, catering and recreational facilities. In remote areas, these can be easily managed with little impact on the few local residents, but cultural problems can arise if the workers are sourced from different countries, ethnic groups or religions. Where construction workers are located close to large residential areas resentment from local residents can occur. Most large construction projects involve significant community relations packages which continue once the plant starts operations.

As stated earlier, there might also be a need to improve existing roads or build new ones to access the site and for future transport requirements. The adequacy of the water supply and power supply (if the site does not have its own generating capacity) will need to be assessed and, if necessary, improvements made.

The costs of any infrastructure improvements and community relations packages will need to be factored into the project economics.

The infrastructure requirements during construction also need to be considered:

1 Where will the site offices be?
2 What fabrication workshops will be needed?
3 How much parking will be needed?
4 What accommodation will be needed for the workforce?
5 What utilities will be required, and how will they be supplied?
6 How will heavy equipment reach the site?
7 What emergency response facilities will be needed during construction?
8 What security arrangements will be necessary?

Many of these, and other, questions will be industry and site specific.

12.6 WHAT CAN GO WRONG?

Even with the best analysis, there are times when mistakes will be made; sometimes these will be technical but economic and political factors will also intervene. Some examples are given in the following table.

Economics	The Trakhill power plant in Kabul has never worked efficiently and is known as the "white elephant" of Kabul.[3] The economics of the plant were incorrectly analysed and fuel costs made power more expensive than that obtained from other sources.
New technology	A new glass-producing plant was successfully completed and commissioned and then shut down and never commercially operated. The process used had been superseded by the float glass process—a much more economical alternative.
	One of the first processes for direct oxidation of ethylene to produce vinyl acetate utilised a nitric acid-based reactor, but it became contaminated by hydrochloric acid used in the catalyst regeneration process. This produced aqua regia (a mixture of nitric and hydrochloric acids) which is highly corrosive; the plant was abandoned after six months.

REFERENCES

1 Bunn A & Hailwood M, Fire and Explosion of LPG Tanks at Feyzin France, *Loss Prevention Bulletin (no. 251)*, IChemE, October 2016, www.icheme.org/media/1278/lpb251_digimag.pdf

2 French Ministry of Sustainable Development, Explosion in the AZF Fertilizer Plant September 21st, *2001 Toulouse France*, July 2013, www.aria.developpement-durable.gouv.fr/wp-content/files_mf/FD_21329_TOULOUSE_DP_JLC_GB_29072013.pdf

3 Chatterjee P, Black & Veatch's Tarakhil Power Plant: White Elephant in Kabul, *Corpwatch*, November 2009, https://corpwatch.org/article/black-veatchs-tarakhil-power-plant-white-elephant-kabul

13 Hazard Study 2—Front End Engineering Design

13.1 INTRODUCTION

The front end engineering design (FEED) stage involves detailed design of the plant and the development of a more accurate cost estimate. This is the second major step in the development of the project, and its place in the overall project life cycle is shown in Figure 13.1.

Until the FEED stage, the project team will have been relatively small and the cost of project abandonment would be low. FEED will involve a significant engineering team and the costs of abandonment increase. If there is any doubt about the project viability at the concept stage, the FEED phase should be paused until there is better knowledge and cost assessments.

When the FEED study is complete, the project is likely to be submitted for approval. Depending on timescales, the long delivery items may be ordered (but subject to subsequent change at the detailed design stage). This may only require a reservation of fabrication time but will also require matching drivers with large rotating machines which may be from different suppliers. Some large heat exchangers and specialised items will also fall into this category. For high-hazard projects preparation to meet regulatory requirements, including a *safety case* will be necessary. During FEED, the major safety and environmental issues will be identified and solutions found. This is one of the key elements of Inherent Safety (see Chapter 11). Errors which aren't identified at the FEED stage will require expensive, and time-consuming, rectification at the detailed design stage and the importance of getting it right first time cannot be over emphasised.

The broader elements of FEED are covered in this chapter with more detailed aspects of design in Chapter 14. In practice much of the detail provided in Chapter 14 could, and often is, undertaken at the FEED stage. Sometimes, the FEED and detailed design stages are combined to shorten the lead time to get the plant on stream. This increases the risk of costly design errors and is inadvisable unless the process technology is well proven.

The amount of documentation produced will increase as the project develops. Documents and drawings are likely to be revised several times, and it is important that the development of the design through the various revisions can be tracked and the reasons for changes clearly understood. The documentation will be required throughout the project life cycle. Further details are given in Section 1.7.

The FEED should provide as much detail on costs as possible to ensure that the decision to sanction the project, or reject it, can be taken with a high degree of confidence. The FEED study may also be used as the initial input to a *safety case*.

13.2 SAFETY STUDIES

A range of safety studies will be performed at the FEED stage. These will be discussed later. These studies will need to be checked or refined at the detailed design stage.

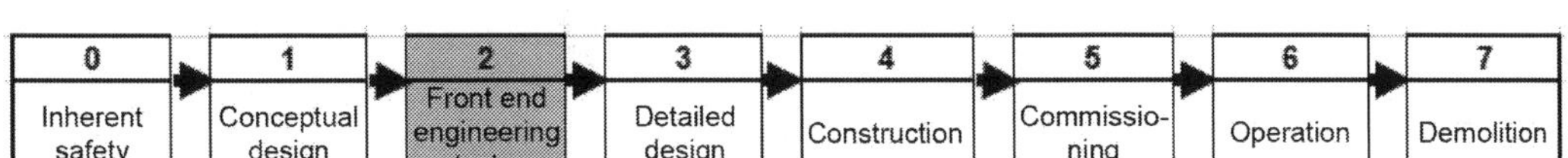

FIGURE 13.1 Project Life Cycle FEED

DOI: 10.1201/9781003510680-16

Studies will be needed to determine the following:

1	Chemical safety data	To cover all chemicals used in the process, including by-products, intermediates, catalysts and additives.
2	Reaction kinetics	Particularly exothermic properties of reactants and reactions and the potential for unwanted side reactions.
3	Preliminary safety studies: HAZID, What-if?, LOPA, HAZOP and others as required.	These should include hazards during startup, shut down and decommissioning as well as normal operation
4	Reliability and availability studies of critical equipment and systems including ESD systems	This applies not only to safety instrumented systems (SISs) but also to systems which affect overall availably and hence project economics.
5	Storage of raw materials, products, intermediates and other chemicals	Monomers may polymerise in storage and require injection of stabilisers. Catalysts may require special storage due to pyrophoric hazards or require dry or refrigerated storage. Reviews of intermediate and off-specification storage requirements should include transient conditions such as startup, maintenance, shut down and decommissioning. Don't underestimate the size, particularly during startup. The sizing of any intermediate storage may rely on a high availability, so this must be an integrated study with 4 discussed earlier.
6	Loading and offloading arrangements	This will need to consider the maximum loading/offloading rates and the number of road, rail or marine vessels that need to be handled simultaneously.
7	Emissions (both routine and following an incident)	Environmental impact studies may be necessary at this stage.
8	Effluent disposal	A dialogue with the environmental regulator will be necessary. This should result in *consent limits* for effluents being defined and any other requirements identified. One consent often forgotten may be the temperature of cooling water discharged to a river.
9	Utilities	Utilities may be shared on existing sites, but new sites, as a minimum, will require water and power.
10	Layout	Site layout is subject to review to ensure that common (and less common) problems are avoided. Prior to the development of computer-based 3D modelling a physical model of the plant was often constructed to ensure that access was adequate and valves could easily be reached. Nowadays virtual models are produced for most significant projects. Some types of risk assessment have been established to check equipment layout, but often a "what-if?" approach is used. Typical questions to be addressed include: 1 Can a fire at one location cause significant plant-wide disruption (for example overhead cabling affecting motors in other parts of the site or a small process area fire causing major damage to utility systems)? 2 Can PSVs be accessed without impacting operating equipment? (For some equipment such as storage spheres and distillation columns a crane or scaffolding might be needed.) 3 Are there adequate means of escape from all parts of the site? 4 Is there adequate vehicle access for maintenance and emergency response activities. 5 Is there any potential for "domino effects"? See Section 13.8.

11	Mechanical Handling	The most obvious aspect of mechanical handling is the transfer of raw materials and products. Where both feedstock and products are fluids, transfers may be by pipework. In other cases, choices may need to be made, for example between hoses or loading arms for fluid transfer. Where solids are handled significant pneumatic conveying or conveyor systems may be necessary. Provisions will also be necessary for the handling of other materials: packaging, spare parts, chemical additives, catalysts and so on. Finally provision for installation of, and removal of, all equipment at the site which might require removal for maintenance must be considered. For larger items, or those above ground level, crane access may be necessary. Lifting beans may be adequate for smaller items, particularly within enclosed areas.
12	Materials of Construction	The choice of material will affect the rate of corrosion and therefore the potential for leakage of process fluids. In particular, galvanic action where piping and other equipment involve changes to metallurgical specifications can be problematic. Where rapid temperature reductions due to depressurising (caused by the *Joule-Thomson Effect*) becomes significant, special low-temperature steels may be required. Material selection must cover all materials and not just those in contact with process and utility fluids and solids. For example, a thermite reaction between aluminium and iron oxide (rust) can cause a spark and should be avoided where flammable materials may be present.
13	Relief and Blowdown study	See Sections 13.10 and 14.8.

Additional studies will often be necessary, depending on the process and chemicals involved.

It might at this stage be appropriate to start the development of a safety case to meet the needs of the Regulatory Authorities.

13.3 CODES AND STANDARDS

Much of the basic design information will be available from easily accessible sources:

1. Materials safety data sheets (MSDS)
2. Regulatory bodies such as the HSE in the UK and OSHA in the USA
3. National, international and corporate standards
4. Trade associations and similar bodies, the most common being:
5. American Society of Mechanical Engineers (ASME)
6. American National Standards Institute (ANSI)
7. Tubular Exchanger Manufacturers Association (TEMA)
8. American Petroleum Institute (API)
9. National Fire Protection Association (NFPA)
10. Engineering Equipment Users Manufacturing Association (EEMUA)

It is sometimes said that "codes and standards are for the guidance of wise men and the blind adherence of fools". There is some truth in this, and blind adherence to a code can led to operating/maintenance difficulties which could have been minimised by a more detailed consideration of the issue. It is essential that codes are used consistently. At no time should there be a "mix and match" approach.

Codes and standards are periodically updated—sometimes at short notice; it is important to ensure that the latest version is used. This is one reason why safety cases are reviewed regularly.

It may be thought that compliance with codes and standards may be a protection in law in the event of an incident. This may be true to some extent, but codes/standards can be misapplied. They represent a minimum standard and over-reliance will not assist in the development of an ALARP approach to risk reduction (see Section 1.5).

13.4 CHEMICAL SAFETY DATA

Chemical safety data should have been reviewed at the conceptual design stage, but continuing research and testing make it essential to review the materials used in the process, most obviously for the following:

1 Reactivity
2 Flammability
3 Toxicity
4 Corrosiveness
5 Sensitisation/irritation

Other hazards may be less obvious or less well reported:

1 Radiological
2 Carcinogenic
3 Teratogenic
4 Narcotic

Specific processes may have additional health hazards; for example biochemical processes could result in exposure to bacteria or viruses.

MSDS are often the primary source of information, backed up by a company's own in-house research or data available from national or international sources such as the European Union (EU) REACH programme (see Section 3.6).

Analysis of hazardous properties will identify storage and handling requirements and safety requirements for normal and emergency situations, for example:

1 What Personal Protective Equipment (PPE) is required for normal activities?
2 Is additional PPE required for specific conditions (such as unloading) or emergency response/escape?
3 Is there a need for a *Safe refuge* in the event of a release?
4 Is special safety equipment (such as safety showers, decontamination chambers) needed?
5 Are there health and safety implications for neighbouring plants and population if a chemical release occurs?

13.5 REACTION KINETICS

Reactor design is a specialist area and will vary significantly depending on the process and scale of throughput. Reactors may vary in size from those producing a few litres per day in a pilot plant to thousands of tonnes per day (2,000 tonnes per day in a large ammonia plant and 5,000 tonnes/day in a methanol plant is common for new plants). Reactor design is continually changing as new catalysts and materials of construction are developed. For new processes, laboratory/pilot plant results might not scale up to a full-scale plant, and unexpected side reactions might take place. Detailed dialogue between chemists and process engineers are necessary to reduce this potential. A detailed, if dated, description of chemical reaction hazards is available from Barton and Rogers.[1]

Reactors come in many forms, the most obvious being batch or continuous. Other classifications are as follows:

1 Exothermic—heat given out by the reaction
2 Endothermic—heat consumed by the reaction
3 Solid bed—usually a catalyst
4 Back-mixed—internally mixed (usually mixed gas and liquid phase)

5 Liquid phase
6 Gas phase

Combinations are possible:

1 Exothermic, solid bed, liquid phase
2 Endothermic, solid bed, liquid phase
3 Exothermic, back-mixed, liquid phase

In general, endothermic reactions are not an issue as they "die" if heat is not added. There may, however, be some issues with by-products. Of greater concern are exothermic reactions. In these, heat is generated, and if not controlled or removed, the reactants heat up and follow the *Arrhenius equation*[2] causing the reaction to accelerate. It is not difficult to see that the loss of temperature control in a reactor will result in a *runaway reaction*. Regardless of the potential for runaway, it is desirable to keep the reaction within a fixed temperature range to maximise yield and minimise unwanted side reactions.

Reliability of the reactor temperature control is fundamental to both operability and safety. Heat exchangers used to cool the reactor should be oversized to account for possible fouling and likewise pumps to allow for fouling or wear. Generally, the power supply to cooling water pumps and the agitator (if installed) can be supplied from the emergency power supply to enhance reliability.

Other issues encountered in reactors include the following:

1 Channelling (where there is maldistribution of fluids in a catalyst bed) and localised hot spots. In a fixed bed hydrogenation process it is normal to fit a number of thermowells to monitor the uniformity of the reaction.
2 By-product formation if operated outside closely defined conditions
3 Reactant "slippage" (incomplete reaction)
4 Catalyst poisoning
5 Explosive decomposition of reactants/products

A change in the throughput of a back-mixed continuously stirred reactor generated dynamic fluid switching of the flow regime. Whilst unusual and unexpected this is an indication of the problems that can occur: (unreported work by one of the authors).

Monitoring and control of a reactor are fundamental to safe operation and special shut down features are imperative to avoid runaways. Shut down could involve arresting the feed of one of the reactants, dumping the reactants, adding a "kill" reagent to arrest the reaction, oversizing coolers to give adequate safety margins, and depressurising the reactor to reduce the reaction rate. The best strategies for dealing with runaway reactions will be based on knowledge of the process (from laboratory and pilot plant studies or experience of similar plants). A significant proportion of process safety incidents involve reactors, and even in common processes, previously unknown side reactions can occur.

As well as meeting the design requirements of conversion, flowrate and minimum by-products, the design should prevent any unwanted events and, if this cannot be totally prevented, the design should be reviewed to reduce their magnitude and frequency.

If all else fails, the reactor must be designed for the maximum pressure that can occur or, more commonly, an adequately sized reactor venting system installed. Calculation of vent sizes is complex as the reaction may be continuing—at an accelerating rate—as the reactor mass is vented, and the discharge will often be either two phase (liquid/gas) or three phase (gas/liquid/solid). Considerable research into reactor venting has been carried out over the years with the Design Institute for Emergency Relief Systems (DIERS)[3] being a good source of information. Chemical manufacturers will generally have a good understanding of the reaction kinetics of their process.

13.6 FLOW SHEETS

The development of *Process Flow Diagrams* PFDs and *process and instrumentation diagrams* (P&IDs) is crucial foundations of the FEED. The PFDs will define flowrates, temperature and pressure, material compositions of the major process lines. Often, there will be several cases, covering different flowrates, products, feed composition and other factors. This information is essential for initial equipment and pipework sizing. The design must also consider the need for buffer storage to cope with variations in flowrates or stoppages in different parts of the process. However, inventories should be minimised, particularly if buffer vessels contain toxic or flammable materials.

Whilst developing P&IDs decisions on the need for duplicating, or triplicating, equipment will be made. Spare equipment is often installed to allow for equipment failures or maintenance during operation. For pumps, it is common to have either two pumps (100% capacity each) or three pumps with 50% capacity. The choice is usually determined by economics or availability of suitable sized units. P&IDs will show the process equipment, valve types, locations and instrumentation. P&IDs need to be developed early in the FEED but are likely to be revised frequently as the design progresses. The initial P&IDs are often marked as *Preliminary* for review. It is essential that all changes are recorded in the drawing revision block and, ideally, as a "cloud" around the individual changes on the drawing. Further information on documentation is given in Section 1.7.

Whilst much of the process equipment will be fully integrated into the design, beware of *vendor packages* (equipment units which are not fully defined on the P&ID). At the FEED stage, these are often "black boxes" on the P&ID with no information on the internal arrangement of equipment. Common examples include nitrogen systems, lube oil systems, air dryers and chemical additive systems.

On an offshore platform a vendor package required a compressed air supply at a defined pressure. The flowsheet therefore included a standalone PCV. Unknown to the FEED engineers, the vendor had also supplied a PCV within the package. The facility was built with this arrangement of two PCVs in series. Surprisingly the vendor package worked well without the two PCVs continually working against each other.

A slop oil tank was not fully defined before the HAZOP took place; the study was therefore superficial. The final design of the tank included nitrogen inerting and an electrical heating element. The atmospheric vent on the tank was fitted with a flame arrestor. This had a very low-pressure drop, and it proved impossible to maintain an inert atmosphere in the tank. During operation, a pressure switch was bypassed. The operators were unaware that bypassing the pressure switch also bypassed the level switch. Inevitably, the day arrived when the level in the tank dropped, and the electric heater came into contact with the flammable atmosphere above the tank. An explosion resulted.

Once the P&IDs are well defined, they can be used for HAZOP, what-if? and HAZAN studies. Any changes to a P&ID will require the study to be reviewed and documentation updated.

13.7 UNIT OPERATIONS AND PROCESS EQUIPMENT

Unit operations are a convenient way of describing the various stages that exist in a chemical plant and can be considered as the individual building blocks of the process connected by piping, valves, instrumentation and items. In simple terms, a unit operation is something that changes the process, for example an increase or reduction in pressure, temperature or composition. There are a relatively small number of unit operations in common use but a much larger number which have limited application. For each type of unit operation, there are huge variations in the types of equipment available with variations in complexity and hazard. A unit operation can be a single item of equipment, for example a pump which changes pressure or a distillation unit which separates the feedstock to two or more outputs with different compositions. The distillation system will essentially consist of process vessels, heat exchangers, a distillation column (or columns) and pumps. Table 13.1 lists the most common unit operations.

TABLE 13.1
Common Unit Operations

Unit Operation	Comments
Compressors	Compressors vary greatly in size ranging from small units (such as instrument air compressors) to massive machines used in gas transmission lines and large process plants such as ethylene crackers and LNG units. Many compressors for low flowrates or very high-pressure duties are positive displacement types. Centrifugal types are generally found on higher volume duties, but axial flow machines are used in some situations. General points to note include the following: 1 They are noisy—particularly if there is a gearbox and high impellor tip speeds. 2 They require special shaft sealing arrangements. In the case of air these can be simple *labyrinth seals*, but toxic and flammable gases generally require sophisticated seal types. Older machines generally have oil seals, but newer machines have dry seals. Pumps for oil seals may be affected by a power failure, and an emergency pump and/or rundown tank (to supply seal oil by gravity) is required. The seal is one of the weak spots of most compressor types. 3 Centrifugal compressors must maintain a "minimum flow" of around 60–65% of design to prevent aerodynamic breakdown (similar to aircraft stall). This can result in reverse flow and may affect overpressure protection of suction vessels. This is often called *surging* or *panting*. 4 Compression generates heat, and aftercooling is generally required. For multistage casing compressors, intercooling between each casing is usually needed. This will also need to address the anti-surge recycle during startup. In the case of refrigeration systems, it is usual to spray condensed liquid refrigerant (also known as quench) into the hot recycle flow upstream of the knock-out vessel. This results in the heat of compression being removed in the condenser. 5 Compressors have NRVs in the discharge to prevent back flow. The ability to overpressure suction vessels has already been mentioned, but, additionally, oil-lubricated bearings are only designed for a unique direction of rotation. 6 In some cases, flow is controlled by running the compressor at a constant throughput but recycling excess gas to the compressor suction. The location of the minimum flow recycle and the location of the NRVs must address the possibility of high reverse flow around the compressor and possible overpressure of the suction vessels. 7 The impellor of a compressor can be damaged by ingesting liquids. A suction vessel (often called a *knock-out drum, suction scrubber* or *interceptor*) is usually essential. On multi-casing compressors, a knock-out drum between each casing will be necessary if there is a possibility of liquid condensation between stages or potential carryover from condensers. 8 Air compressors must be either of an oil-free design or fitted with an oil demister to avoid oil mist explosions. Air for use in instrument air system must be dried to avoid damage to instrumentation. In 1965 the whole instrument air system of a large chemical complex in the UK was shut down by an oil mist/air explosion in the discharge manifold. The ignition was the result of a slow exotherm in oil-soaked mill scale in the manifold. 9 Long shafts on centrifugal compressors have a number of potential critical speeds which can result in damage. The unit must not linger within 5% of these during startup and shut down. 10 Positive displacement compressors require overpressure protection at the discharge (if the discharge is blocked pressure will continue to rise until something ruptures). 11 Consideration should be given to overpressure protection of the compressor suction if the discharge pressure is high and discharge valve leakage is credible when the compressor is isolated. 12 Small piping close to high-pressure reciprocating compressors will be subject to induced vibration and, if inadequately designed or supported, may fail due to fatigue.

(*Continued*)

TABLE 13.1 ***(Continued)***
Common Unit Operations

Unit Operation	Comments
Pumps	Pumps, like compressors, are either centrifugal or positive displacement. 1 Pumps also require shaft seals. For cooling water, this could be as simple as a packed seal, but for other duties there may be a single or double mechanical seals. In some cases, the interface between the two seals has a pressurised oil supply, with an alarm to indicate loss of pressurisation. The pump can then be shut down when loss of pressure indicates a seal failure. This is a relatively common occurrence in pumps and the consequences of a failure should be considered as part of the plant hazard studies. 2 In the case of slurry systems, a secondary barrier on the pump side of the seal fitted with a compatible clean flushing fluid must be used. 3 Pumps can foul with debris carried in the process fluids, and suction strainers are often installed. They may also foul due to particulates in the process fluid. This problem can be overcome using open impellor designs. 4 Suction filters are sometimes removed by operations teams to reduce the pressure drop and hence the potential for cavitation. Frequently, however, this has resulted in debris damaging the pump impellor. 5 Pumps require a minimum flow of 5–10% design flow to remove the *heat of inefficiency*. Several pumps have ruptured when operated between closed suction and discharge valves.[4] 6 Non-return valves are essential at the discharge to prevent back-flow when the pump is not operating. 7 If pump suction pressure falls below the *net positive suction head* (NPSH) cavitation can occur, causing severe damage. 8 Centrifugal pumps can generally be specified to avoid the discharge pressure exceeding the design pressure of the downstream plant and equipment. However, the pressure downstream of a positive displacement pump will continue to increase until there is a rupture—either of the pump or downstream equipment. A PSV must therefore be fitted on the outlet of a positive displacement pump.
Pressure Vessels	Pressure vessels are found in numerous places in process units with uses ranging from charge vessels for batch reactors to stabilise varying flows through the process, for two- or three-phase separation, reflux drums in distillation systems and knockout drums in compressor and gas turbine systems. They are often used for medium-scale storage of liquefied gases. Debris may collect in pressure vessels. Generally, process pressure vessels would contain two phases (vapour and liquid) with the level control (and a safety trip) to prevent a *blow-by* or *blow-though* (where the liquid level is lost and vapour enters a downstream liquid system). Often the liquid system is to a lower design pressure than the vapour system, and vapour entering the system can result in rupture. For similar reasons, a vortex breaker is frequently installed to prevent a two-phase mixture draining into the liquid system. On 22 March 1987, a major explosion occurred on the hydrocracker at BP's Grangemouth, UK refinery. The explosion originated in the low-pressure separator of a hydrocracker and was the result of poor operation of the level control valve for a high-pressure separator. This led to the liquid level in the high-pressure separator being lost and gas flowing into the low-pressure downstream separator—which ruptured.[5] Projectiles from the low-pressure vessel were flung considerable distances, a significant fire followed, and there was one casualty. A low-pressure trip to close the valve on low level had been disconnected—but that's another story. Many pressure vessels have complex internals (such as weirs and baffles in separators, trays or packing in columns). Access for maintenance needs to be considered with suitably positioned manholes to allow the vessel to be *sweetened* (purged of harmful materials and possibly "steamed out" before entry). The location should also address "emergency escape/evacuation" and address natural air flows A scrubber fitted with "*disc and donut*" trays required internal inspection. The gases entered at "A" and flowed downwards in a co-flow mode and then out up the central column (Figure 13.2).

(*Continued*)

TABLE 13.1 *(Continued)*
Common Unit Operations

Unit Operation | **Comments**

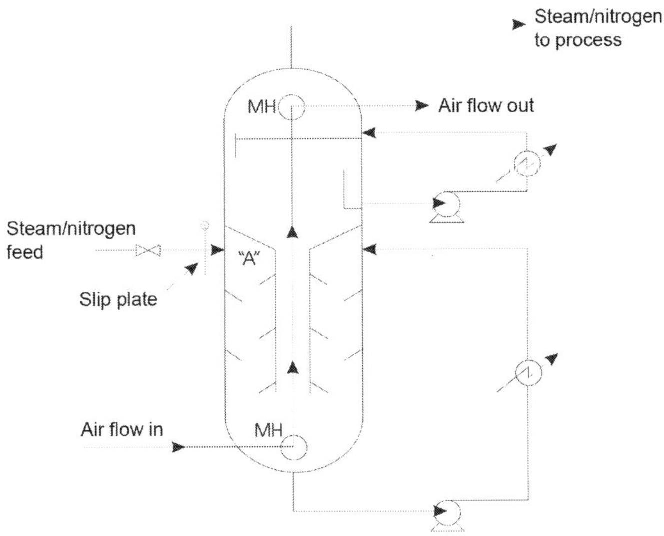

Scrubber Internal Baffles

In preparation for entry and inspection residual gas was displaced by nitrogen from point "A" following the process route. The nitrogen was then displaced with air via the two open manholes (MH). This left a dead-end at "A" which may still have contained nitrogen. Anyone entering this area could be asphyxiated.

A pressure relief system is required on all pressure vessels. Even if the process cannot generate pressures in excess of the vessel design pressure, heat from an external fire could cause boiling and an increase in vapour pressure which will ultimately exceed the design pressure. Note that a pressure relief system will not protect against direct flame impingement which could cause a BLEVE (see Section 9.8). Pressure relief is discussed in more detail in Sections 13.10 and 14.8.

Columns

Most columns will be pressure (or in a few cases, vacuum) vessels, and the comments made earlier will also be applicable.

For all kinds of large columns, there can be a significant time lag when control parameters are varied before operation is stabilised. This can create control issues and can affect the hold-up in the base of the column. These are issues which must be considered during FEED.

A specific issue arises with crude distillation columns where pyrophoric iron sulphite may ignite when exposed to air.

Trayed Columns

The internal design of distillation and other contactor columns fitted with trays will depend on the fluid handled.

Whilst the bulk of the liquid hold-up in a vessel will be at the base, trays will often have a depth of 2–3 cm of liquid, with a further 10 cm in the *downcomers*. For a large column with numerous trays this can add significantly to the vessel inventory and will need to be considered in risk assessments. This is of particular importance for columns handling liquefied gas. The potential size of the gas cloud following a release must include the liquid in the trays/downcomers as well as in the base of the column.

Attention must be paid to the possibility of *jet flooding* at startup. If the reflux is started late, and the gas phase is at or above about 50% of normal flow, it is virtually impossible to establish downward liquid flows. Increasing flow slowly is the only solution.

As with pressure vessels, access for maintenance and arrangements for sweetening must be considered in detail, as well as arrangements for removing and inserting trays. Whilst there are no absolute rules, access ways are usually provided every 10–15 trays.

(Continued)

TABLE 13.1 *(Continued)*
Common Unit Operations

Unit Operation	Comments
	One column was fitted with 150 trays. Any change in reflux rate took 10 minutes to affect the level in the column base due to the velocity time lapse between the trays. This is not instantaneous and makes control of the specification very difficult. Changes had to be made slowly over a period of time.
	Packed Columns Again, access for maintenance and sweetening arrangements are a key concern, as well as arrangements to remove the packing for cleaning or replacement. In general, packed columns are not recommended for fouling duties.
Heat Exchangers	There are a variety of heat exchanger types available, but the majority are either shell and tube or plate types. The former are more widely used in the process industries, but plate heater exchanges are frequently used for low-pressure duties where space is at a premium (in offshore and marine applications) or where frequent cleaning is necessary (food and pharmaceutical industries).
	Shell and Tube (S&T) Exchangers There are a number of different variations in S&T design. There are codified by the Tubular Exchangers Manufacturers Association's (TEMA's) code which defines the baffle arrangement, flow regime and other design feature. Several problems are commonly encountered:
	1 Tube vibration, created by *Von Karman vortex shedding*[6] of the tube, results in tube/tube impact, tube/baffle impact or fatigue of the tube at the tube sheet joint. Design codes will assess the likelihood of vibration based on the flow velocities across the tube and the spans between tube sheet and baffle or baffle to baffle. 2 Tubes are normally expanded into the tube sheet (rolled by an internal tool, creating plastic distortion). There may be a groove in the tube sheet which acts as a secondary seal, but these may not seal properly and are a potential source of leakage and/or become corrosion sites. 3 Impingement plates are often installed to prevent erosion and forced vibration of tubes. 4 The design of a heat exchanger usually has a *fouling factor* to allow for the build-up of a film on the exchanger surfaces which reduced the efficiency of heat transfer. Tubes within the heat exchangers are on either a square or triangular pitch. Cleaning the outsides of the tubes is easier with square pitch tubes—triangular pitching tube bundles only permit the cleaning of about two rows of tubes. However, triangular pitching gives a better shell-side heat transfer coefficient—hence a smaller (and cheaper) heat exchanger by comparison with a square pitch exchanger. Ease of cleaning must be balanced against increased capital cost. Triangular pitch exchangers are generally used on clean shell-side duties. 5 Tube-side cleaning will be facilitated by a "*floating head*" or second tube sheet that is smaller than the shell diameter with a removable return head (TEMA code A, E or S). However, there is a potential leak route from the tube side to the shell side. This will allow the removal of the whole tube bundle for clearing both the shell and tube sides and is essential if there is a significant *thermal growth* or expansion as the exchanger reaches operating temperature. Long heat exchangers on clean duty can have fixed tube sheets which allow the tubes to "expand" in the form of a "wiggle".
	Except for reboilers, the higher pressure or more hazardous fluid is in the tubes, and the less hazardous fluid is on the shell side. This reduces capital cost of the shell and reduces the inventory of the more hazardous fluid. However, in the event of a tube rupture high-pressure fluid is discharged into the low-pressure shell. The shell design must cater for this possibility by either increasing the design pressure or sizing the pressure relief system to cover this situation. Instantaneous rupture is unusual and is virtually impossible to size relief protection for instantaneous tube rupture and release of gas due to the dynamics of displacing liquids at the same rate as gases enter the shell side. Practical guidance is given in API 521.[7]
	Plate Exchangers Plate heat exchangers are only suitable for low-pressure duties due to the vulnerability of the plate/sealing arrangement. They must be specified to avoid corrosion. As well as correct specification of the plates the elastomer used for sealing must be suitable for the fluids being handled. They are frequently used for water cooling duties, and water trapped can freeze at low temperatures. There are numerous other kinds of exchangers for specific duties such as fin fan coolers, spiral exchangers and rotating exchangers. An overview is given by Ronquillo.[8]

(*Continued*)

TABLE 13.1 ***(Continued)***
Common Unit Operations

Unit Operation	Comments
Fired Heaters	Fired heaters and boilers are fired with a variety of different fuels. Within the process industries, most are either gases or liquids. Prior to starting a fired heater, a purge cycle must take place to ensure that any unburnt fuel in the firebox has been removed. Failure to do this will result in an explosion when the heater is ignited—a common cause of fired heater/boiler accidents. All but the oldest and smallest units have flame failure detection (commonly called "fire eyes" or "purple peepers") which will isolate the fuel supply if the flame is lost. These sometimes fail if the burners are not burning fuel efficiently and a film of soot builds up over the lens. Valves on the fuel lines do not always close tightly, and there may be fuel leakage into the firebox—hence the importance of purging before startup.
Reactors	Most reactors are either *continuous stirred tank reactors* (CSTR) or some form of linear flow reactor with continuous flow through the reactor. As a rule CSTR types are for smaller scale duties with flow-type reactors for larger more complex duties. Reactor kinetics have been outlined in Section 13.5, but some key points of reactor design are considered in the following: A weak point in CSTRs is the agitator (or stirrer) with several issues: 1 Agitators are often fabricated from several components fixed together. If the fastenings fail, mixing will be impaired, affecting the reaction kinetics—possibly resulting in a reaction runaway. 2 The mechanical seal around the agitator may fail releasing reactants and finished products. 3 If the reaction process does not proceed as desired, high viscosity products may be produced. In extreme cases this can result in failure of the agitator or the agitator motor burning out. 4 There is a general rule (called the "dead cat" rule for obvious reasons) that shear pins may be required so that the shaft will separate from the motor if there is excessive viscosity in the reactor. 5 The necessity of a reliable cooling water supply has previously been discussed. A similar high reliability is needed for the power supply to ensure that the agitator can operate. Linear flow reactors, such as ammonia converters, often consist of several fixed beds with baffle arrangements to guide the flow through beds in turn. These need to be assembled carefully to prevent leaks through the baffles which will reduce the reaction efficiency. The response of operators in this case might be to increase pressure (possibly close to the design pressure) or increase flowrate. Maintenance requirements will be similar to pressure vessels, but there may be additional requirements such as ensuring that an inert atmosphere is maintained over the catalyst during shut downs as well as the provision of facilities for activating/deactivating catalysts. Where an agitator is fitted, this must be "locked off" to prevent accidental operation (see Section 16.8). Fines from catalyst systems may be a respiratory hazard and some (such as finally divided carbon) could spontaneously ignite. A particular problem occurs in polymer reactors with blockage of the pipework leading to the pressure relief systems. As there is normally no flow through these lines, polymer can build up and cause a restriction in flow if a runaway reaction occurs. For this reason, bursting discs are often installed upstream of PSVs.
Filters/Strainers	Filters in process systems with the design intent of stopping fine particles are generally duplicated (known as duplex filters) so that one can be removed for cleaning without affecting production. They are one of the few items of process equipment which is intended to be opened frequently whilst the plant is in operation. The changeover system must address the possibility of one filter being isolated before the spare is brought online. Usually, the valves are interlocked to prevent this, but the potential for leaking valves must be considered. If the filter is used with flammable materials the hazardous area (ATEX) classification zone must address the likely frequency and duration (see Section 14.6).
Dryers	Dryers come in many design types, shapes and sizes. Large dryers are often used for powder drying and incorrect operation may cause a dust explosion.[9] Many dryers are desiccant types: molecular sieve (beads) or alumina (chips) and, like filters, generally need to be duplicated to ensure that one can be regenerated whilst the other is in use. In some cases, a "lead dryer" unit is installed which absorbs the moisture followed by a "guard dryer" which is left online during the changeover. The guard becomes the lead and a new guard is put in place. The valve sequencing requires detailed examination. "Slippage" or moisture breakthrough is detected by a humidity/dryness meter. This then becomes a key process parameter. Other types of dryer available include condensing dryers and liquid contact using a drying fluid which has to be regenerated continuously

(*Continued*)

TABLE 13.1 *(Continued)*
Common Unit Operations

Unit Operation	Comments
Crystallisers	Crystallisers are relatively uncommon but have two main forms: 1 Batch—not of use in a continuous process without buffer storage. This will result in a cake with a high spread of crystal sizes. 2 Continuous classifying, which produces a more uniform crystal size due to the use of Stoke's law. These must have oversized crystal destruction loops. Care must be taken to extract the crystals at the same rate as they are grown.
Centrifuges	There are numerous centrifuge types: 1 Solid bowl decanters (continuous) 2 Pushers (semi-continuous) 3 Peelers (semi-batch) 4 Vertical—bottom discharge (batch) 5 Decanters In some cases, a pressure filter with a pre-coat may be necessary. Rotary vacuum filters can also be considered as a type of centrifuge. All centrifuges will require a cake wash to remove the traces of "mother liquor". The wash then has to be further processed. The solid bowl decanters develop a slight pressure at the discharge point; it may be necessary to fit flexible seals and a closure at the discharge. Where flammable solvents are present the vapour must be contained. Centrifuges can and do become unbalanced when laying down the cake, washing the cake and ploughing out the cake. Operators may attempt to set up hoses to wash the cake; this is unlikely to be helpful and is likely to increase unbalance. The centrifuge design must reflect the nature and morphology of the crystals. At high acceleration forces, the cake becomes compressed—hindering the removal of mother liquor. Common mechanical failures[10] include the following: 1 Out of balance 2 Bearing failure 3 Incorrect assembly 4 Incompatibility between process materials and materials of construction 5 Unauthorised operator intervention when the centrifuge is in operation 6 Incomplete or inappropriate replacement of wash nozzles, creating an imbalance due to disturbance of the cake Flammable mother liquors have caused explosions in centrifuges and some form of inerting is necessary.

13.8 LAYOUT

Like many aspects of engineering the layout of a process plant involves compromise. Short pipe runs reduce pressure drop between equipment items but increase the amount of damage a pool or jet fire might cause. The designer will want to reduce overall cost whilst juggling with conflicting requirements. Layout is a complex issue, and experience plays a significant part in the design process. The information in the sections later are generalisations, there is an engineering limit to the spacing of the equipment due to increasing capital and operational costs. Often local building codes will provide detailed guidance on many aspects of layout and construction.

13.8.1 Spacing

As a rule of thumb, the process area of the plant should be approximately 20 times the footprint area of all the major pieces of equipment. This should allow access for maintenance and also give adequate separation between equipment items. Other factors may influence spacing, particularly in offshore platforms. Availability of land may also result in undesirably close spacing. In other locations—particularly deserts—the availability of land will allow generous spacing. A major explosion will cause blast damage to all the units within the overpressure radius. Therefore, it is desirable to arrange process units in a row rather than a grid. This is, however, not always possible and may increase construction costs as a consequence of longer pipe and cabling runs.

Generally, there will be spacing between the site facilities and the boundary to minimise the effects of an onsite incident on property outside the boundary. The overall layout should minimise the need for personnel and visitors to enter hazardous areas. Offices, workshops and similar facilities should be located close to the main entrance. For all but the smallest of sites, a minimum of two well-separated, entrances will be required to ensure access for emergency services regardless of wind direction. There are often dedicated entrances for trucks supplying raw materials or exporting products. This allows administration and control of trucks away from the main entrance used by employees and visitors.

The designer should separate potentially hazardous parts of the plant from each other. Fired heaters and boilers should not be located where they could provide a source of ignition for escaping flammable gases and are usually located upwind of potential leak sources. Roadways must be suitable for access by maintenance vehicles (including cranes) and emergency vehicles. Walkways must be provided for access and emergency escape.

Confinement and congestion in the process areas are undesirable but, to a degree, unavailable. They will cause air turbulence and increase the velocity of the flame front in the event of an explosion, leading to higher overpressures (see Chapter 10). A degree of openness is, in any event, desirable to allow rapid dispersion of gas leaks which could explode in a confined area. There are however cases where enclosure is necessary and alternative mitigations will be necessary. Typical examples are compressor houses in gas transmission systems in cold climates or close to residential areas (where noise problems could occur). Toxic liquefied gases are sometimes stored within enclosed buildings to provide secondary containment in the event of a leak.

Individual process units are often surrounded by an open area with a roadway between units. Generally free passage of vehicles and personnel along roadways is possible but, in some cases, where space is limited roadways may encroach into hazardous areas (see Section 14.6). This is not generally a problem. If access is required, it can be done under PTW control (see Section 16.6). If required, access for emergency services will also be under supervised control. There should however be a barrier and warning notices.

Spacing must also be adequate for emergency response and escape from a hazardous event. Spacing between individual equipment items has often been based on prescribed separation distances, but in recent years a more risk-based approach[11] has come into use.

Process plants are three-dimensional, and consideration of the third dimension is sometimes forgotten. Typically, pumps will be offset from the base of distillation columns to reduce the overall height of the structure, and, where practical, pump seals and flanges should not be located where a fire from a seal or flange leak could impinge on structural steelwork or process facilities. The area under fin fans should be free of pumps on flammable duties to avoid the situation where a seal failure results in a pool fire. The updraft from the fin fans will exacerbate the fire, increasing the level of damage. At a more mundane level, instrument and electrical cabling should be routed away from pumps on flammable duties, again to avoid damage from fire. On multi-level pipe racks, corrosive chemicals should be in the lowest level of pipe racks so that leakage will not corrode piping at lower levels. For similar reasons, cabling is generally run above piping.

Layout also needs to consider sloping of ground to ensure that any spillages flow towards drains and pools do not form where pool fires could occur. A convex camber or graded/sloping concrete is often constructed under pipe racks to divert liquids to the sides of the rack where drains are installed.

Where practical, electrical cabling should be run underground where it will be less exposed to fire or accidental damage. However, it is generally more economical to run cabling above ground and many newer plants are constructed on a modular basis with all cabling above ground. Fire protection can be supplied by fire-resistant enclosures or passive fire protection (PFP) (see Section 9.12). Some cabling will, of course, need to be located above ground, for example, for lighting, motor terminations and electrical facilities.

Within process areas, it is also becoming increasingly common for low containment walls to be installed around likely leak sources, such as lube oil systems and the seals of hazardous pumps.

In the case of processes handling toxic or corrosive fluids it may be desirable to forbid access to certain areas. In this case the design may have to cater for remote valve operation and instruments, other than sensors, may have to be located out of the restricted area. Other areas where access should be restricted include areas with automatic CO_2 fire protection and areas where flammable gas sources could be present (for example, analyser houses). Access to electrical switchgear is also often restricted.

Whilst attention will be focused on the process areas, other parts of the site—utilities, storage and feedstock/product handling—all require consideration. Access in utility areas will be similar to process areas, but hazards will generally be more limited (although it is not uncommon for hazardous materials to be used in utility systems). Atmospheric storage tanks containing flammable liquids should be separated by a minimum of half the diameter of the largest tank to minimise the risk of flame spread between tanks. Secondary containment (bunds) should also be provided in case of tank leakage and are generally a design code requirement. Pumps to fill and empty the tanks are generally located away from the tanks. This will provide the ability to pump out of a tank on fire.

Pressurised storage of flammable liquids in "bullets" or spheres is generally not fully bunded to prevent a spill of liquid causing a BLEVE. The ground underneath storage vessels should be sloped to a collection point which drains to a remote impounding basis where a fire could do no harm. Small walls may be installed around storage vessels to help channel liquid towards the impounding basin. A significant leak of LPG or other liquefied gas stored under pressure will initially have a large degree of *flash* but will contain an increasing proportion of liquid as the ground underneath the leak area cools as a result of heat transfer to vaporise liquid (see Chapter 7). For cryogenic storage, the design must address *frost heave* which can "spring" joints resulting in additional leakage. This is usually addressed by subsoil heating and location on raised platforms.

Road tanker loading/unloading areas should avoid the need for vehicles to run close to process areas, and drainage/containment areas should be installed to contain spillage during loading/unloading. A one-way system through the loading rack is preferred to one where tankers need to reverse into position. Properly designed earthing systems must be installed in each bay if flammable materials are handled.

13.8.2 Access

13.8.2.1 Maintenance

Access to all equipment for maintenance is essential. Smaller items, which may require removal to workshops for repair, will require space for a crane or the installation of lifting beams above the equipment to be moved plus space for a trolley for transport.

Within individual process and utility areas spacing should be adequate to allow access for maintenance (for example: removal of S&T bundles, catalyst replacement and removal of trays or packing from columns). In the case of tube bundles there may have to be an anchor point for a ratchet-operated device with pulleys and ropes. It may be necessary to have an overhead runway or cradle to support the bundle using strops to avoid damage through bending. Consideration must be given to the route by which the bundle could be removed from the process area for inspection or cleaning.

The use of cranes within process units should be avoided where possible but may be necessary, under controlled conditions (see Section 16.6). This should be considered at the design stage and be part of the *mechanical handling study*. Equipment which may require the use of cranes should be located with good access and should avoid the need to lift over pipe racks, cable ways and other equipment. Access routes for cranes should avoid underground services, especially drains which could suffer compression damage. This is also a useful consideration for construction where there will be widespread use of cranes and other equipment.

Whilst most equipment will require maintenance on an occasional basis, there are a few items which may be opened on a regular basis. Filters are the most common example but plate heat exchangers on fouling duties, and pig traps may also require frequent opening. The area around such equipment needs to allow good access for personnel access and escape. Hazardous area classification around these areas needs to be considered to avoid electrical equipment which could be a source of ignition. Pig traps can be opened under pressure and it is becoming common practice to install a substantial concrete barrier to block a pig released under pressure from escaping and causing impact damage. Whilst many pig traps have interlocks to prevent accidental opening, there are still cases of traps being opened under pressure. The area classification of the pig trap also needs to be considered.

Many maintenance activities involve working at height, and space should be available for scaffolding to be installed without the use of piping or equipment as anchoring points.

13.8.2.2 Human Activities

Access must avoid the need for excessive bending/reaching to avoid back injury or overbalancing. The location and orientation of valves, instruments and access structures therefore require detailed assessment. Walkways should be sufficiently wide and high that operators can walk without snagging valves, rotating equipment (which should have guards covering rotating elements) and hot/cold piping. Valves spindles should not impinge on walkways. Even if not required for process purposes, insulation or a protective guard is often added to sections of hot or cold piping close to walkways to avoid injury if they are touched accidently.

All valves regularly used should be accessible from ground level or via a fixed platform. Ideally, they should be installed vertically without the need to twist whilst opening/closing. There should be no requirement for the use of temporary steps or other makeshift arrangements (unfortunately, it is common to find scaffolding installed where access to valves had not been identified at the design stage). Valves requiring routine operation should not be located in pits or other inaccessible areas.

13.8.2.3 Emergency Escape

Consideration must be given to emergency access/escape. Single walkways should be an absolute minimum of 1 m, and preferably 1.5 m, wide. Platforms and walkways above ground level generally require two means of access or escape at opposite ends. This can be either two stairways or a stairway for normal use and a vertical ladder for emergency escape. The ladder should have a safety cage until 2 m above ground level and a safety gate at the top. Whilst this is the ideal, it may not always be practical, for example on distillation columns.

Escape routes or routes where injured personnel may require stretchers must be at least 1.5 m wide and have sufficient access on landings to rotate a stretcher. Headroom in all cases should be at least 2.25 m.

> During an emergency exercise, a dummy was being evacuated from an elevated platform down a stairway with a bend in the middle. The landing at the bend was undersized which created difficulties in manoeuvring the stretcher. After several attempts, it was decided that the only way to rotate the stretcher was to have part of it overhanging the guard rail. Whilst doing this, control of the stretcher was lost, and both the stretcher and dummy fell to the ground: (first hand experience of author).

The guiding question must be "Can I get into and out of the area in an emergency and can I assist an injured person to leave the area?"

Confined spaces must only be entered under permit control (see Section 16.6). Two, or more routes are preferred for both *gas freeing* the space and continuous ventilation. In some cases, this may not be possible due to other design constraints.

13.8.2.4 Access—Emergency Services

The need for access to rescue personnel is obvious. Firefighters have different needs. An aspect that is often overlooked is the requirement for a clear view between a fire hose nozzle or monitor and the source of fire. There may be a need to set up firewater nozzles for cooling through 360° which can be hindered by walkways and structural steelwork. Emergency services may also require *hard standing* for fire engines (or ambulances) and easy access to hydrants, monitors and fixed systems. A review of the layout by emergency responders at an early stage will prove useful (see Chapter 22). Conversely, fire hydrants and monitors should not be located too close to the source of a fire where high levels of thermal radiation could prevent access.

Most large plants will have significant water storage facilities with fire water pumps supplying a ring main around the site. If there are additional water sources nearby clear access to these would be useful so they can supplement the main supply if necessary.

13.8.2.5 Lighting

Most plants will be in darkness for part of each day, and adequate lighting is necessary (see Section 3.10). The location of light fittings needs to be considered with care. Walkways, stairs and field instruments need to be clearly illuminated without shadows.

13.8.2.6 Social Impact

Although the plant designers will want to optimise the layout, the plant will impact the surrounding area. In some locations there will be limited concern, but in other cases, particularly where the site is near built-up areas or in open countryside, the concerns of local residents and others need addressing. Considerations can be summarised as shown in the following table:

Visual	Attention should be paid to the location of tall pieces of equipment—these should be as far away from the public as is practicable. Distillation columns can be split into two shorter sections, but the penalty will be increased capital cost, reduced reliability (due to the need for additional pumps) and greater inventories of hazardous process materials. Cooling towers should also be located as far away from the public as is practicable—not just due to the impact of the towers themselves but also from the plume of steam which may occur in some weather conditions.
Noise	Power plant, compressors and other noisy equipment should be located as far as is practicable from the public and where possible fitted in acoustic booths around machinery whilst still maximising the potential for natural ventilation of gas leaks.
Odour	Many processes—not only chemical plant but foodstuffs, pharmaceuticals and effluent treatment—can, and often do, produce odours.
Traffic	Additional traffic on local roads will occur. Access for the emergency services to the site must also be allowed.

Further information is available in Chapter 4.

13.9 CONTROL AND SHUT DOWN SYSTEMS

Control engineering is another specialist activity and is not discussed in detail. A basic overview of process control has been provided by King,[12] and there are a number of readily available texts covering the design of control systems which should be referred to for detailed knowledge. Process control, automation, optimisation and digitalisation are areas of rapid development. This section only covers aspects of control that relate to safety.

13.9.1 Control

There are five main parameters which must be controlled:

1 Composition
2 Flow
3 Level (including liquid/liquid interfaces)
4 Pressure
5 Temperature

Composition may be a control parameter but is often an *inferred* parameter. For example, the inflexion point for temperature in a distillation column may be used as a partial control parameter. Pressure (due to vapour/liquid equilibrium) is a secondary composition parameter. Increasing use is being made of fast analysers for composition control.

Whilst there are still numerous analogue control systems almost all modern plants utilise computer control. Both are based on control utilising:

1 *Proportional band* (P). The control signal is proportional to the offset from the control setting.
2 *Derivative action* (D). The control signal is a function of the rate of change in the parameter.
3 *Integral* (I). The control signal is proportional to the time away from the set point. Beware of reset or integral saturation. The control will direct a full-scale movement and if the valve is "sticky" it moves suddenly.

Two extra controls, often called *safety instrumented systems* (SIS), are required:[13,14]

1 *Alarms* to warn of a significant deviation from the desired parameter.
2 *Trips* (sometimes called *interlocks*) to put the process into a safe but shut down condition.

These will be covered in more detail in Section 14.7.

Prior to the widespread use of computer control, most plants had large analogue panels, often stretching 20–30 m in length, A few still exist. However almost all modern plant uses computer-based systems. The layout of the screens is a major part of the control system design. Whilst some screens will show an entire part of the process, others may show only a single item of process equipment. To allow the operator to quickly analyse process upsets key parameters may be displayed on more than one screen. In addition, there will be screens to show process performance and an *alarm log* to show alarm and trip readings. The selection of parameters requires detailed discussion with the operations group. This has become a variation of HAZOP, generally called a Control Systems Hazard and Operability Study (CHAZOP). A version of HAZOP covering electrical systems (EHAZOP) has also been developed.

Computer control allows some internal checking of the validity or the *drift* in a parameter. However, it is often useful to have a second set of measurements as a verification of the parameter. This is one reason why readings are generally required by field operators at regular intervals (another benefit is to encourage regular patrols around the plant so that leaks, fires and other faults can be discovered at an early stage).

Whilst the details of the control system will require detailed work by specialists there are a few control room basics which should be given consideration by the design team:

1 Control Room Design and Location	The control room should, ideally, be located away from hazardous areas, but if within the blast radius of a potential explosion, a degree of blast resistance needs to be built into the design. Control rooms are often pressurised to prevent the ingress of flammable and toxic gas. The air inlet for the HVAC should be located where it is least likely to encounter flammable or toxic gas, for example, at an elevated location if heavier-than-air gases could be present. Often a gas detector is installed on the air inlet which closes dampers in the HVAC system.

2 Number of Screens and Level of Redundancy	An operating system will generally have several screens and have an UPS. The number of screens, level of equipment redundancy and duration of the UPS all need to be specified.
3 Standard or Touch Screens	Whilst touch screens may seem desirable, there is a risk of accidental contact causing unwanted changes to the control parameters.
4 Use of Colour	Red and green are often used to indicate if valves are open or closed, or pumps are running or in standby. This will be problematic if panel operators are colour blind. Whilst they will often be tested for colour blindness as part of an employment medical, this may not always be the case, and the use of words "open/closed or "off/on" should also be considered in addition to colours.
5 Illumination	Screens must be arranged so that the natural or artificial light does not cause glare.
6 Operating History	Most modern control systems record operating data to allow for future analysis of upsets. Often the storage capacity is limited, and data is downloaded on a regular basis. Downloaded data should, ideally, be located in two separate locations so that damage (such as fire, flood or impact damage) at one location will not result in the loss of data.
7 Trip Bypass Functions	It is often necessary to bypass trip functions during startup (for example, if a vessel is empty a low-level trip will usually prevent the flow of liquid out of the vessel). To allow startup, a critical *trip bypass procedure* has traditionally been used, but some modern systems now allow trips to be bypassed for a limited time period. Even where this is the case, it is advisable to still use a bypass procedure as an additional safeguard (see Section 16.7).
8 Access	Screens and terminals are generally installed in a long desk containing several screens. Each screen will be monitored by a panel operator and will cover a different section of the plant. There can be a temptation, particularly during upsets, for managers, field operators and others to look over the shoulders of the operators and point at the screens. At best this causes a distraction and, possible, worse—especially with touch screens. Some control desks have a line marked on the floor with only authorised personnel able to cross and approach the screens. There is also a trend to install a large wall-mounted screen which can be viewed away from the control screens. In an emergency, the control room should not be used as a coordination centre. Some control rooms may be in a hazardous area, and emergency managers would increase the number of personnel at risk. Even if the control room is in a safe location or is designed to be occupied during an emergency, staff entering or leaving would be at risk. Equally importantly, additional staff in the control room would disturb the concentration of operators trying to bring the plant to a safe condition (see Section 22.3).
9 Cybersecurity	There have been examples of cyber-attacks either to extort a ransom or to inflict damage on process facilities. So far, the number of incidents has been small but this will be a growing problem.[15] An obvious way to reduce potential for cyber-attack is to isolate the control system from the internet, but this may be difficult if, for example, remote monitoring of equipment is to take place (see Section 19.5). Access to the control system via USB ports needs to be strictly controlled, but there may be occasions when this is necessary for example when downloading data or uploading software updates. Cybercriminals and terrorists are constantly looking for loopholes in security, and this is an area where continuous scrutiny and improvements will be necessary.
10 Control Cabling	Most control systems use a *control* bus to send and receive signals and process readings; this avoids the need for individual control cables to each instrument from the control room. But it does mean that damage to a cable can effectively shut down the plant. To minimise this potential, it is common to install more than one bus with separate routing to avoid a single incident damaging both buses—an example of *common-mode failure*. Critical shut down systems may individually be hardwired back to the shut down device, which is generally an *Emergency Shut Down Valve* ESDV.

13.9.2 Alarms and Emergency Shut Down

The requirements for alarms and trips will be determined during design. The trips and their actions are generally shown in a *shut down matrix*. This lists each instrument which has a trip function and the effects of its activation in tabular format. Most alarms and trips are part of a digital system.

However, it is still common to have *hardwired* ESD buttons on critical systems in case "all else fails".

Early digital computer systems suffered from *alarm flooding*.[16,17] As the inclusion of alarms into control software was cheap and easy, many control system designers added numerous alarms to help identify upset conditions. This created control systems where an upset could affect several process parameters and, in the worst cases, could generate thousands of alarms per hour. Such a high number of alarms made a quick analysis of the problem and corrective analysis difficult—with the operator going into *alarm saturation* mode. This has been a contributory cause in several serious accidents where the relatively few critical alarms have been hidden by the flood of second- and third-order alarms with the operator suffering *information overload* and hindering him/her from keeping the control parameters within the operating envelope. Once the envelope is breached the shut down systems should safely shut the plant down. Even if there is no physical damage or injury there could be significant costs in lost production and increased maintenance.

In 1994 an explosion and fire at the Texaco Milford Haven refinery injured twenty-six people and caused damage of around £48 million plus a significant production loss.[18] The incident was caused by a process vessel that was overfilled with a volatile hydrocarbon with the exit route isolated. Eventually the vessel's safety relief valve opened and discharged the hydrocarbon liquids into a flare main which was only designed to handle vapour. Eventually the flare main collapsed under the weight of liquid, ruptured, and there was an explosion. (There were other contributory causes which need not be discussed in this example.)

Key factors that emerged from the health and safety executive's (HSE's) investigation were the following:

- There were too many alarms, and they were poorly prioritised.
- The control room displays did not help the operators understand what was happening.
- There had been inadequate training for dealing with a stressful and sustained plant upset.

In the last 11 minutes before the explosion the two operators had to recognise, acknowledge and act on 275 alarms.

Ideally trip systems should be independent from the control system, although there will be obvious links between the two. However, control and trip systems are increasingly becoming integrated. Trip systems need to have a high level of reliability, and some components are often duplicated or triplicated, but poor design can negate this advantage. For example, low-level trips, like many others, are often triplicated and work on a two-out-of-three (2-o-o-3) voting system. If all three trips are installed on the same impulse line, a blockage would prevent any of them from operating.

Most trip systems are now assigned a SIL rating to specify their reliability, which is often a mixture of hardware design and maintenance regime. Safety integrity levels are further discussed in Section 14.7. The process is specified in "Functional Safety of Electrical/Electronic/Programmable Electronic Safety-related Systems (IEC 61508)"; this is an umbrella safety standard. A related document for the process industries is "Functional Safety—Safety Instrumented Systems for the Process Industry Sector (IEC 6151)". There are several guides giving an overview of the assessment process.[19]

13.10 PRESSURE RELIEF AND BLOWDOWN

The various process safety systems installed should prevent excessive pressures from being generated, but overpressuring of equipment can always occur. This has been a common problem from the early days of the Industrial Revolution when boiler explosions were common.[20] Overpressure can be mitigated by some form of pressure relief system—generally a PSV or bursting disc (BD), sometimes called a rupture disc. A blowdown valve (BDV) is sometimes installed; this is a control valve which opens in an emergency (or on instrument failure) to reduce the pressure in a vessel or system. Both busting discs and blowdown valves will reduce the pressure, but a PSV will maintain the pressure to the set pressure plus additional tolerance. Until the 1980s, it was common practice to consider the process equipment as the important components of a system with the piping being installed to connect the equipment together. This led to some anomalies in design, and it is now common practice to think of *pressure systems* as incorporating all the piping and equipment in a given loop.[21]

In some cases, fluids discharged from the pressure relief system can be safely released to atmosphere, but many process materials will require some form of treatment. The most common is burning in a flare system for the disposal of flammable gases. Toxic or corrosive materials may require to be *scrubbed* before discharge (this was the design intent at Bhopal, but the system failed). Discharge of gases from a vent stack is also a common means of disposal. The discharge from individual vessels or pressure systems will be connected into a header. The design of the header is complex as several relief systems might operate at the same time. It is impracticable to design the header to assume that all relief systems operate at the same time—especially as some systems may relieve for more than one reason. A HAZOP (see Section 5.3) should identify the causes of overpressure and where several systems will be affected simultaneously.

The sizing of PSVs or BDs for any item of equipment should address all the upset conditions which might occur. The most common causes of overpressure are shown in Table 13.2.

Table 13.2 covers overpressure resulting from process upsets. Explosions can also occur in pressurised systems. Whilst oil mist in compressed air systems can be a problem, more common

TABLE 13.2
Common Causes of Overpressure

Power Failure	This could allow liquid levels to build up and possibly cause flooding. Localised failure of power may also result in an obstruction to flow at some point in the process system. The design should be for full heat input with allowances made for elevation of the boiling point at the PSV set pressure. Failure of the power supply to a reactor could lead to runaway reaction (see Section 13.5). Power failure could result in the relief systems in several vessels operating simultaneously.
Failure of Cooling Systems	Condensers will cease to operate and vapour pressure increase until the PSV set pressure is reached. This, again, could result in several relief systems operating simultaneously
Failure of Heating Systems	This could result in high viscosity and restricted flow in pipework, leading to increased upstream pressures.
Localised Instrument Failure	The consequences should be analysed in the HAZOP study, but a typical example is the liquid flow into or out of a pressure vessel. This may cause a control valve to open or close. An opening control valve may result in a high-pressure to low-pressure blow through with high-pressure fluids entering a system designed for low-pressure operation.
Total Instrument Air Supply Failure	In generally valves should go to a safe position in the event of air failure. The effects of instrument air failure are again best analysed by HAZOP but often cause BDVs to open to reduce system pressure.
Pump Failure	This is likely to result in high liquid levels in upstream vessels or loss of cooling in cooling water systems.
Compressor Failure	Failure of a refrigeration compressor will result in temperature and generally pressure in the refrigerated system to increase.
"Dead Heading" of a Pump or Compressor	This is the discharge of a pump or compressor against a closed valve or other blockage. In some cases, the characteristic curve of the machine will be below rupture pressure of the system. However positive displacement machines will always increase downstream pressure until either the machine or the pressure system fails.

(Continued)

TABLE 13.2 *(Continued)*
Common Causes of Overpressure

Heat Exchanger	In most cases failure of a heat exchange tube, with leakage of fluids from the high- to the low-pressure side will be as a result of a longitudinal split in the tube. The worst case is a guillotine fracture where the tube splits across its diameter with flow coming from both sides of the fracture. The dynamics following the transient generation of a gas channel to the PSV through the liquids can cause the vessel to rupture if the tube split is "sudden". The worst-case event would be with high-pressure gas on tube side and cooling water on shell side. Fortunately, sudden total severance is rare as impending rupture can be indicated by leakage. The type and sizing of the pressure relief device will depend on a number of factors and will require detailed assessment[22]. Damage to a heat exchanger can also occur if the cold side is blocked in with heat transfer from the hot side continuing. Thermal expansion of the cold side will continue until the temperature difference is reduced to zero.
Interconnections (sometimes called HP–LP interfaces)	This will arise where a higher-pressure system flows into a lower-pressure system with pressure controlled by a regulator of PCV. A typical example would be a pressure control valve to reduce steam pressure. A failure in the PCV would allow high-pressure steam into the low-pressure piping. For this reason, it is common for PSVs to be installed downstream of PCVs. Another example would be the use of nitrogen for purging of plant before or after maintenance. Nitrogen from a high-pressure source without overpressure protection can be particularly destructive—especially with atmospheric storage tanks which are designed for only a few inches, or centimetres, Water Gauge (WG) of pressure. Note that bypasses around control valves can create a problem for PSV sizing which must be adequate to discharge the combined flow of a failed open control valve, plus the flow through the bypass. Bypasses may be installed after commissioning, and the need to consider the combined flow is often forgotten but should be identified by the MOC procedure (see Section 16.9).
Blockage	Blockages can occur for a number of reasons including accidental closure of an isolation valve, control valve failing closed, blockage by process material such as a build-up of polymer or hydrate formation. The effect will be similar to the build-up of pressure from dead heading a pump or compressor.
Hydraulic Expansion	Liquid in a confined space will expand when heated causing a significant rise in pressure. Where this may occur (often in sections of pipe exposed to ambient or radiant heat) the line might rupture but more commonly a flange leak occurs. This situation can be prevented by the installation of a small thermal relief valve *discharging* to a safe location.
Fire	Heat input to vessels or piping from an external fire will raise the temperature of vessel contents and could result in boiling. It is possible that several systems could be affected by the same event. Calculation of required relieving capacity for external fire is generally based on API 521 pressure relieving and depressurizing systems.[23] It is important to note that a PSV will not protect a gas-filled vessel (or a vessel where the vapour space above liquid is exposed to thermal radiation) against rupture. The vessel will heat up to a point where the ultimate tensile strength of the vessel is reached and the vessel will rupture. If the vessel contains flammable material, a BLEVE will result (see Section 9.8).
Chemical Reaction	Specifying the relieving requirement for a runaway chemical reaction is complex and best left to experts in the process kinetics. The material to be released may be gaseous, liquid, solid or a mixture of all three. Reaction may be continuing as the reaction contents are being released. As pressure rise during a runaway can be extremely rapid and the relieved material viscous, BDs are a better solution than PSVs.
Storage Tanks	Atmospheric storage tanks are surprisingly fragile. Although of solid construction, they are generally designed for only a few inches WG. In fixed-roof tanks problems can arise if the tanks are over filled, tanks with internal heating coils operating above 100°C will rupture if the heating coil bursts or water is inadvertently added to the tank. There is no easy way to provide pressure relief in this situation, and tanks are generally constructed with a weak seam between the wall and roof so that the tank contents will be contained even if the tank is overpressured. Refrigerated storage tanks can be subject to *rollover*[24]—where a warmer liquid layer is covered by a colder heavier layer. The layers can then roll over with the warmer layer now above the colder layer with rapid vaporisation.

explosions are in pneumatic systems handling powders—grain and sugar handling being prime examples. Explosion venting is briefly covered in Section 10.7.

Calculations for sizing PSVs (the orifice size) and BDs are readily available from a variety of sources. Whilst the same valves can be used for both liquid and gas duties, the flow-rates will be significantly different. Sizing for two phase flow (including flashing flow) through the valve needs special consideration. The former practice of specifying the PSV based on the orifice area required for liquid and vapour flows then adding them together has been superseded by more sophisticated methods.

13.10.1 Vacuum

Whilst more emphasis is placed on overpressure of systems, there are cases where a vacuum can be generated. Common examples are the following:

1. The draining down or pumping out of a vessel.
2. Cooling a vessel with a cooling coil.
3. Condensing steam or vapour in a vessel when the weather changes or cold fluids are put into the vessel.[25]

Most pressurised systems will have adequate strength to tolerate vacuum, but this should be checked. This may not be the case for storage tanks.

13.10.2 Factors Affecting Release Rates

Having identified the source of overpressure the designer must now determine the mass of fluid to be discharged to avoid overpressure (or under pressure). The designer must decide, for each system, which condition produces the highest release rate and under what conditions. This can be a complex task and requires a systematic approach examining all possible causes for all pressurised systems in the plant. The physical properties of the discharged fluid could change over time, for example, being initially liquid but then becoming a two-phase flow mixture. The HAZOP study should identify the worst-case scenario for each vessel or system if this has not been done previously.

After the pressure relief requirements for each system have been identified, the number of systems which can relieve simultaneously needs to be established. In the case of a fire or power failure, numerous PSVs may operate and discharge into a common header. This may require the completion of a fire study to determine which pressure vessels will be affected by each fire scenario. The worst-case condition will form the basis of sizing the flare header.

PSVs will be sized for the worst-case event. However, in some cases, there may be significant differences in the discharge rates for different events, and more than one PSV may be required with staged settings; there are no definitive rules for this.

The detailed specification of pressure relief devices is covered in Section 14.8.

13.10.3 Relief Headers and Flare Stacks

With the exception of non-hazardous materials, PSVs, BDs and BDVs discharge into a closed header leading to a disposal system, generally, a vent, scrubber or flare. Key factors in the design are as follows.

1 Minimising pressure drop	When the pressure falls after discharge through the relief device, the physical properties of the material may change. The most dramatic cases involve high-pressure gases where drastic falls in temperature occur due to adiabatic expansion, and liquid fallout takes place. The header must be designed for the minimum temperature that may occur and must continuously slope in the downstream direction. Back pressures in the header may be significant in some cases and possibly prevent PSVs or BDs on low-pressure equipment operating. High-pressure drop in discharge piping can also affect PSV performance (see Section 14.8).

2 Mixing	The header may be required to handle several different materials at the same time, and the mixing of cold and wet fluids can result in ice or hydrate formation which will block or restrict flow. Where this can occur, it is common practice to have separate cold and wet flares. Common examples of this are ethylene crackers and LNG plants. Sometimes a separate toxic (or sour) gas flare is also installed—typically where hydrogen sulphide is discharged. These materials might require different materials of construction and disposal methods.
3 Discharge rate	The discharge rates from relief devices will vary, particularly for BDVs and BDs. The upstream pressure (and hence discharge rate) will decline as pressure in the system is reduced. For this reason, where several BDVs discharge into the same system they are often designed to discharge in sequence. The alternative would be a very large-diameter header.
4 Drainage	The header must slope in the direction of travel to ensure that any liquids which enter the header (or condense out of vapour streams) do not accumulate and potentially cause a restriction or blockage.
5 Valves	The inclusion of insolation valves should be avoided if at all possible. Where valves are absolutely essential, a gate valve is preferable with the valve stem installed in the 6 *o'clock* position. In the unlikely (but possible) event of corrosion of the valve steam the valve will fall into the open and not the closed position.

Ultimate disposal from the relief header is generally via one of the following:

1 Flare	This is the most common disposal mechanism in hydrocarbon plants. The flare has a continuous pilot flame and will often discharge small releases from valve leaks but must be capable of handling the much larger flows from process upsets. Between the header and the flare, a knock-out drum is generally installed to prevent flammable liquids from being carried into the flare and falling as *burning* or *flaming rain*. Some form of sealing arrangement, often a *molecular seal*, is also installed to prevent air from being sucked into the flare system when the flow is low. Alternatively, a gas purge may be used. If air enters the flare system, an explosion is likely. It is sometimes tempting to split the flow from a header to feed two flares. This is highly undesirable, as differential pressure between the two flare types can result in air entering the system. If flows are sufficiently large that two flares on the same header are essential, great care to provide adequate sealing is necessary. During normal operation the flame at the flare tip should be small, but when the relief systems are operating, the flame length may be over 100 m long and impose considerable thermal radiation levels around the flare stack. The distance to a tolerable level of thermal radiation must be determined, and a barrier installed to prevent entry into this *sterile area*.
2 Vent	This is the simplest method of disposal and will require turbulent mixing. Venting can be noisy and, if flammable gases are vented, they may ignite and impose significant thermal radiation levels in the surrounding area. Comments regarding flares are also relevant here. Venting is generally used for the discharge of non-toxic materials such as air, steam and nitrogen. Note that fixed-roof tanks are also fitted with vents. These are effectively *breathers* to allow air to enter when the tank is emptied and discharge air when the tank is filled.
3 Scrubber	Mixtures of gases with toxic components generally pass through a scrubber to remove them from the discharge flow by washing in a suitable liquid. Most systems involve counter-current systems and to ensure reliability, the recirculating pumps should be powered from an emergency power supply or a gravity head tank used.

Sizing of the relief piping is complex. It is likely that all PSVs will not lift simultaneously. This then requires the designation of various "zones". The pipe sizing must allow for the worst credible combinations of discharge rates. Where BDVs are incorporated in the design, they are usually sized to reduce pressure by 50% within a 15-minute period.[26] High-pressure venting can give rise to significant noise generation at the peak flow.

Depending on the plant type and the characteristics of the relieved materials, it may be necessary to segregate discharges from PSVs, BDs and BDVs into several different headers. The following are commonly encountered:

1. Low-pressure headers—separate from the main process vessel headers
2. High-pressure headers
3. Dry (or cold) headers where the expansion of gases creates a low-temperature (metallurgical) problem)
4. Wet headers where there is the potential for ice formation which might create blockages
5. Toxic or corrosive fluid headers (again, special materials of construction may be necessary.

Where possible the lowest pressure system should enter the header last in the sequence to avoid reverse flows induced by pressure drops.

13.11 DRAIN SYSTEMS

Drains are often considered as a "black hole" into which anything can disappear but which often cause problems with blockages or chemical reaction between incompatible chemicals. These issues should be identified during the HAZOP and may require separate drain systems for incompatible chemicals such as acids and alkalis. The routing of drains needs to be considered with care. As with vent headers, drain lines should continuously slope in the direction of flow.

> Due to a "piping clash" at the design stage, a low point was installed in an above-ground drain line. Due to maloperation, gas entered the drain line and travelled in the direction of flow. The low point was full of liquid which was accelerated by the gas flow. When the slug of liquid reached a bend, it causes a pipe rupture.
>
> Similar situations have occurred in steam headers where drainage of condensate was inadequate.

As well as the necessary slope drains will need to be designed to withstand external loads. In some areas, such as under roadways, the need is obvious; in other cases the need for reinforcement may not be recognised. For example, underground drains will be constructed early in the project, and site areas may be designated as temporary laydown areas or fabrication areas with heavy traffic after installation of the drain system. Surface drains may be either buried or run in an open channel; the choice will depend on economics and factors such as odour and the potential for blockage.

Most facilities will have two or three drainage systems:

1. Open process drains from process vessels and secondary containment which discharge into local containment systems or the stormwater drains.
2. Closed drains which connect to a closed drains vessel. Disposal may be by removal for treatment elsewhere, recycling through the process system or to a burn pit or ground flare.
3. Stormwater (often called *open*) drains will be situated throughout the plant and the ground should be sloped to the grates covering the drains. The grates should be easily accessed for *rodding* in the event of blockage. They should be covered during *hot work* (see Section 16.6) if it is taking place close by—both to prevent gases from the drain being ignited and to prevent hot welding slag from entering the drain. Outside the process areas, drainage is often provided by open channels, but within process areas, drains are underground and should be sealed by water traps to prevent flammable or toxic vapours from freely travelling through the underground system and venting from grates in hazardous areas. Some older facilities still have drains without seals.

A spillage of light hydrocarbons occurred at an elderly topping refinery where the stormwater drains were not sealed, and light hydrocarbons were therefore free to travel through the drain system and could exit through the grated openings throughout the process area. One such point was close to a fired heater which acted as an ignition source. A serious flash fire then occurred.

Rainfall will often be the design basis for sizing surface drains, but runoff water from firefighting should also be considered. There have been cases where undersized drains have "backed up" with water spreading across surfaces where a pool fire is burning. As hydrocarbons are lighter than water, they will float, and the pool fire may spread to engulf a larger area than would otherwise be the case.

Whilst water will usually be the predominant liquid in the surface drain system, it is likely to be contaminated and some form of treatment system will be necessary to capture impurities before discharge. The level of treatment varies depending on the likely contaminants and local environmental regulations but, at a minimum, there should be a collector for floating impurities such as oil.

With very few exceptions, some further form of clean-up system will be necessary. For small, simple installations it may be possible to discharge directly into the local authority drainage system. But for many locations, some form of onsite clean-up will be necessary. This can vary from open pits where contaminated water is allowed to evaporate and the residue collected for disposal to complex systems involving skimmers to remove oil contamination, filtration and then biological treatment. In some cases, the products of effluent treatment may provide a raw material such as methane which can be used for power generation. Incineration is a common method of disposing of solids provided that the products of combustion are safe.

One plant combined the open and closed process drain systems into a common tank with weirs separating both sides. Oil was skimmed from both sides into a middle compartment which was vented to atmosphere. The water was discharged under level control. Unfortunately, one drain became blocked and then suddenly freed itself such that the flammable gases from the closed drain system overloaded the vent and gases entered a non-process area where they ignited with the loss of three lives (Confidential Information) (57) & (58) (59).

Effluent treatment facilities are often located close to the site boundary and factors which may affect the surrounding area need to be considered, such as odour, effect on wildlife and light pollution.

13.12 STARTUP, SHUT DOWN AND OPERATING INSTRUCTIONS

Whilst detailed operating instructions will often not be developed at this stage, some consideration of operating practices is essential, for example, ensuring a safe, and consistent, level of isolation (see Section 16.8) and ensuring there is adequate access for maintenance activities which may require dismantling of equipment within process areas.

It would also be prudent to consider startup and shut down. Often plant startup features a number of stages. If there are problems moving to the next stage, it might be possible to hold the startup at the previous stage until the problem is fixed. Similarly, it might be able to partially shut down and restart if a problem occurs. Reverting to a "cold start" is often unproductive and may create other issues such leaks of process materials during thermal expansion and contraction of pipework. Startup and shut down can be analysed using HAZOP techniques with suitable "hold points" identified where the process is controlled and can then move forward. This analysis should dictate the sizing of any "off-spec" storage.

13.13 WHAT CAN GO WRONG?

A number of incidents have been given in previous sections, and there are numerous examples where design failures have resulted in serious loss or major operational problems. Sometimes, the transition from FEED to detailed design is inadequately considered with the assumption that the design is correct at FEED and the detailed design is only a matter of crossing the t's and dotting the i's". This is not the case. Any mistakes in the FEED will then be carried through to the operational stage and only make themselves known once an incident has occurred.

Issues sometimes occur where the project envelope is not adequately considered. Whilst it would be expected that everything within the plant site is reviewed at this stage, interfaces penetrating the envelope may not be adequately considered. The interfaces will generally be for the supply of raw materials, export of products or the provision of services and utilities. These may impose limitations on operations.

A waste treatment was constructed in a remote location with external power supplied by low-voltage three-phase feeders. The treatment plant was based around a large compressor. At the commissioning stage, it was determined that the starting current for the compressor was so high that it could only be started during the early hours of the morning.

REFERENCES

1 Barton J & Rogers R, *Chemical Reaction Hazards* (2nd edition), IChemE, 1997
2 Anon, *Rate Constants and the Arrhenius Equation*, www.chemguide.co.uk/physical/basicrates/arrhenius.html
3 Design Institute for Emergency Relief Systems (DIERS), www.aiche.org/diers
4 Anon, *A Pump Explosion, Loss Prevention Bulletin (No. 147)*, IChemE, 1999, www.icheme.org/media/5147/lpb_issue147p017.pdf
5 Anon, *Hydrocracker Accident, Loss Prevention Bulletin (No. 89)*, IChemE, October 1989, www.icheme.org/media/6049/lpb_issue089p013.pdf
6 Younis B A, Banica M C & Weigand B, Prediction of Vortex Shedding with Heat Transfer, *Numerical Heat Transfer, Part A: Applications*, 48, 2005, https://doi.org/10.1080/10407780590929892
7 American Petroleum Institute (API), *Pressure-Relieving and Depressurising Systems Standard 521* (6th edition), latest edition
8 Ronquillo R, *Understanding Heat Exchangers, June 2023, A Risk-Based Approach to Safety Distance Determination in the Process Industry*, IChemE, www.thomasnet.com/articles/process-equipment/understanding-heat-exchangers/
9 Kong D, How to Avoid Dryer Fires and Explosions, *Hydrocarbon Processing*, April 2022, www.processingmagazine.com/material-handling-dry-wet/powder-bulk-solids/article/21261846/how-to-avoid-dryer-fires-and-explosions
10 Lindley J (ed), *User Guide for the Safe Operation of Centrifuges* (2nd edition), ICHEME, 1987
11 Benintendi R, Deisy A, Guio R & Marsh S, A Risk-Based Approach to Safety Distance Determination in the Process Industry, *IChemE Symposium Series No. 159*, IChemE, May 2014, www.icheme.org/media/8916/xxiv-paper-20.pdf
12 Myke King, Practical Process Control Part 1 Introduction, *The Chemical Engineer*, IChemE, January 2023 (this is the first in a multipart series of articles)
13 International Electrotechnical Commission (IEC), *Functional Safety of Electrical/Electronic/Programmable Electronic Safety-Related Systems—Part 1: General Requirements* (see Functional Safety and IEC 61508)61508–1:2010, IEC, 2010
14 International Electrotechnical Commission (IEC), *Functional Safety—Safety Instrumented Systems for the Process Industry Sector—Part 1: Framework, Definitions, System, Hardware and Application Programming Requirements IEC 61511–1:2016*, IEC, 2016

15 Ayral T & O'Donnell J, Minimize Industrial Cyber Security Risk in Plants in 12 Steps, *Process Control and Instrumentation*, July 2016, www.hydrocarbonprocessing.com/magazine/2016/july-2016/process-control-and-instrumentation/minimize-industrial-cyber-security-risk-in-plants-in-12-steps
16 UK Health and Safety Executive, *Better Alarm Handling (Chemicals Sheet 6)*, HSE, www.hse.gov.uk/pubns/chis6.pdfwww.hse.gov.uk/pubns/chis6.pdf
17 UK Health and Safety Executive, *Human Factors: Alarm Management*, www.hse.gov.uk/humanfactors/topics/alarm-management.htm
18 Newsholme G, Too Much Data Too Little Information Some Thoughts on the Explosion at Milford Haven in 1994, *Loss Prevention Bulletin (No. 193)*, IChemE, 2007, www.icheme.org/media/4598/lpb193pp15-16.pdf
19 Foord A G, Gulland W G, Howard C R, Kellacher T & Smith W H, Applying the Latest Standard for Functional Safety—IEC 61511, *IChemE Symposium Series No. 150*, IChemE, 2004, www.icheme.org/media/9906/xviii-paper-23.pdf
20 C H Hewison, *Locomotive Boiler Explosions*, David and Charles, 1983
21 UKHealth&SafetyExecutive,*SafetyofPressureSystemsPressureSystemsSafetyRegulations2000Approved Code of Practice*, HSE, 2009, www.hseni.gov.uk/sites/hseni.gov.uk/files/publications/%5Bcurrent-domain%3Amachine-name%5D/l122-safety-of-pressure-systems.pdf
22 Short L & Lindsay F, Screening Heat Exchangers for High Pressure Differential Relief, *The Chemical Engineer*, IChemE, March 2008, www.thechemicalengineer.com/features/screening-heat-exchangers/
23 American Petroleum Institute (API), *Pressure-Relieving and Depressuring Systems Standard 521* (6th edition), latest edition
24 Davis L N, *Frozen Fire*, Friends of the Earth, 1979
25 Anon, *Storage Tanks Safety Safety Considerations for Atmospheric Storage Tanks*, www.wermac.org/equipment/storage_tanks_safety.html
26 American Petroleum Institute (API), *Pressure-Relieving and Depressuring Systems Standard 521* (6th edition), latest edition

14 Hazard Study 3—Detailed Design

14.1 INTRODUCTION

Chapter 13 has focused on the selection of the major process equipment, unit operations plus the design of some critical systems. This is appropriate as the selected equipment and its ancillaries must be incorporated into the PDFs and P&IDs. Equipment selection will influence plant layout and civil engineering aspects. This chapter adds detail to the design. The position of detailed design within the overall project lifecycle is illustrated in **Figure 14.1**. Unless mistakes were made at the FEED stage, the key aspects of the design will be unaltered, but detailed changes are likely.

Cost estimates will become more accurate as project definition improves and more detail is available. Cost estimates should have an accuracy of +/−10%. The "product" of FEED will be drawings and specifications marked as approved for design (AFD), and the drawings and specifications available at the end of detailed design will be marked approved for construction (AFC).

Financial sanction will be obtained at the end of FEED or at the beginning of detailed design. Deviations from the FEED will incur project delays and/or cost increases. It is important that the structure of the project is well defined during FEED with additional detail being added during the detailed design stage. There have been cases where FEED and detailed design have been combined into a single phase to compress project delivery times. Whilst this may look desirable and economic, it reduces the amount of time spent on the design process, and there is considerably greater scope for error.

At the start of detailed design a dedicated design team will normally be established in a dedicated project office. An initial task will be to confirm the basis of the FEED design with a review of the mass and energy balances, PFDs, layout, pressure relief and blowdown demands. As these aspects interact with each other reviews must be performed in parallel.

Detailed design (and in some cases FEED) may not be complete when civil construction starts and can impose limitations on the design if civil engineers make incorrect assumptions. It is important that the layout is fixed early in the detailed design process so that basic civil work can be started and expensive, time-consuming changes are not necessary. This does not mean that all foundation designs are complete, but key features such as road layout and location of major equipment items are fixed. The design should also consider access for equipment storage, laydown and installation. Long-delivery items might be located in areas where construction is almost complete and provision must be made for this.

> Expansion of an existing offshore complex involved the addition of new equipment. The design was successfully completed and onshore module construction started. Arrangements to ship and install the modules offshore then began. At this stage, it was discovered that the only way of moving the equipment into position was by using a crane barge to lift modules over equipment operating at high pressure.

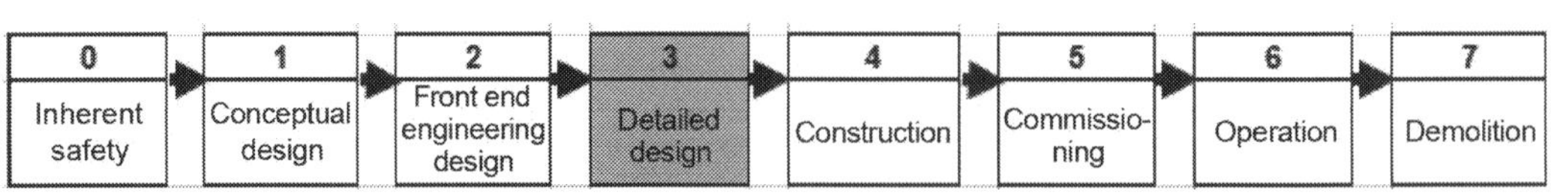

FIGURE 14.1 Project Lifecycle Detailed Design

DOI: 10.1201/9781003510680-17

Detailed design should not be considered a "rubber-stamping" exercise. Original design assumptions might be incorrect, new technology might have become available or there might be a change in regulations as a consequence of a major incident elsewhere.

> During the detailed design of crude unit expansion, the HAZOP study picked up the fact that river water (used for cooling) temperatures used in FEED were out of date, and river water temperatures were warmer than those used in the FEED.

Operating instructions will be developed during detailed design. These are often basic and will need rewriting in more detail as plant commissioning approaches—often incorporating issues identified in the HAZOP study. However, it is important that the main features of operation and commissioning are considered at this stage to ensure they can be incorporated in the design. For example, to ensure that space and lifting equipment are available for equipment removal for maintenance and catalyst charging/changing. Such needs should be identified in the materials handling study. Operating requirements will influence many other factors, including the control system configuration and piping isolation requirements.

> The design of an offshore platform included piping arrangements which operations staff felt would be difficult to operate. No agreement could be reached between the design and operations team. To avoid project delays the operations team eventually concurred with the original design but intended to rip it out and replace with their preferred option as soon as possible after commissioning.

14.2 SAFETY STUDIES

A wide range of safety studies will be required during the detailed design stage, and there will probably be a series of iterations as design concepts are refined and studies need to be updated. Once the P&IDs have been completed, they will require HAZOP study, and this will inevitably highlight issues that require changes to drawings and possibly the addition of new equipment or changes to existing equipment. Equally important, the HAZOP team will probably make recommendations which should be incorporated into the commissioning or operating procedures.

> The HAZOP of a cryogenic gas plant identified the possibility of low concentrations of mercury in the feed gas. If present, mercury would concentrate in the aluminium cold box in the cryogenic system and cause liquid metal embrittlement (LME).[1] Whilst the limited analysis of gas performed prior to detailed design showed no indication of mercury, the potential for LME was considered unacceptable and a guard bed was installed upstream of process equipment.

It is sometimes difficult to remember that the purpose of HAZOP (see Section 5.3) is to identify hazards and operability issues (the latter are sometimes forgotten)—not to resolve them. The resolution of HAZOP findings is the task of the design team. Any changes to the design or procedures should be subject to a further HAZOP review. Whilst normal operation is inevitably part of the HAZOP, commissioning and startup/shut down are sometimes overlooked. A procedural HAZOP which looks at an activity stage by stage might be necessary for some critical tasks such as startup of critical equipment or frequent, but hazardous tasks, such as filter cleaning.

Other studies will need to be completed for either the first time or previous versions revised. These will include the following.

1	Layer of Protection Analysis (LOPA)	LOPA is a useful method for the initial screening of the performance of SIS. It is likely that it will be necessary to carry out a full risk assessment for complex systems. LOPA is not necessarily the final study method (see Section 6.6).
2	Relief and Blowdown Study	This will have been performed at FEED (see Section 13.10) but will need to be reviewed and updated (see Section 14.8).
3	Emergency Shut Down (ESD)	The ESD system will have been designed at the FEED stage (see Section 13.9), but again changes are likely during detailed design. Whilst the system may have been specified in detail during FEED some details cannot be completed until the design is fixed—for example the location of field-mounted ESD buttons.
4	Reliability and Availability Studies of Critical Equipment and Systems	Whilst studies of some equipment may have been undertaken at FEED, these will need to be reviewed. In some cases, decisions might have been made which did not undergo critical analysis at an earlier stage of the project but could have profound consequences. During the detailed design of an offshore platform, a decision was made to provide seawater cooling to the emergency generators. This was considered cheaper than a large air cooler when weight and size factors were included. Water cooling was more compact and required less equipment with valve executive actions being made through the existing control system. A reliability assessment concluded that 12 valves (butterfly type)—many of which were normally open—would need to operate for adequate cooling water to be available. The probability of all valves operating as designed was considered low, and the air-cooling option was eventually chosen.
5	Quantitative Risk Assessment (QRA)	Assessments may have been performed earlier in the project, but a more accurate assessment will be possible as operating conditions, inventory sizes, layout and pipe routings are finalised. See Chapter 6. The level of detail required for the QRA differs greatly between countries, and there is currently no requirement for its use in some parts of the world, but many companies follow international practices regardless of plant location. In other parts of the world, the QRA would form a key part of the safety case (see Section 1.6), which must be approved by regulatory authorities before the plant can operate.
6	Environmental Risk Assessment (ERA)	At the FEED stage the focus is likely to have been on emissions during normal operation and emergencies. At the detailed design stage, approvals with local authorities will be required, and environmental issues during construction will also need to be considered (see Chapter 4).
7	Fire Hazard Analysis	Again, work will have been performed during FEED but final analysis cannot be completed until all details of the design are available, such as the size of storage tanks and pressure vessels where the required water application rate will depend on the surface area. Once complete, the study will provide the basis for final design of the Fire and Gas (F&G) system, fire water system and other active or passive fire protection.
8	Effluent Disposal	Amounts of effluent and disposal methods will have been developed during FEED (see Section 13.11) but will need to be confirmed during detailed design. All effluents must be approved by the regulator who will specify *consent limits*. Disposal of solid wastes will also need to be considered with approval by an approved contractor to a registered disposal location.

14.3 CODES AND STANDARDS

The use of codes and standards will be similar to those assumed in FEED (see Section 13.3), but checks should be made to ensure that there have been no revisions since the completion of FEED and that they have been correctly interpreted by the FEED engineering team, which is not always the case. Due recognition should be paid to the potential extremes of both temperature and pressure. For example, chilling down a refrigeration system or when depressurising a high-pressure system where transient temperatures well outside the normal operating range may be experienced.

The plant should be built to a consistent set of engineering standards. Internationally, US standards are widely used, but many countries, and some companies, have their own corporate standards. It is not unusual to have an overall set of standards but for some components, for example pressure vessels, to be built to a different standard, possibly due to price or delivery time. Ideally this should not happen but where it is unavoidable care must be taken to ensure there are no differences between the standards which could cause problems.

14.4 FLOWSHEETS

After FEED, the PFDs will contain significant detail which will need to be confirmed: stream compositions, pressures, temperatures and flowrates; these will influence piping and equipment sizing as well as heat balances which will influence the design of heat exchangers. It is also useful to determine if a slight change in design parameters (particularly pressure) might reduce or increase the size of a PSV or even remove the need for a PSV except for the fire case (see Section 13.10).

Once the process PFDs are complete, utility requirements can be confirmed to ensure adequate supplies of water (for potable, utility, cooling and firefighting) are available. Similarly, the adequacy of the power supply must be determined—note that the starting current of motors is around four to five times higher than normal running loads and may be a decisive factor in determining cable sizes.

> The design of a refinery included a large electrically driven fire pump. Only after the refinery was commissioned was it realised that the starting current of the pump was so large that it would cause some refinery process units to trip. As a consequence, the pump was not regularly started and its reliability proved difficult to determine.

At the end of FEED, the P&IDs will have been reviewed a number of times by experienced engineers. The level of detail will generally increase as the design is finalised. The P&IDs are used for piping design, and the piping designer will require them to contain all relative information, for example the slope of the line, requirements for high-point vents/low-point drains, plus detailed instructions for the location of valves. The P&IDs will also give every pipe section a specific reference number which includes information on the contents of the pipe, pressure rating plus its diameter and pipe schedule (wall thickness). Each line will have a unique reference. Whilst individual organisations have different notations, most follow a similar system:

Duty-Line No.-Diameter-Material of Construction-Flange Rating-Insulation

For example, the designation PR-1001–10-SS318–150-IN2 would indicate the following:

Process Duty:	PR = Propylene
Line number:	1001
Diameter:	10 inches
Material of Construction:	SS318 = Stainless steel 318
Flange rating:	ASA 150
Insulation:	2 inches

Piping specifications often use imperial units, but sometimes an untidy mixture of metric and imperial units is found. To avoid confusion, the first P&ID is generally an index sheet defining all the symbols used and a key to the pipe numbering system. Although symbols and line numbering will vary from company to company, they should be consistent within a project.

Determining the piping specification is the job of the piping designers, but a "sanity check" during the development of P&IDs is desirable. For example, a change in conditions might make a specification change necessary. A typical situation would be a pressure reducing valve (PRV) in a light hydrocarbon line which would lead to low temperature on the downstream side of the valve due to the Joule-Thomson (JT) effect. A change to a low-temperature steel may be necessary. Similarly, a specification change might be necessary once a flow stream is heated—the mechanical strength of metals falls as the temperature rises.

Hazard studies will often proceed in parallel with the development of flow sheets. Once the P&IDs have been developed, they will become "Approved for HAZOP". Feedback and recommendations can then be incorporated into later versions of the P&IDs with the minimum of delay. After all comments and checks have been completed, these then become "approved for construction" (AFC).

14.5 PIPING SYSTEMS

Pipe sizing, pipe stressing and routing would have been performed as part of FEED but will need to be rechecked and refined for detailed design. It is very likely that changes will have occurred to pipe routes and pipe sizes. Valve selection may also have altered. "Line clashes" (where two pipes pass through the same point) sometimes occur, but with the use of CAD, clashes have become far fewer and more quickly identified.

> Prior to the widespread use of CAD, a line clash was identified between a liquid drain line and a depressurising line. As operation of the drain line could result in "hydraulic hammer", simply reorienting the drain line to avoid the depressurising line was not possible. The drain system had to be extensively redesigned and run at a lower elevation.

14.5.1 Valve Types

Valves installed will be a mixture of control, actuated and manual valves. The design of control systems is a specialist activity and is not considered here. Whilst there are relatively few valve types, there are numerous variations available for different fluids, operating pressures and temperatures and duties. Valve selection will initially be based on pressure rating and the properties of the fluid in the pipeline. The most common types of valve are listed in Table 14.1.

There are numerous guides to valve selection available,[2] many of these are issued by valve manufacturers and may be biased towards the designs they offer. Note that valves required to operate in a fire situation will need to be designed for that purpose and specified to an appropriate standard. If the valve is actuated, the actuator and cabling within the fire zone will need to be suitably specified.

Control valves usually operate with a plug-and-seat arrangement thus adjusting the flow area in the valve. However, many other types of valve have been adapted for control duties. Control valves may not offer a "tight shut-off" and should not be used on their own for isolation duties.

A specific valve application worthy of mention is a *remotely operated isolation valve* (ROIV). These are located in areas which may be inaccessible in an emergency but where isolation is required. A common example is on the upstream side of a pump handling flammable liquid. If the pump seal fails, liquid will leak and potentially start a pool fire. Closing a valve on the pump suction will stop the flow of fuel to the fire, but flames will generally prevent access to manual isolation valves. An ROIV is frequently installed in these situations. The valve and electrical/instrument cabling in the fire-exposed area should be protected with PFP. This is not to suggest that an ROIV should be installed on every flammable pump suction, but if there is a significant inventory of liquid which could escape, for example on the bottom pumps of a distillation column, it is common practice.

14.5.2 Isolation Standards

Most valves are used for various types of isolation: isolation of offline equipment, isolation for maintenance or for isolation during an emergency. The required reliability of the isolation system

TABLE 14.1
Common Valve Types

Type	Details
Gate	A gate valve simply consists of a *gate* which blocks flow through the pipe when the valve is fully closed. Opening the valve will allow the fluid to flow past the gate. Most gate valves are of the rising stem type, and when fully open, the stem (spindle) can extend a metre or so outside the valve body, depending on valve diameter. This has the advantage that the position of the valves can usually be easily determined (in some cases, there is a cover over the stem). However, the stem can become an obstruction and should not block walkways. Slurries may become trapped in the seal of a gate valve and inhibit complete closure. They work poorly as control devices as there is not a linear relationship between the position of the steam and the proportion of the valve area that is open.
Globe	In a globe valve fluid enters the valve and flows through an internal orifice to the discharge side of the valve. A spindle connects the valve wheel to a plug (generally cone shaped) which closes the orifice as the wheel is rotated. Globe valves control flow better than other types of valve and are often used on bypasses around control valves.
Ball and Tapered Plug	Whilst both gate and globe valves take multiple rotations to open and close, ball and tapered plug valves go from fully open to fully closed in 90 degrees. They are often used as isolation valves, particularly for automated duties such as air dryers which automatically regenerate at regular intervals. Whilst these types of valve can be used for LPG duties, they must have a vented ball or plug. Without the vent, liquefied gas can be trapped and, if the temperature increases the fluid will expand—ultimately rupturing the ball/plug and the valve body. At least one example of this has occurred when a ball valve had been removed to a workshop for repair. The seats around the ball or plug are "soft" and may not seal if the valve is exposed to a fire.
Diaphragm	A diaphragm valve has a membrane which closes over the valve inlet and outlet ports as the valve is closed. They are suitable for low-temperature duties and are often used for corrosive fluids or slurries. Diaphragm valves are also unlikely to seal tightly in a fire. Some types are suitable for control duties.
Wafer (or Butterfly Valve)	Butterfly valves are installed between pipe flanges and consist of a simple baffle plate which in the open position is "edge on" to the flow of fluid; when closed the baffle blocks the flow of liquid. They will not give as tight shut-off as a globe or gate valve. They should not be used with flammable fluids as they are prone to leakage when exposed to a fire.

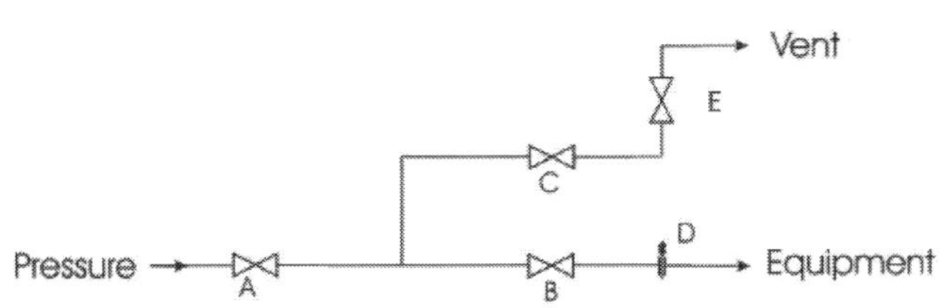

FIGURE 14.2 Double Block and Bleed Isolation

will vary depending on the hazards involved. For example, isolation of a low-pressure water pump will require only a single valve to be closed. For high-pressure toxic gas isolation, a double isolation will be required involving a *double block* or a *double block and bleed*. See Figure 14.2. The former consists of two valves in series (ideally different types of valve), and the latter is similar but has a small-bore valve venting the space between the two main valves to a safe location. There is often a slip plate installed to ensure a positive isolation. The slip plate is a "solid flange" rated to the piping specification which fits between two flanges. Any leakage from the "pressure" side cannot enter the

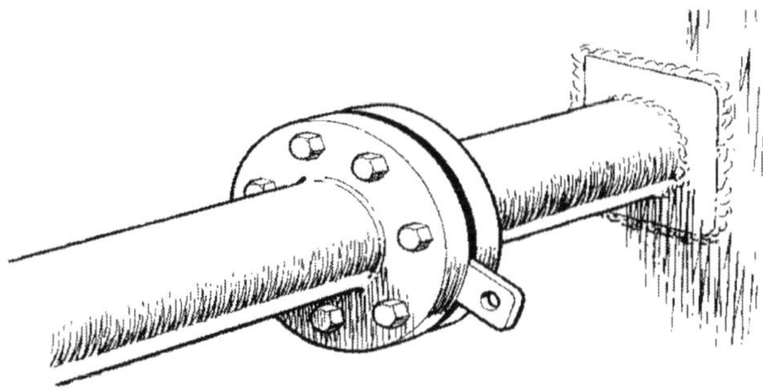

FIGURE 14.3 Typical slip plate (reproduced courtesy of the Kletz Estate)

"safe" side. The typical slip plate illustrated in Figure 14.3 represents a positive isolation and its use is best isolation for maintenance and, particularly, vessel entry. In this situation fluid from a leaking valve will not pass through the second isolation valve.

It is now common practice for an *isolation matrix* to be developed for each plant listing each fluid and defining the isolation requirements for different temperatures and pressures. Alternatively, an algorithm may be used.[3]

Shut-off systems for boilers and fired heaters are generally double-block isolation for liquid fuels and double block and bleed for gaseous fuels. Many ESDVs consist of a single valve which shuts rapidly when activated. The speed of closure, particularly in pipelines, needs to be checked to ensure that *hydraulic hammer* does not occur.[4] As a valve cannot be guaranteed not to leak, it is good practice after an ESD to close additional valves manually until the plant is restarted.

14.5.3 Some Common Issues

Regrettably, some piping problems are experienced by each generation of plant designers with little learnt from the previous generation. Some of the more common are listed here:

1 Sampling and blowdown of cryogenic fluids will result in cooling of the pipework and ice formation on the outside. This can result in the isolation valve in the pipeline freezing in the open position. To counter this, it is common practice to install two valves separated by around 1.5 m of piping. The upstream valve is generally a 90-degree turn valve with low-pressure drop, and the downstream one is a globe-type valve to control flow. The 90° turn valve will be opened and flow controlled by the globe valve. If the downstream valve becomes frozen, the upstream valve can be closed to stop the flow. Although this feature has been common in some companies since the 1980s it is not included in a depressingly large number of new plants.

On 4 January 1966 at Feyzin, France,[5] operators were taking a sample from the base of an LPG sphere. The sampling procedure was not followed correctly, and LPG escaped. The adiabatic cooling effect of the released LPG resulted in the sampling system valves freezing open—the leak could not be stopped. The ensuing gas cloud spread beyond the site boundary and was ignited by a passing car. The fire spread back to the LPG area and engulfed the sphere. Local authority firefighters arrived and started tackling the blaze. They were unaware of the potential of BLEVE, and when the sphere ruptured and a BLEVE occurred several firefighters were killed.

2 Fluid taken from a sample point on the pipe wall will often not be a representative sample of the piping contents. It is therefore necessary to flow fluid through the sample point for a period to ensure that a sample is representative. At the design stage a pipe loop from the pipeline to the sample point, returning to the pipeline, can easily be installed with a connection for a *sample bomb*. This situation is often missed by piping designers, requiring on-site changes or, very often, the discharge of process fluids to atmosphere. A better alternative is to obtain the sample from the centre of the pipe, known as *isokinetic sampling*.
3 Piping engineers will provide adequate bracing and support for valves—which, for larger diameters, can weigh up to a tonne or more—but may ignore the weight of valves, steam traps and other fittings installed on small-bore piping. Even the weight of a small valve on a horizontal small-bore pipe a metre or so in length can result in fatigue failure, especially where the pipe is subject to vibration.
4 Design reviews should ensure that valves are accessible but, as discussed in Section 13.8, it is common on new plant to find temporary scaffolding to provide access to valves where access has not been considered. Good communication between the piping engineers, the plant designer responsible for layout and plant operators is essential.
5 Most piping once installed will, apart from inspection, not be considered further until demolition. However, some piping may need to be frequently accessed. Common examples are systems which require *cleaning in place* (CIP)—where high standards of hygiene are required—or piping which may be subject to blockage—low-pressure pneumatic systems are a good example. Flanged connections for easy disassembly should be provided; otherwise operators will inevitably hit the outside of the low-pressure piping (often called *ducting*) with a wooden post, scaffolding pole, hammer or some or other implement causing damage.
6 Piping systems must not block access to other parts of the plant—particularly if run at ground level. Where this happens, inevitably people will walk over the piping. At best there is a tripping hazard (with the possibility of burns if the piping is hot). Insulated piping will be damaged by footsteps, possibly allowing water ingress and *corrosion under insulation* (CUI) (see Section 20.5). This can occur on cold fluids due to condensation or on other piping due to rainwater ingress.

14.6 HAZARDOUS AREA CLASSIFICATION

Electrical equipment can pose an ignition hazard in both dusty environments and areas where flammable gases may be present. It is therefore specified for the potential hazard using hazardous area classification methods. Recently the term ATEX (ATmosphere EXplosible) has often been applied as a short form for regulations developed by the European Union. A useful overview of ATEX has been produced by Neleman.[6]

The classification identifies sources of leakage and likelihood to determine the appropriate classification. For liquids, a flammable atmosphere is likely in or around:

1 Bund or diked areas
2 Sample and drain points
3 Near pump seals
4 Tanker loading points
5 Pig traps
6. Vents and "breathers" on storage tanks and other low-pressure equipment
7 Analyser houses
8 Filters opened frequently for cleaning
9 Vents and drains in frequent use

In some places, leaks of flammable gases might be present but would not be expected during normal operation.

1 Flanges
2 Blinded vents and drains

3 Compressor seals (away from the immediate area)
4 Filters which are opened very infrequently

Distances from the leak source to the perimeter of the classified area are calculated from international standards[7,8] and apply both horizontally and vertically. The specification of the electrical equipment is based on the frequency with which a flammable atmosphere may be expected and the ignition energy of the material.

14.6.1 Classification of Zones

Classification is a pragmatic approach. If there are many flanges in an area the judgement may be that, overall, some leakage could be expected during normal operation. The durations have no scientific basis, other than they are based on *engineering judgement* and experience, but they work.

1 Zone 0—flammable gas is expected to occur frequently, for long periods or continuously
2 Zone 1—flammable gas is expected occasionally (cross hatch in Figure 14.5)
3 Zone 2—flammable gas is not expected but may occasionally occur for short periods (single hatch in the Figure 14.5)

Areas that are not covered by Zones 0–2 are classified as non-hazardous; this does not mean they are safe—only that hazards are not expected.

14.6.2 Extent of the Hazardous Zone

The extent of each zone depends on the following:

1 The type of hazard
2 The effectiveness of ventilation
3 Characteristics of the released flammable liquid, gas or vapour, particularly whether it is lighter or heavier than air
4 The layout of equipment

The extent of the zone is available from standard references:

Energy Institute (Formerly Institution of Petroleum)	EI model code of safe practice Part 15: Area classification for installations handling flammable fluids[9]
International Electrotechnical Commission (IEC)	IEC 60079 explosive atmospheres. IEC 60079–10.1 classification of areas—explosive gas atmospheres EC 60079–10.2 classification of areas—combustible dust atmospheres[10]
UK Health and Safety Executive (HSE)	Hazardous area classification and control of ignition sources[11]
American Petroleum Institute (API)	API RP 500: Recommended practice for classification of locations for electrical installations at petroleum facilities classified as class I, division 1 and division 2.[12] Note that API classification of zones differs from that used elsewhere.

Different codes have different classifications and distances for the extent of the three zones; the code used should be defined in the project standards. Only one code should be used for any project, and the temptation to mix and match codes must be avoided. Many codes now quote distances in SI units which are obtained by converting feet into metres. Distances which are integers in imperial units will include *rounding* and a decimal point when converted to SI units. A decimal point does not imply an increase in accuracy. The codes are revised at intervals, and the latest version should be used.

14.6.3 Reducing the Risk of Ignition

The following are common strategies to reduce the risk of ignition.

1. Use a relevant code to define the design requirements for those pieces of equipment which may be used in the appropriate areas.
2. Draw a *petal diagram*. This is a series of intersecting arcs taken from each leak source (see Figure 14.4).
3. Rearrange the layout as necessary.
4. Install only appropriately classified equipment within defined zones.

No doors or windows facing the plant are permitted in the switch room. Ventilation is supplied from a safe location.

The distances around equipment are based on judgement—no one measures them with a tape measure, but some classification methods do attempt to be more analytical. Each classification method, be it corporate or national, will define different distances and shapes around potential leak sources where gas may be present. It is convention, and convenient, to extend the classified area to the boundary of the plot area.

When all of the subsets of classification are applied (drains, sumps and so on) the overall classification can be determined. An example is shown in Figure 14.5 following which is an illustration of the final classification diagram.

As described in Section 13.8 the layout does not follow codes but is "risk based". A full classification would involve three dimensions which is especially important when considering equipment such as sumps and compressor houses. Note the fire points (FPs) in the diagram; these are located behind a concrete wall (for protection against thermal radiation) and allow firefighters to open a valve and inject foam solution into the rim seal of a floating roof tank—a relatively frequent source of fire. Fire protection and foam are discussed in Section 9.12.

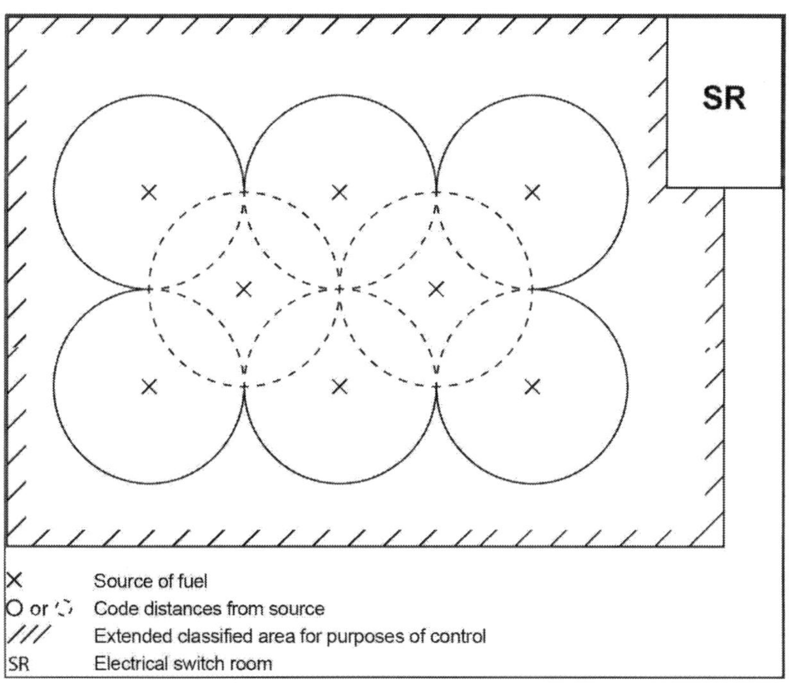

FIGURE 14.4 Example of a Petal Diagram

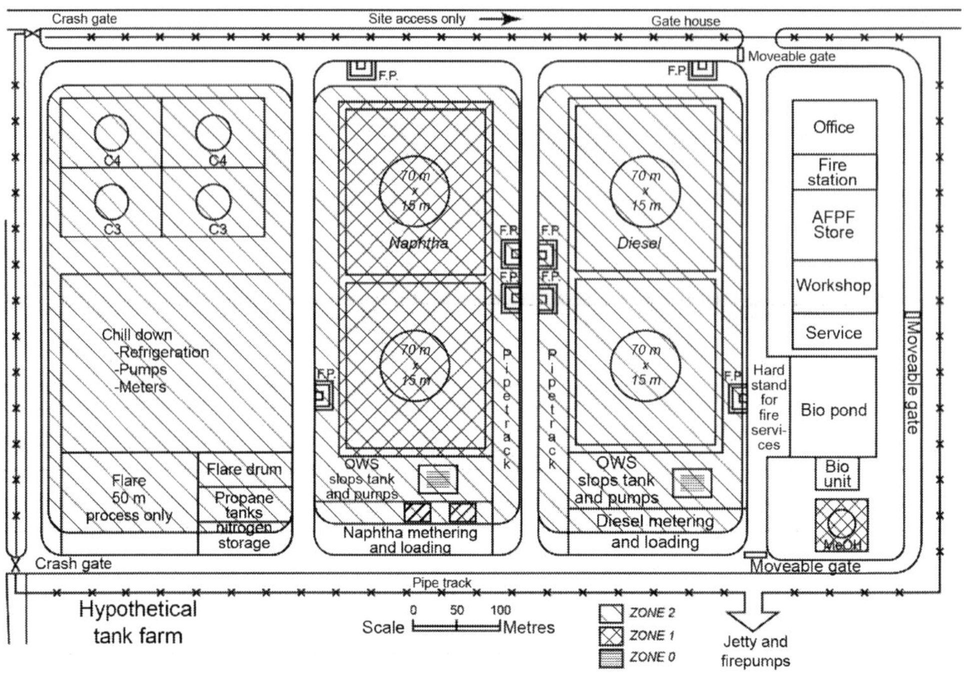

FIGURE 14.5 Example of Electrical Area Classification

14.6.4 Electrical Standards and "Fitness" for the Zone

Electrical equipment must be matched to the likelihood of flammable gas being present. In the case of zone 0 the equipment must be *intrinsically safe* (IS) which will ensure that there is insufficient energy present to ignite a cloud of flammable gas.

IS equipment is labelled as Exia, Exib or Exic, the difference being in the design of the system and, hence, its level of reliability which reduces from Exia to Exic.

In zone 1 areas there are two types of protection, both located with enclosures:

1 Exp—Pressurised equipment with interlock to shut down if pressurisation fails
2 EXd—Flameproof—flanges are designed to quench flames that occur within the enclosure.

Note: If anyone disturbs the interlock on Exp-rated equipment or interferes with the flanges on Exd equipment, its integrity is lost. Operators, thinking they are being helpful, often install a gasket between the flanges of a flameproof fitting—rendering it ineffective. Exd equipment is expensive and should be inspected for integrity on a regular basis. To minimise cost, electrical equipment is not usually installed in zone 1 areas if it cannot easily be located elsewhere.

In zone 1 areas sometimes equipment with increased safety features and special internal clearances is used and is designed Exe. There is debate regarding the use of Exe equipment in zone 1 locations. In zone 2 areas *non-sparking* equipment may be used and is designed Exn. Note that non-sparking does not mean never non-sparking, and adequate maintenance of the equipment is essential.

14.6.5 Classification of Gases

Gases are categorised into groups according to the ignition energy (see Table 14.2).

TABLE 14.2
Classification of Gases

Group I	Contains methane
Group IIA	Contains saturated gases such as propane plus volatile materials such as acetone
Group IIB	Includes unsaturated gases such as ethylene or propylene
Group IIC	Includes Hydrogen and acetylene

TABLE 14.3
Classification of Auto-Ignition Temperatures[13]

T6	Temperature will not exceed 85°C
T5	Temperature will not exceed 100°C
T4	Temperature will not exceed 135°C
T3	Temperature will not exceed 200°C
T2	Temperature T2 will not exceed 300°C
T1	Temperature will not exceed 450°C

Source: UK HSE

TABLE 14.4
Classification of Dusts

Group IIIA	Combustible "flyings"
Group IIIB	Non-conductive dusts
Group IIIC	Conductive dusts

TABLE 14.5
Electrical Area Classification of Gases and Dusts

Gas	Dust
0	20
1	21
2	22

Gases are further categorised according to their auto-ignition temperature (AIT) as shown in Table 14.3. Ignition may occur below a material's published AIT depending on a number of factors, such as pressure and the presence of welding slag or rust.

In summary, electrical equipment must not only satisfy the demands of spark frequency but also match the demands of ignition energy as well as surface temperature.

14.6.6 Classification of Dusts

Dusts are classified in a similar way to gases, see Table 14.4.

Electrical area classification of dusts is categorised into three zones,[14] similar to that for gases (see Table 14.5).

As equipment specified for use in electrically classified areas is more expensive than standard electrical equipment, the designer will attempt to minimise the size of the zone 0/20, 2/21 and 2/22 areas by considering engineered protections such as locating equipment within pressurised cubicles to prevent ingress of flammable gases, the use of suitable rated containment and location away from potential sources of release. In the latter case the following must be considered:

1 Release rate
2 Release velocity
3 Angle of the release relative to likely wind direction
4 Wind speed and stability
5 Weather
6 Distance between release and equipment/piping
7 Physical properties of the gas
8 The height of the release
9 The relative elevation of the release point and receiver

14.7 CONTROL AND SHUT DOWN SYSTEMS

All processes normally operate within defined parameters, often called the *operating envelope*, with no real need for intervention. Basic control systems are increasingly being optimised by the use of advanced computer systems to maximise output and minimise energy use. Parameters such as flow, pressure, level and temperature are required to maintain the product specification and plant throughput. The basics of control and shut down systems are covered in Section 13.9. Occasionally, there are process upsets, generally caused by human, mechanical, control or computer faults where parameters move out of the defined range. Under these circumstances the operator must take control of the affected part of the process and return it to within the operating envelope. The operator will be alerted to deviations from operating parameters by audible or visual alarms.

A fixed hierarchy for alarms will be built into the design so that the operator can differentiate between trivial and significant alarms. The development of computerised control systems has made it simple and cheap to add additional alarms into control systems and resulted in cases of alarm flooding (see Section 13.9). Once the tolerance within the envelope is breached the *safety instrumented systems* (SIS) should safely shut the plant down. The final element of an SIS is usually an ESDV or blowdown valve but can occasionally be something different. For example, in polymer plants a *reaction stopper* can be added to control a runaway reaction.

The reliability of each SIS is assessed, most frequently using LOPA (see Section 6.6), to define its reliability based on the required SIL. This, in turn, is based on the *probability of failure to danger on demand* (PFD) and the *risk reduction factor* (RRF) as shown in Table 14.6.

TABLE 14.6
Safety Integrity Level (SIL) Classification

SIL	PFD	PFD (power)	RRF
1	0.1–0.01	10^{-1}–10^{-2}	10–100
2	0.01–0.001	10^{-2}–10^{-3}	100–1000
3	0.001–0.0001	10^{-3}–10^{-4}	1000–10,000
4	0.0001–0.00001	10^{-4}–10^{-5}	10,000–100,000

Source: Adapted from IChemE, Layer of Protection Analysis (LOPA): Simplified Risk Assessment, 2nd Ed., 20XX, which reproduces IEC 61508:2010.

The SIL level will, in part, be defined by the potential hazard; the consequence of overfilling a tank of cold water is different from rupturing a high-pressure boiler and the latter would require a higher SIL. There can be a tendency to over-specify SIL levels and the higher the level, the more difficult the design and maintenance of the system become. To ensure that the required level of reliability is met, most SIS systems have three independent sensors with a "voting system" which will initiate an executive action when two of the three sensors detect that the alarm set point has been reached. SIL 4 systems are seldom found in the process industries and would require detailed assessment of reliability, including a review of maintenance/inspection procedures.

The individual components of the loop are selected and the maintenance programme is defined to meet the required SIL level. There is a common misconception that the controller of the loop, generally a *programmable logic controller* (PLC), defines the SIL level. The instrument loop in its entirety must meet the required SIL.

Most modern SISs are PLC based and have self-checking features for the logic but not necessarily for the status of the detection system and the final shut-off system. For example, self-checking is unlikely to identify a blockage in an instrument impulse line. However, it may be possible to track the output of the primary control system and compare it to the output of the shut down system. If the SIS fails to danger, the final line of defence will be a PSV, BD or, in some boilers and fired heaters, a fusible plug. Additionally, most plants have manually operated *hard wired* ESD buttons which can be activated in the control room or at strategic locations around the plant.

In a few cases a very high level of reliability for an SIS is required and a *high integrity protective system* (HIPS)[15] is installed. These are very highly specified with comprehensive maintenance requirements.[16] In general, a SIS will only operate on one item whilst a HIPS may operate on several items or the whole plant.

The design of a SIS requires an estimate of its reliability. Whilst complex systems are available, a simple rule of thumb is the concept of FDT (see Section 6.2). In the simplest shut down—a *simplex system*—where no elements are duplicated but are separated from the control system, the FDT (sometimes called probability of fail to danger (PFD), which can be confused with process flow diagram) is made up from three elements:

1 The failure of the shut down system itself
2 The time when the shut down is defeated (or overridden) whilst being tested
3 Human factors failures where the test was carried out imperfectly and/or some part of the system is not returned to the correct state after use

Whilst SISs should not be isolated from the process (which will increase the FDT), this frequently happens for a variety of reasons and a procedure to cover this situation is required (see Section 16.7).

FDT was defined in Section 6.2 and is repeated in a slightly modified form:

$$FDT = \left[\left[\frac{FT}{2} + 0.005 + \frac{t}{(8760T)}\right]\right] \tag{14.1}$$

where F = Sum of all of the failure rates of all elements in the ESD system from initiator through the comparison circuit to the shut down valve and initiator—failures to danger/year
T = The time interval between tests—fraction of a year
t = The test time when the shut down system will be isolated—hours/test
0.005 = Typical human failure rate

This equation has a minimum value for the FDT where t approximates to once per week but is fairly constant up to t = once per quarter. In simple terms a *simplex system* has a 95% to 98% availability or probability of operating successfully on demand and a FDT of 2% to 5%. A common practice for reducing the FDT is to duplicate or triplicate the sensing devices.

Testing should be possible without bypassing the trip system. For example, for a 1-o-o-2 system, the FDT is reduced to

$$FDT = \tfrac{1}{2}t^2T^2 \tag{14.2}$$

This reduces the FDT to around 0.1% or less. Note that the definition of failure needs to be considered carefully.

In the case of a HIPS, all elements are redundant and, as a minimum, duplicated with different types of equipment and with maintenance by different instrument personnel. They are often *diverse* (achieving the same function by utilising different techniques). For example, level detection by using a float switch and a differential pressure cell or radar gauge.

Sharing of components between the control loop and SIS can lead to a process upset not being detected. For example, if a level controller activates a shut down on high level, failure of the controller will also prevent the valve operating to prevent high level. Poorly designed protective systems which use elements of the primary control loop reduce the performance of the SIS by perhaps an order of magnitude. The use of elements of the control system introduces a *CMF.* Other common causes are as follows:

1 Using the same instrument tapping for all the sensors in the SIS. This is a CMF as the tapping may block.
2 Dirt in the process fouling both control and shut down valves (note that in some cases a control valve acts together with an ESDV to provide a double block but as stated previously, control valves should not be considered as providing a tight shut-off.)
3 The same maintenance group making the same mistake on all sensors during routine tests.
4 The same designer/manufacturer creating the same fault in both the control and protective system.

CCFs can also occur, such as a set of conditions which affect all elements of control and shut down systems in the same manner at the same time, such as voltage surge, dirt in the instrument air system and high vibration. Common mode or common cause failures reduce the reliability of duplex or triplex SISs to that of a simplex system.

> Damage to a boiler in a refinery occurred when the level control in the steam drum was lost and the boiler "ran dry" and overheated. Although the drum was fitted with three independent level switches to prevent loss of level none of them activated. The investigation discovered that there was no inspection of the level switches as the site management could not conceive of a situation where all three would fail.
>
> One of the uses of steam from the boiler was for heating heavy fuel oil. A leak had occurred in one of the heat exchangers, allowing oil to enter the steam/condensate system. As the temperature in the condensate reduced, the oil became more viscous and blocked the impulse lines to the level controller and the SIS level switches. When level control was lost, the SIS was unable to detect low level and activate corrective action.
>
> This situation could have existed for a significant period before the event occurred and is an example of both common cause failure and an unrevealed fault.

There are various failure modes for each component. If a component fails and puts the plant into a safe condition it is *revealed* or *fail-safe.* Where the failure of a protective system is not evident

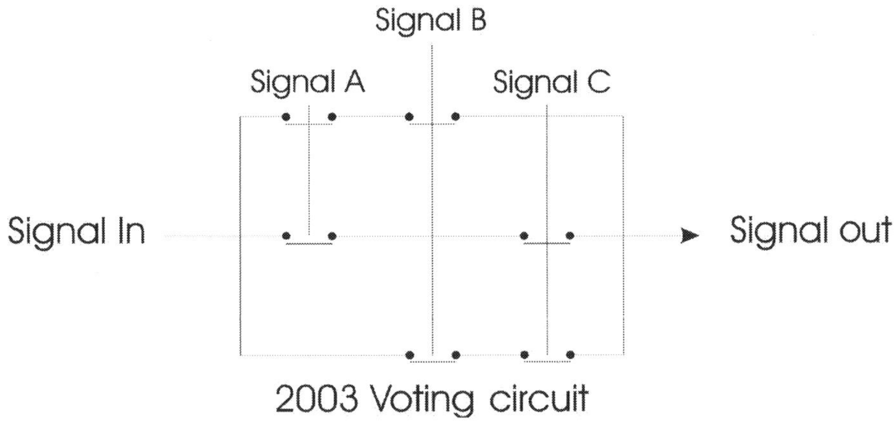

FIGURE 14.6 Simple 2 out of 3 Voting System

or will prevent a shut down, it is *unrevealed* or *failed to danger.* There are variations on this such as *degraded* where the system will only partially work. Sometimes a failure which is revealed and initiates a shut down may actually be a danger or a spurious failure resulting in loss of production. Therefore, majority voting systems are used with *two-out-of-three voting (2-o-o-3)* being the most common. See Figure 14.6.

In some situations, SISs need to be bypassed. In hard-wired systems, this is often in the form of a key lock, but electronic systems often used a password system. The latter has the advantage that all bypasses, in a modern plant, will be recorded by the control system *historian*. Regardless of how the bypass is achieved, a comprehensive procedure will be required to ensure bypasses are as infrequent and as short as possible whilst ensuring that adequate testing can be performed. This is further discussed in Section 16.7.

Valves used in ESD duties must fulfil three criteria:

1 There should be negligible leakage past the seat when closed (a control valve does not always have a tight shut-off; it has leakage of a few per cent of full flow). This is particularly important with HIPS which are designed to protect against overpressure. In this case it will almost certainly be required to have a venting system downstream of the valve and probably double block and bleed isolation.
2 The valve should be driven to the "safe" position *(fail-safe)*, open or closed by a spring or equivalent. It must not be driven by a system or mechanism which could fail on demand.
3 If the valve is intended to arrest flammable fluids in the closed position the seat arrangement should be *fire-safe* so that even when exposed to fire the seat integrity is maintained.

The ESD must operate on test and on demand. It is therefore necessary to consider the need to protect the system—not only the valve but also the signals from the plant and to the valve against damage from fire, impact and even explosion.

On large, complex plant there will be a hierarchy of shut downs ranging from the isolation of a single piece of equipment or process section to isolation of the entire plant. If the plant is operating under pressure, it will generally be isolated but maintained at pressure, but in extreme emergencies depressurising (blowdown) will occur. This hierarchy of shut downs is particularly common in gas processing plants.

14.8 PRESSURE RELIEF AND BLOWDOWN

Pressure relief will have been considered as part of FEED (see Chapter 13), but the calculations will need to be repeated to ensure accuracy and allow for any variations in design since FEED. For example, inlet and outlet piping runs might be longer than assumed in the FEED thus changing PSV specification. Similarly changes to piping and control valve sizes (and bypasses) will affect relief requirements.

A very large gas processing plant had progressed from FEED to detailed design, and construction had started. Well into mechanical erection, it was noted that flow-rates had changed between FEED and detailed design, but the PSV capacities had not been rechecked.

Fortunately, after rechecking the capacity of the PSVs was found to be adequate.

The basics for sizing relief headers were covered in Section 13.10 and a good introduction to the subject is available from Parry.[17] It is now appropriate to consider the different types of relief device in more detail.

There is a belief that overpressure protection can be provided by instrumental systems. Whilst this may, in theory, be possible, such systems will be complex and require significant maintenance to ensure reliability. The standard options of PSVs and BDs are cheaper and more practical. Instrumented pressure relief should not be confused with blowdown or depressurising systems which are designed to reduce pressure rather than preventing excessive pressures from being reached.

14.8.1 Pressure Safety Valve

The most common type of device is the PSV which is defined in API 520 (Part 1) Sizing, Selection, and Installation of Pressure-Relieving Devices in Refineries[18] as a spring-loaded pressure relief valve actuated by the static pressure upstream of the valve.

PSVs for incompressible (liquid) streams are generally called pressure relief valves and open in proportion to the increase in pressure above the set point. Pressure safety valves are used on compressible (vapour) duties where the valve "pops" open at the set pressure.

An important point to recognise is that even after the PSV has opened the pressure in the system will rise, generally by 10% for compressible vapours but possibly up to 25% for non-compressible fluids. Formulas for sizing are available in manufactures' literature and API 521.

Correct sizing is important, and the temptation to oversize and "err on the side of caution" will not work. Where relieving requirements are overestimated, the PSV will *chatter* (open and close rapidly) as excess pressure is discharged, followed by drop in pressure causing the valve to close, followed by a build-up of pressure with the PSV reopening. This repeated rapid opening and closing damages the valve which may not tightly reseat.

Upstream and downstream pressure drop can also affect the performance of a PSV, and normal limits for conventional PSVs are 3% of set pressure in upstream piping and 10% in downstream piping. Higher backpressures, up to 40%, are possible using *balanced pressure*-type valves. To work at higher backpressures, the PSV bonnet is vented to atmosphere. If the valve is handling a hazardous material, the vent will need to be piped to a safe location. The vent most not be blocked, or the valve will not operate correctly. Note that a plastic packaging piece is generally installed in the vent during transit and is frequently not removed when the valve is installed—again this will prevent correct operation. Alternatively pilot operated valves (where the set pressure is not affected by back pressure) can be used. These have an on/off snap action which makes them particularly useful for assisting atmospheric dispersion of relieved gases.

API 520 requires all PSVs to have a plate displaying the following information:

1 Size
2 Set pressure
3 Type
4 Back pressure
5 Capacity at overpressure
6 Cold differential test pressure
7 Serial number

The terminology for describing a PSV is in three parts: upstream pipe diameter, orifice designation and downstream pipe diameter. For example, a PSV designated 4P6 has a 4" inlet, 6" outlet and a "P" orifice (orifice sizes range from "D" with a cross-sectional area of 71 mm^2 through to "T" with a cross-sectional area of 16,774 mm^2). The set pressure of a PSV may need corrections for operation at high temperature. The pressure setting at ambient pressure is called the *cold differential set pressure*.

Detailed mechanical specification of PSVs, including orifice sizes, is given in API Standard 526.[19]

Crowl and Tipler[20] provide a good introduction to the selection and sizing of PSVs. Similar, but less complete, information is available from API 520 Part 1.[21] On new plants, it is common practice to install two or more PSVs for each process duty. This is to allow a PSV to be removed for inspection whilst the plant remains in operation, protected by the second PSV. Occasionally the relieving duty is so large that a single PSV will not have adequate capacity, multiple PSVs are then required, sometimes set at different pressures. This is particularly common on large steam boilers. There is often a temptation to leave both PSVs online "to be on the safe side". This should be avoided. If the PSVs are correctly sized, leaving both PSVs online will provide excess relief capacity and cause the valves to *chatter*, potentially leading to damage and a failure to reseat properly.

Pressure relief valves have been developed for a number of specific duties such as water hammer and explosion relief. These are specialist applications and are not discussed further.

Whilst the sizing of PSVs is relatively straightforward, there are a few common pitfalls:

1 A pressure relief valve will not protect a gas-filled vessel from rupture in a fire. The PSV maintains the pressure in the vessel. However, heating of the metal wall which loses strength, softens and eventually ruptures. The same situation occurs in the vapour spaces of vessels. Protection in these circumstances will require a blowdown system (see Sections 9.4 and 9.8).
2 A pressure relief valve opens relatively slowly due to inertial effects and will not necessarily protect a vessel against rupture of a high-pressure gas tube in an S&T exchanger. Bursting discs are more effective but have the disadvantage of not shutting off the gas release as the pressure falls.
3 A relief valve sized to handle a specific volumetric flow of gas will only handle a fraction of the flow as liquid. Mixed (two phase) flow is complex and special design case.
4 A control valve designed to pass liquid will pass a significant volume of gas—a downstream PSV sized for liquid could be overloaded in this situation (commonly called *blowby*).
5 Where a PSV is installed to protect an upstream PRV or PCV and a bypass is installed, the PSV must be sized to cope with flow through the PRV/PCV and bypass.

14.8.2 Bursting Discs

Bursting discs, sometimes known as rupture discs, fail at a predetermined pressure to provide an almost full-bore vent area. They are used either where pressure rise is rapid, such as a runaway reaction, or in combination with a PSV, in which case the capacity of the PSVs must be derated.

There are several different kinds of bursting disc; they are all installed in special holders located between a pair of flanges. The most common types of disc are as follows:

1 Conventional	This is a simple domed disc, fitted into a holder with the dome "pointing" in the direction of flow. These discs can fail prematurely, particularly when subject to pressure and temperature cycles which causes metal fatigue. Conventional discs are thin and easily damaged. A particular problem is *pinholing* where a small hole in the disc occurs allowing the pressure to equalise on both sides of the disc—rendering it ineffective. A variation of the simple conventional disc is a "scored disc" which is of thicker material but is scored so that when the rupture pressure is reached the disc opens along the line of the score with the "petals" between the score lines being forced downstream. Care must be exercised when inserting the disc and holder between flanges if it is installed the "wrong way round" (a frequent occurrence) it will not work.
2 Reverse Buckling Disc	These are installed with the dome facing the pressure source—hence the name. When the rupture pressure is reached, the dome is reversed, the force of rupture pushes the disc against sharp blades which cut the disc, giving almost the full pipe bore available for discharge. The effectiveness of this design depends on the blades being sharp, which may not be the case after a long period in service. Alternative designs have been developed where the disc reverses and is pushed out of its holder and folds itself on a downstream capture bar. Reverse buckling discs are thicker than conventional discs and therefore less prone to accidental damage. However, they can fail spuriously, often as a result of being incorrectly installed in their holder or the holder being installed "back to front" in the pipework.
3 Graphite	Where metal discs cannot be installed, for example due to corrosion or low rupture pressure, graphite discs are used. Unlike metal discs, graphite fragments and, when installed in combination with a PSV, particles can be trapped in the PSV preventing it from reseating. Some form of trap for the fragments is therefore necessary. This is usually achieved in by the piping configuration.

14.8.3 Rupture Pins

A third type of relief device is a rupture pin which holds a flap closed but fails at a present pressure. These are rarely encountered and are not discussed further.

14.8.4 Relief System Piping

The basic sizing of the relief header will have been determined during FEED (see Section 13.10) but will need to be reconfirmed. At this stage full details of the system must be marked on the P&IDs; these include:

1 Slope: Relief headers must slop in the direction of flow to avoid accumulation of liquid at low points. Without an adequate and continuous slope, at best the flowrate will be affected and at worst *hydraulic surge* (also known as *water hammer*) can occur.
2 Isolation: It is generally considered bad practice to install isolation valves in a relief header. However, this is often necessary upstream and downstream of PSVs and BDs to allow removal for replacement or calibration. Some form of interlock and/or procedure is required to ensure that a relief route is always open.
3 BD/PSV Combinations: Where process materials could block the PSV (such as in polymer reactors) or cause corrosion, it is common to install a BD upstream of the PSV. Where this occurs, a pressure indicator (PI) must be installed to monitor the interspace between the BD and PSV. Ideally this should signal an alarm if activated but, at the least, should be monitored on a shift basis. Whilst the BD will rupture if there is a sudden rise in pressure, if a *pinhole* occurs, pressure will equalise across the BD, which will then not actuate

until the rupture pressure plus the pressure in the interspace is reached. This is likely to be greater than the rupture pressure of the system. This simple but important fact is often not understood, even by senior personnel.[22] Occasionally a design with two BDs in series is encountered, and again the interspace needs to be monitored.

A dated, but still useful, overview of common shortcomings in pressure relief system design and operation has been provided by Scott.[23]

14.9 VENT AND DRAIN SYSTEMS

Vent and drain systems are often treated as an afterthought. They may connect many pieces of equipment through a manifold, and any misuse of isolations may introduce high pressure into a low-pressure system, with the potential for serious overpressure. This is especially important if there is also potential for freezing (choking) in any part of the line.

> An ice plug formed in the drain line of a process vessel on an offshore platform. The operator was not aware of the plug which melted and was eventually blown out due to upstream pressure in the system. Flammable gas was released and entered the open drains system which was connected to numerous areas around the platform. Some gas was discharged from the open drain system into a non-hazardous area where an ignition source was present. An explosion resulted[24].

Ideally drains from hazardous and non-hazardous areas should not connect but this is sometimes difficult, especially in offshore installations. Drains from chemical storage tanks often feed to a common treatment point, but the consequences of reaction of incompatible chemicals in the drains are not always considered. A comprehensive HAZOP study should identify problems of the type identified earlier.

High-point vents and low-point drains are often required in process and utility piping for commissioning and maintenance. These remove gases from high points and liquids from low points. When not in use they should be capped or blinded to avoid being accidently opened. In the past, it was common practice to allow small releases of flammable materials to atmosphere but regulations are becoming ever more restrictive and discharge of any fluids to atmosphere other than air, nitrogen and uncontaminated water should be discouraged. On hazardous services, such as liquefied gases, it is common practice to have two isolation valves, separated by approximately 1.5 m to prevent a valve from freezing due to the JT effect. This is a similar situation to sampling of liquefied gases discussed in Section 14.5.

> A maintenance technician was walking through a hydrocarbon plant with a tool bag over his shoulder. At one point the strap of the tool bag caught on the handle of a ball valve on the drain line. The valve partially opened and released flammable liquid. Fortunately, the leak was quickly stopped but if a plug or blind had been installed—as should have been the case—there would have been no release.

14.10 WHAT CAN GO WRONG?

Good design practices and rigorous assessment techniques should prevent, or at least minimise, the number of issues that will occur as the project changes from calculations, drawings and specification to concrete, steelwork, piping and cabling. Problems that still exist are unlikely to be discovered until partway through construction or commissioning, at which point they may become time-consuming and expensive to correct.

> One of the plants within an existing complex was to be expanded as the market for products was growing. As is common each of the plants within the complex had a large degree of autonomy—as did the plant utilities function with its own separate management. The plant expansion was designed and construction started. Whilst the increased utility requirements had been identified, it was assumed that the utilities' function would be able to supply these without difficult. It therefore came as a shock to discover that the cooling water system was undersized to meet full capacity from the expanded plant. There followed a frantic race to provide additional cooling towers before the plant was commissioned.
>
> The design of a steelworks included a large overhead gas line supported on steel trestles. Whilst the steelwork was suitable for the weight of the piping and its contents, it was inadequate to allow hydrotesting prior to commissioning.
>
> The design of the process system in an offshore platform included stainless steel pipework where low temperatures were anticipated. At the design stage, no consideration had been given to pressure testing of the pipework offshore. The usual medium for testing was seawater which would not be suitable for use with stainless steel. This issue was identified during HAZOP, and provision for the use of treated water was made.

REFERENCES

1 UK Health and Safety Executive, *Liquid Metal Embrittlement by Mercury in the Refining and Allied Process Industries*, HSE, February 2014, www.hse.gov.uk/safetybulletins/lme-by-mercury.htm

2 Smith P & Zappe R W, *Valve Selection Handbook Engineering Fundamentals for Selecting the Right Valve Design for Every Industrial Flow Application Book Fifth Edition*, Elsevier, 2004

3 UK Health and Safety Executive, *The Safe Isolation of Plant and Equipment (HSG 253)*, 2006, www.hse.gov.uk/pubns/books/hsg253.htm

4 Wood D J, Calculation of Water Hammer Pressure Due to Valve Closure, *Journal AWWA*, 60(11), Wiley, November 1968, www.jstor.org/stable/41265482

5 Anon, *The Feyzin Disaster, Loss Prevention Disaster (No. 77)*, IChemE, October 1987, www.icheme.org/media/5935/lpb_issue077p001.pdf

6 Neleman K, *Overview of Explosion Protection Techniques*, IECEx 2018 International Conference, 8 and 9 August 2018, www.iecex.com/assets/Uploads/D2S2-Ex-Protection-Techniques-BARTEC.pdf

7 American Petroleum Institute (API), *Recommended Practice for Classification of Locations for Electrical Installations at Petroleum Facilities Classified as Class I, Division 1 and Division 2 API RP 500:2023*, API, latest edition

8 International Electrotechnical Commission (IEC), *Classification of Areas Various Parts IEC 60079-10-1*, IEC, latest edition

9 Energy Institute (EI), *Code of Safe Practice Part 15: Area Classification for Installations Handling Flammable Fluids*, EI, latest edition, https://publishing.energyinst.org/topics/asset-integrity/ei-model-code-of-safe-practice-part-15-area-classification-for-installations-handling-flammable-fluids#:~:text=Status%3A%20New-,EI%20Model%20code%20of%20safe%20practice%20Part%2015%3A%20Area%20classification,areas%20in%20the%20petroleum%20industry

10 International Electrotechnical Commission. IEC 60079-10-2:2015 – Explosive atmospheres – Part 10-2: Classification of areas – Explosive dust atmospheres. Geneva: IEC; 2015.

11 UK Health and Safety Executive, *Hazardous Area Classification and Control of Ignition Sources*, www.hse.gov.uk/comah/sragtech/techmeasareaclas.htm

12 American Petroleum Institute (API), *Recommended Practice for Classification of Locations for Electrical Installations at Petroleum Facilities Classified as Class I, Division 1 and Division 2 API RP 500:2023*, API, latest edition,
13 *UK Health and Safety Executive*, www.hse.gov.uk/comah/sragtech/techmeasareaclas.htm
14 Dust Explosion Info, *Dust Explosion Info*, www.dustexplosion.info/atex%20zones%20for%20dust.htm
15 Many HIPS are for protection against overpressure and the abbreviation HIPPS (high integrity pressure protective system) is sometimes encountered
16 International Association of Oil and Gas Producers (IOPG), *High Integrity Protection Systems—Recommended Practice*, IOGP Report 443, IOGP
17 Perry C E, *Relief Systems Handbook*, ICHEME, 1992
18 American Petroleum Institute (API) Sizing, *Selection, and Installation of Pressure-relieving Devices (API Standard 520) Parts I & II*, API, latest edition
19 American Petroleum Institute (API), *Flanged Steel Pressure Relief Valves (API Standard 526)*, API, latest edition
20 Crowl D A & Tipler S A, *Sizing Pressure Relief Devices, Chemical Engineering Progress*, AChemE, October 2013, www.aiche.org/sites/default/files/cep/20131068_r.pdf
21 American Petroleum Institute (API) Sizing, *Selection, and Installation of Pressure-relieving Devices (API Standard 520) Parts I & II*, API, latest edition
22 Baker J A (chair), *The Report of the BP US Refineries Independent Safety Review Panel (p 134)*, January 2007
23 Scott D S, Some Seldom Considered Aspects of Pressure Relief Systems, *Loss Prevention Bulletin (No. 34)*, ICHEME, August 1980, www.icheme.org/media/5630/lpb_issue034p007.pdf
24 Personal Communication

15 Hazard Studies 4 and 5—Construction and Commissioning

15.1 INTRODUCTION

Whilst design followed by construction and commissioning would be logical and desirable, the phases often overlap with design being completed as construction mobilisation and civil work start. The construction and commissioning phases of the project lifecycle are illustrated in Figure 15.1. Construction of some parts of the plant will often still be continuing as commissioning starts elsewhere, this reduces the overall project timescale, allowing the plant to enter revenue earning service in a shorter period.

Preliminary arrangements for construction will include the provision of temporary offices, laydown and storage areas, fabrication workshops and vehicle parking plus a secure perimeter (with a minimum of two entrances to ensure access for the emergency services).

Where possible, utility systems are commissioned in advance of process facilities to ensure the availability of water, power, steam and other utilities for activities such as refractory drying, line blowing and hydrotesting. Fire detection and protection systems should be amongst the first equipment to be commissioned. Where this is not possible, temporary arrangements must be made. If a project consists of two or more process units, one might be in commercial operation whilst the other is still under construction. The juxtaposition of operating and construction facilities can create problems, and a procedure to control simultaneous operations (often called SIMOPS) will be necessary. The hazards associated with SIMOPS should be identified by HAZID (see Section 5.4). Strict controls will also be necessary to ensure that incompatible activities are not taking place simultaneously. For example, construction work involving welding adjacent to an area involving line breaks which could release flammable gas. An overview of construction safety is given by Davies and Tomasi.[1] Construction and process safety are two separate specialisms, and different safety specialists should be employed during construction and commissioning.

Construction management varies from project to project, but for large projects a main contractor will be appointed with numerous subcontractors for individual activities such as civil work, pipework, structural steel erection and so on. For a large project the number of subcontractors can run into hundreds. It would be normal practice for subcontractors' safety performance to be one aspect of the bid analysis but problems of unsafe working often still occur, particular for low skill activities. In principle each subcontractor will follow the safety procedures defined by the main contractor and be subject to audits and penalties for unsafe working. Most subcontractors will employ their own safety personnel, but these can be powerless and may be bullied into submission by aggressive construction managers. It is therefore essential that the main contractor has good oversight of the subcontractor's safety personnel and can intervene directly with the subcontractor's management

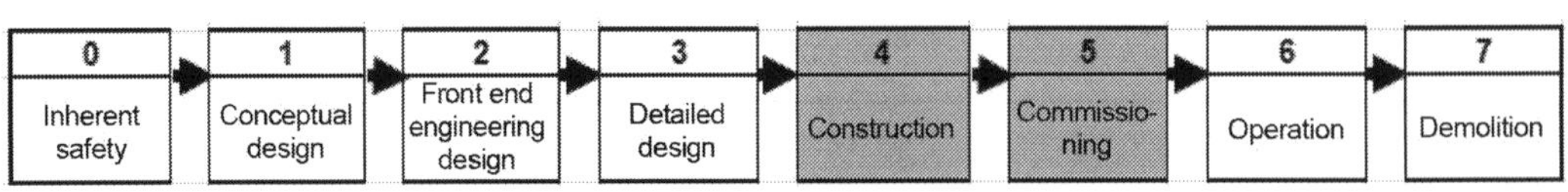

FIGURE 15.1 Project Lifecycle Construction and Commissioning

DOI: 10.1201/9781003510680-18

where necessary. It is common practice that all staff and visitors to the construction site receive a safety induction.

A critical path analysis (CPA) should be available at the start of the construction phase to identify bottlenecks in the construction schedule. This will be updated at regular intervals with construction management taking steps to minimise the impact of any delays in construction, equipment delivery or other factors. Long-delivery equipment, such as compressors and other complex mechanical machinery, may have been ordered prior to formal project sanction to minimise lead time.

Well before the advent of modern computerised methods of project management/control, a major construction project (estimated construction time of approximately 2.5 years) was suffering serious delay due to poor planning. A CPA expert was jointly appointed by the contractor and plant owner with daily meetings to update progress. The use of CPA allowed a three-month improvement in the completion date.

At some point during construction a permit to work (PTW) procedure, will be initiated (see Section 16.6). In some cases, a simple PTW will be used almost from the start of the project. However, a more comprehensive PTW will be required as mechanical completion and commissioning approach. The requirements for a construction PTW frequently differ from those for operational PTW, and care is necessary to avoid confusion if two types of PTW are in use at the same site.

15.2 SAFETY STUDIES

When construction starts plant design should be virtually complete with only minor work to be completed. The standard safety studies such as HAZOP, LOPA and so on should be complete (see Chapters 5 and 6). Environmental impact studies should also be available (see Chapter 4). Any studies required for regulatory approvals, including the safety case, should be largely complete. In addition to studies to ensure safety operation, additional studies will be required specifically for construction. Requirements will be location dependent. Further studies to determine specific conditions at the site will often be necessary, particularly if archaeological remains (which will delay construction) could be present. Construction in former war zones might identify UneXploded Ordnance (UXO), and brownfield sites may have hidden contamination or worse from previous occupation.

Development adjacent to an existing chemical plant was thought to be close to unrecorded mine workings. As a precautionary method a ground survey was undertaken and three mineshafts identified.

Typically, an EIA, covering construction activities will be necessary to ensure, for example, that increases in traffic will not result in excessive noise, fumes or congestion. During construction there may be a requirement for a wide range of chemicals on-site, such as concrete additives, gases for welding, acids (for *pickling* of pipework) and fuel for vehicles. The potential for spills from construction vehicles, temporary storage tanks and other sources must be considered.

Waste materials, including packaging and "used" chemicals, must be disposed of in an environmentally acceptable manner. Radioactive sources are likely to be required for NDT weld inspection. These will be used and stored under specific procedures for radioactive materials.

It would be unusual for a project not to require modifications as construction progresses. Changes may be minor or major. For example, soil studies may have underestimated ground load-bearing strength, changes in technology, market conditions or cost overruns might require changes in construction techniques. All changes must be reviewed from a safety viewpoint by a MOC procedure (see Section 16.9). Changes and actions must be fully documented in the same manner as changes at the design stage. Design changes are generally agreed under a *change order* procedure which not only records the change but also ensures necessary safety studies are performed and documents updated. Changes to cost and scheduling are agreed between the contractor and the eventual plant owner. Contractors are generally happy to agree change orders as they are charged to the plant owner and will increase the contractor's income.

It is now common practice for projects to prepare a *risk register* which identifies risks which could affect the project. Often registers include only items which affect project cost or schedule but they should also cover safety-related risks. Typically, the register is in the form of a table which lists the event, probability, magnitude and mitigations for each event. It is undesirable for the risk register to ignore safety and environmental issues, except those which affect cost/schedule but this frequently occurs and may be a yardstick of the contractor's commitment to safe working.

The construction team will produce numerous *method statements* to ensure that each task is done economically, effectively and safely.

15.3 MOBILISATION

The first step in construction is *mobilisation* to provide temporary infrastructure. This will include storage (warehouses and laydown areas), fabrication facilities, temporary office facilities plus temporary utility systems. On larger projects, or those in remote locations, temporary fire and medical facilities will also be required. Regardless of location, liaison with government authorities will be necessary to ensure emergency response is adequate and, often, to review security. More remote locations will also require accommodation and catering facilities. A large construction site may have a workforce of thousands and resemble a small town, but entertainment facilities will be limited and, usually, alcohol will be banned. To service the construction workforce, teams of support workers such as cooks, cleaners, drivers and guards will be needed. The contractor will need to consider public safety issues such as infections and food poisoning as well as construction and process safety.

Demobilisation will be required at the end of construction work and needs handling sensitively to avoid unrest amongst the workforce who might find themselves jobless. Industrial action may occur, in the worst case, resulting in sabotage.

> Final inspection of an offshore module identified that nails had been driven into electrical cables, requiring significant delays whilst cabling was removed and replaced.

15.4 CONSTRUCTION

The first stage of construction will be civil work, which often produces an unsightly, muddy area (and significant noise) whilst work takes place.

The initial civil works of ground levelling and compaction can be performed whilst design details are being finalised. High levels of vibration can occur during dynamic compaction, pile driving or demolishing of existing buildings. Considerable earth movement may be necessary and the import or export of material might be necessary to provide a flat area at the correct elevation. This will be followed by piling, the excavation of trenches and casting of foundations. Much of the civil work

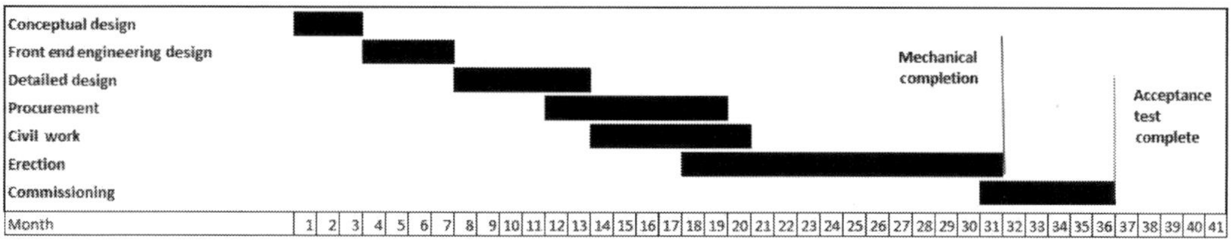

FIGURE 15.2 Simple Construction Bar Chart

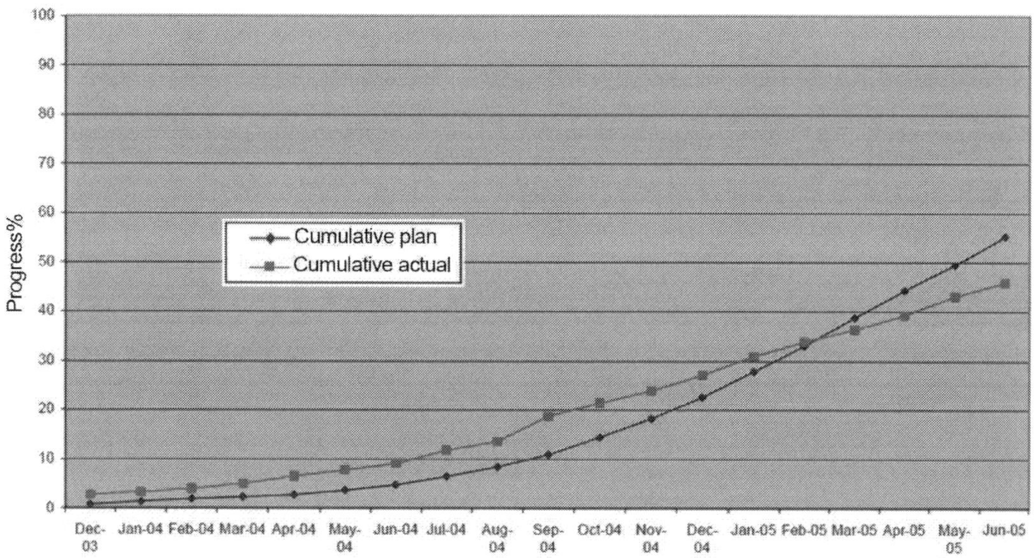

FIGURE 15.3 Typical "S" Curve

will be complete before the installation of steel work, equipment, piping cabling and utilities. On large projects, civil work may be completed in one area before moving onto a new area with erection starting where civil work is complete. This will have the advantage of requiring a smaller workforce and maintaining continuity of employment.

At the start of construction, a time schedule in the form of a bar chart will have been prepared showing the various construction activities and the start and finish times plus the *float* (the delay that can be tolerated without delaying the project). There are generally various versions of the schedule showing increasing levels of detail. Progress against the bar charts is monitored constantly, and resources of manpower and equipment are moved between activities to minimise delays—particularly those which could affect the *critical path* for project completion. A typical bar chart is shown in Figure 15.2.

Frequently, there will be an overall project bar chart, with separate bar charts for different project activities. Another common method for monitoring progress is the "S" curve which monitors actual progress over time against planned progress. A typical S curve is shown in Figure 15.3. The name "S" comes from the shape of the curve. There will be limited spending at the start and end of the project but an increasing rate of expenditure once procurement and construction are in "full swing". Normally there are separate curves for different activities such as engineering, procurement, construction and commissioning which are combined into an overall curve. Control of the schedule and costs are complex and challenging activities but essential to successive project completion.

The storage of equipment and components arriving at site is critical. There will be an initial examination to ensure that the items are to the correct specification, undamaged and the correct number has been supplied. Bulk items will then go into storage in a warehouse or laydown area. It is important that all items entering are correctly recorded, along with their location. Projects have been delayed when critical components could not be identified. The equipment will be delivered in a "workshop" condition—clean, pressure-tested and drained of all fluids.

Rotating equipment should be stored in a vibration-free environment and rotated regularly. Spare rotating compressor shafts should be stored in a workshop or storeroom environment in the vertical position so as to prevent sagging and later unbalance unless the manufacturer's packaging is designed to prevent damage during extended periods of storage.

Construction sites are hazardous workplaces and with relatively high accident rates. Scott[2] has provided a simple overview of construction hazards. Until commissioning, accidents are not generally process related and are summarised in Table 15.1. Detailed planning of construction activities is necessary to minimise movement around the site particularly of heavy loads and critical equipment. The impact of an incident in one area on another needs to be considered.

TABLE 15.1
Common Construction Hazards

Exposure	Specific Hazards	Comments
Interface with Existing Equipment	Conflict between safety precautions required for operating equipment and hazards caused by construction work	Will only apply where operational and construction work are taking place on the same site and can be controlled by SIMOPS. Good planning should minimise the extent of construction work in operational areas, but there are likely to be times when it is unavoidable, such as *tie-ins* to existing pipework. Such activities represent high-hazard situations.
	Barriers between operational and construction areas hindering the access of emergency services or evacuation of workers	Construction personnel are unlikely to be knowledgeable about process hazards. Similarly operational staff might have limited knowledge of construction safety. It is desirable that the workforces are separated, which is usually achieved by a fence between construction and operational areas. However, there must be some form of controlled access between areas to escape in an emergency.
	Poor mechanical, electrical or instrument isolation practices	Isolation practices should be defined at an early stage (see Section 14.5).
Storage	Damage in transit between laydown/warehouse	The more times equipment is handled, the more likely it is to suffer damage. As far as possible, major equipment items are located on their foundations when they arrive on-site. "Consumable items" arriving at site need to be checked, stored and released for use at the appropriate time. A computerised inventory system is generally used.
	Equipment in a laydown area need to be moved to the construction site which was separated from the site by overhead power lines. The practice was to use a small crane to lift the equipment onto a truck at the laydown area; both crane and truck are then moved to the construction site where the item was unloaded. Eventually (and probably inevitably) the crane boom was not sufficiently lowered when moving under the power cables, and the boom touched a live cable destroying the crane and injuring the occupants. This type of incident is not unusual, and it is common practice to install "goal posts" on either side of overhead cabling (and piperacks) to prevent injury/damage[3].	
	Large values of construction materials in one location.	Equipment is often stored in specific areas, and multiple items can become damaged by the same event such as fire, flooding or vehicle collision.

(*Continued*)

TABLE 15.1 *(Continued)*
Common Construction Hazards

Exposure	Specific Hazards	Comments
	Poor storage and preservation	Storage facilities need to be adequately prepared and maintained. Muddy/wet areas can cause corrosion to equipment. Pipe sections and fittings are generally supplied with end caps fitted but should still be stored on sleepers above ground level to prevent earth from entering the pipe if the end cap is lost. The same is true of pipe spools after fabrication. Pressure vessels are usually nitrogen-filled under a slight positive pressure. All nozzles on vessels must be blanked off to avoid ingress of debris or water and filled with an inert gas. Whilst preservation should be in line with manufacturer's guidelines, it is common practice for pumps and motors with ball or roller bearings to be turned once or twice on a weekly routine to avoid false brinelling[4] on the bearings and also corrosion at the contact point. Grease-filled bearings should be turned monthly. Tarpaulins are frequently used to protect equipment from high temperatures and storms (wind, rain or sand). However, the possibility of condensation and possible damage underneath the tarpaulin should not be overlooked. Some components are sensitive to temperature and humidity and will need to be kept in a controlled environment This particularly applies to electrical and electronic equipment. Water may be trapped externally and result in corrosion. In a cold environment, freezing could occur.
Hazardous Materials	Radioactive sources for weld inspection	These need to be stored in a secure location with radiation warning signs clearly displayed. They must be handled only by suitably qualified personnel.
	Fuels	Small storage tanks with fuel for site vehicles are often installed. These should have secondary containment and spill protection at the filling points.
	Gas cylinders	Gas cylinders should be stored in a secure outdoor location. Flammable gases and oxygen should be separated by a firewall or a distance of several metres.[5]
	Chemicals (These may range from acids or pickling to chlorine for water treatment. Some chemicals may be harmless to personnel and the environment but others will not.)	Chemicals must have spill protection and be stored in such a way that mutually reactive chemicals are not located adjacent to each other. Material safety data sheets (MSDS) should be readily available—but not stored next to the chemical containers where they will be inaccessible during a leak or fire.
	Lube and transformer oil.	These are usually supplied in drums and should ideally be stored in some form of secondary containment.
Fire and Explosion	Unexploded Ordnance (UXO)	Site surveys and clearance should be undertaken before civil work starts. In some cases, partial clearance of the site may take place prior to the start of construction, and personnel must strictly follow instructions from the clearance team. Normal clearance techniques will identify UXO up to a depth of several metres, but air-dropped weapons may have penetrated beyond that limit. Deep excavations need to be partially dug and a second level of scanning for UXO should be undertaken.

(Continued)

TABLE 15.1 *(Continued)*
Common Construction Hazards

Exposure	Specific Hazards	Comments
	Smoking	Smoking is a less common habit than in the past and is more common in some countries than others. Whilst a blanket ban on smoking might appear a good solution, this seldom works as smokers will find a "den" somewhere on the site—possibly in a dangerous location. It is better to have designated smoking locations in well-ventilated areas with fire extinguishers nearby.
	Explosives for civil work/ well stimulation work	Again, explosives will need to be stored in a secure containment and handled by suitably qualified personnel.
	Welding	There are obvious hazards associated with welding (and grinding). Most piping spools are now constructed in fabrication (fab) shops with site welding minimised. Often there are several fab shops for welding different materials of construction. This reduces the potential for the accidental use of incorrect materials. Note that welding sets can transmit vibration to the ground—particularly if the drawbar rests on the ground, which can affect rolling contact bearings.
	Hot tapping (welding a branch onto a pipe section which is in use)	There are specialist companies who undertake this specialist activity. Hot tapping should only be undertaken when no other alternative is available
Pressure Testing	High-pressure leak or pressure testing	Hydraulic testing is preferable to the more dangerous practice of pneumatic testing. Special precautions are necessary including the creation of a safety exclusion zone around the area where testing is to take place.
	On 7 February 2009, an explosion occurred during pneumatic pressure testing at an LNG import terminal in China. One person was killed and 15 injured[6].	
Commissioning (SIMOPS)	Inadequate isolations	Isolation standards were covered in Section 14.5. Removal of electrical or mechanical isolations must be performed with the knowledge and approval of both construction and operations teams.
	Instrument bypassing	Critical trip bypassing may be necessary for a variety of reasons during commissioning but should only be done in compliance with a formal procedure (see Section 16.7).
	Communication with existing plant	Good communications must exist between operating facilities and nearby construction work. Incidents in one location can negatively affect other areas. A gas escape in an operating plant when hot work is taking place in a construction area is one example. Other examples include ensuring easy access for emergency services across the site, warning of routine testing of operational alarms or the potential for operations such as piling to affect sensitive instruments.
	Impairment of fire protection systems.	Impairment of fire protection systems is covered in Section 22.5.
Civil Works	Site preparation	Impact damage from earth moving equipment.
	Buried services	Damage to existing underground services ether by excavation or movement of heavy equipment over utility systems.
	Trench collapse	Large construction projects require foundations to support structures and heavy equipment. The first stage in constructing foundations is to excavate a trench or pit. Depending on soil type and weather conditions excavations can collapse. This will cause injury to personnel working in the excavation. The collapse can also affect the stability of nearby plant and equipment.

(Continued)

TABLE 15.1 ***(Continued)***
Common Construction Hazards

Exposure	Specific Hazards	Comments
	Flooding of trenches	Flooded trenches represent a drowning hazard and inadequately supported trenches may collapse.
	Piling	Vibrations from piling can damage nearby equipment, and incorrectly located piles could damage underground services.
Cranes	Collapse	Large cranes of various types are used in major projects. A crane collapse can result in severe damage to the crane itself and also structures and equipment close to the crane.
	Dropped object	A poorly secured load or a load being raised under adverse weather conditions can fall, causing damage not only to the object itself but also to other equipment in the area where it falls. A particularly hazardous situation occurs if a load has to be lifted over an operating section of plant. If this cannot be avoided a detailed risk assessment is needed before the lift takes place.
Weather	High winds	These will delay work, particularly crane operations
(careful scheduling might limit the potential for weather-related issues)	Rain	This can cause flooding and water ingress into sensitive areas—possibly causing work to stop. Electrical commissioning may be impossible during and immediately after periods of heavy rain.
	Heat	Heat exhaustion and dehydration can be serious problems and may require changes to working patterns with overnight rather than daytime working taking place. Some activities may be limited by high temperature, such as concrete mixing which will have to be performed during lower night-time temperatures or by using an ice water mix.
Others	Food poisoning	Many sites will either have accommodation for the workforce or provide food during the working day. Poor hygiene could lead to mass outbreaks of food poisoning. At best this will affect construction progress and, in a remote location or offshore overwhelm the available medical resources.
	Pandemics	The same issues apply as with food poisoning. In severe cases quarantining of the workforce may be necessary. Problems can also arise with the movement of technical specialists, such as vendor engineers who will need to be present for the commissioning of many equipment items.

Scheduling of construction activities should ensure that foundation and supports for equipment are complete before the equipment arrives on-site. Delivery should be part of the Critical Path and may require prior sanction to avoid construction delays.

To make the best use of the construction labour force, work is often staggered across the site. Once civil work is complete in one area, steel erection starts; the civil construction team move onto a separate area. Whilst it would be desirable to keep each area of the plant dedicated to a particular construction discipline at any point in time, this is often not possible, particularly at the end of a project and especially where small-bore piping is site run. It will be necessary to apply different safety rules in different parts of the site and to keep a close watch to ensure that incompatible activities are not taking place anywhere around the site.

Integrity checks such as concrete testing and weld inspection will take place throughout the construction process, and comprehensive *quality assurance* and *quality control* procedures are required with regular audits and corrective actions specified where necessary. This process is guided by the project *quality plan*. As the project nears completion checks will be made throughout the entire site to ensure that the design has been correctly constructed. To simplify this process the plant is usually considered in three parts:

1 Mechanical completion
2 Pre-commissioning
3 Commissioning

The boundaries between these three activities vary between projects, and each is explored in more detail later. All three activities will take place over a period of time and some parts of the plant may be commissioned before others are mechanically complete. Temporary measures are often necessary at the pre-commissioning and commissioning stages. For example, power may need to be supplied from temporary generators if the permanent power supplies are not ready and fire marshals may need to patrol the site if fire detection systems are not operational. A basic guide to the mechanical completion, pre-commissioning and commissioning process has been produced by Horsley and Parkinson.[7]

15.5 MECHANICAL COMPLETION

Prior to any part of the plant being considered mechanical complete, it will be split into sections and subsections for review to ensure that the plant has been built exactly as designed and has no unauthorised changes or additions. This particularly applies to piping, electrical and electronic systems. The review of systems/subsystems will include the use of detailed checklists covering all civil, mechanical, electrical and instrument components and will be "walked through" by a multi-disciplinary team with, as a minimum, representatives from the contractor (and subcontractor) and plant owner. Vendor representatives for major equipment may also be involved to ensure the equipment is undamaged and there are no issues which might affect equipment warranties.

After confirmation that each subsection is mechanically complete and fit for purpose. pressurised systems will be pressure tested. Hydraulic testing is much safer than pneumatic testing, which has resulted in some serious accidents.[8] Particular attention must be paid to the solids dissolved in the water. Using seawater to hydrotest stainless-steel lines is unacceptable because of the potential for *stress corrosion cracking* (SCC). As stated in Table 15.1 stringent safety precautions are necessary for pneumatic testing if this is unavoidable.

The details to be reviewed for each system/subsystem will vary depending on the type of equipment involved and should be defined in a test procedure. In specific cases such as high temperature or low-temperature steels it may be necessary to carry out metallurgical tests. Rotating equipment will need particular care to ensure that it is correctly levelled, with correct piping alignment, lubrication and direction of rotation checked. Table 15.2 gives examples of typical checks prior to mechanical completion but should not be considered exhaustive.

Incomplete items will be noted and added to a *punch list* (sometimes called a reservations or snagging list). This itemises all defects and incomplete work. The simplest form of punch list has two categories:

1 Defects which must be corrected before pre-commissioning can proceed (for example: installation of safety systems such as fire detection equipment and safety guards on equipment). These should be recorded and signed off by the operations personnel when the defect is rectified. If a modification is necessary, the P&IDs should be modified and a revision number shown. A safety review will also be necessary.

TABLE 15.2
Common Piping and Equipment Checks

Piping	It is important that piping is installed exactly as designed with correct sequencing and spacing of valves and orientation of pipework. Even minor deviations from the P&ID can cause problems. A series of pumps on a fuel forwarding system for a gas turbine power plant fed into a common manifold. The piping arrangement consisted of an upstream and downstream isolation with a NRV located between the pump and the downstream isolation valve. The design considered that as the NRV was a low reliability item, installation between the isolation valves would simplify maintenance. However, the NRV on each pump was installed downstream of the isolation valve and was not discovered during the pre-commissioning piping checks. The only way of removing and repairing the NRVs would involve dismantling part of the manifold—with a requirement to take the turbines out of service. Where lines need to be sloped, for example on steam and flare lines the slope must be checked. Flare header performance will be seriously impaired if liquids build up on inadequately sloped lines and reaction forces at bends have caused damage to both steam and flare headers. Process equipment will have a hydrotest certificate from the manufacturer. Piping on the other hand must be hydrotested on-site. Hydrotesting can impose large static loads on supporting structures and foundations. This should have been recognised at the design stage. There may however be cases where design for hydraulic loading is not possible and pneumatic testing may be the only option. It might be necessary to test large-bore gas piping on the ground before fitting in position and installing a *golden weld*[9] (full NDT) or a flange. This will apply also to tall vessels where the loads on the foundations might be excessive. Pipework, especially large bore pipework, should be thoroughly drained before blowing (to remove dirt and scale). Any slugs of liquid would be pushed through the system and possibly rupture pipework at a bend due to reactive forces. Debris blown out of piping or vessels will often be at high velocity and could cause injury—particularly eye damage.
Vessels	All vessels should be inspected for cleanliness. In the case of columns this will including checking behind weirs and downcomers for construction debris. Checks for residual mill scale in large vessels should also take place. This should have been done by the fabricators, but it might be necessary to remove remaining scale using a stiff wire brush. Distillation columns should be checked to ensure trays are perfectly level. Manways will be left open during construction for access and must be fitted after inspection. There have been examples of manways not being reinstalled after inspection with embarrassing (and dangerous) results.
P&IDs	A thorough review of the P&IDs is necessary to ensure the plant, as constructed reflects the design intent in the P&IDs. Common checks include the following: • Compliance with the P&IDs both in detail and in line sequence • Orientation of control valves (are they in the wrong way around?) • Materials of construction of piping, fittings, nuts and bolts • In specific cases such as high temperature or low-temperature steels it may be necessary to carry out spot metallurgical tests. • Orifice plates for flow measurement and bursting discs are fitted the correct way round. • High point vents are capped off or blanked • Low point drains are capped off or blanked • Plugs on PSV bonnets are removed • Springs supporting piping have their pins removed The P&IDs should be marked up with a highlighter to ensure that everything has been checked. Any deviations should be noted in a contrasting colour and investigated. The accuracy of the P&IDs is paramount as they reflect the "skeleton" of the plant and will be used as soon as pre-commissioning starts to determine isolations for maintenance and as a basis for design of modifications.
Atmospheric Tanks	Weld inspections should have taken place as the tank was being erected but a hydraulic test will take place before the tank enters service. Water is generally use for hydrotesting and will normally take place to the maximum working level. Many tanks will have contents with a significantly lower density than water and, if the tank was fully filled with water, this may overstress the base of the tank. In this case the tank can only be partially filled with water and the weld integrity of the top few levels of strakes cannot be confirmed.

(*Continued*)

TABLE 15.2 *(Continued)*
Common Piping and Equipment Checks

Materials of Construction	Whilst materials will be specified, it is not uncommon for materials to be supplied to an incorrect specification. A *mill certificate* will be supplied with the material, but it is not always possible to trace individual components, for example small sections of pipe cut from a longer length. Mill certificates are not always correct, and an alloy analyser is commonly used to check stainless and alloy materials. Checking carbon steel is more difficult but may be achieved by hardness testing or other means. Nuts and bolts may have a code stamped on them. Inspection is covered in more depth in Chapter 20.

2 Defects which can be corrected later include the following common examples:
 1 Safety gates at the top of ladders are not fitted
 2 Handrails are incomplete
 3 The bottom step of a ladder is the short step
 4 Painting or insulation is incomplete
 5 Access to valves is difficult
 6 Incorrect valve orientation which could cause injury

Temporary solutions must be in place for safety-related items such as scaffolding in place of handrails. "Corrected" later does not mean forgotten which is often the case, particularly with valve access.

Whilst category 2 items will not affect the start of pre-commissioning, long-term delay could affect plant operation. For example, performance could be impacted if thermal insulation is not applied and piping heat losses become accepted or, worse, cause burns (hot or cold). Care is also necessary to ensure that plant commissioning does not make it impossible for minor items to be completed with the plant in operation—it would be hazardous to install insulation on high temperature pipework for example.

The timing of the walk-throughs is critical. Too early and the hundreds, or thousands, of punch list items are generated because construction—which would have corrected the items—is not complete. Tracking hundreds of punch list items, whilst construction is continuing is also a mind-numbingly difficult task. If the walk-throughs are too late, time is wasted before pre-commissioning can start.

A *mechanical completion* certificate for each system/subsystem is issued when category 1 defects are corrected, and eventually an overall certificate is issued for the whole site.

15.6 PREPARATION FOR COMMISSIONING

Once each section of the plant is confirmed as being mechanically complete, pre-commissioning activities can take place. The work necessary should have been fully defined and sequenced in advance. Often PSVs, control valves and sensitive instruments, such as orifice plates and turbine meters, will have been removed whilst hydrotesting or cleaning takes place and will be reinstalled later. A list of removed and refitted items should be established and crosschecked to ensure completion. For low-pressure systems one PSV should be left in place as overpressure protection. After fittings are reinstalled, a leak test is generally required. Each pressurised system is slightly pressurised with a safe gas containing a *tracer* (usually helium) and leakage is detected by an analyser. As the system has been pressure tested the only likely leak sources are from flanged and screwed connections. If tracer gas cannot be detected, this demonstrates there is no leakage. An alternative is a "pressure drop test" of 1% operating pressure per hour. Soapy water is applied over leak sources and the presence of bubbles indicates the location of the leak. Valves will be stroked to ensure they are not stuck and three-phase motors checked for rotation. Function testing of the control and ESD systems and the F & G system plus fire protection systems will be required. These are only part of the complex process of necessary checks.

Different systems will be mechanically complete and ready for pre-commissioning at different times and some temporary arrangements may be required. F & G detection and protection systems should be amongst the first items to be commissioned. This may be an insurance requirement but in some cases it is impractical. For example: fire alarm systems will obviously not be operable until power is supplied to the system. Whilst this should have been identified during construction planning, changes to the schedule may have to be made. Temporary systems may be needed such as the use of temporary heating, ventilating and cooling (HVAC) systems, particularly in hot climates. Where temporary arrangements are in place, these should be subjected to a risk assessment and suitable precautions provided.

Centrifugal (and axial) compressors can be test-run "open circuit" on air. The pressure differential must be low as the ratio of specific heat for air ($\gamma = 1.41$) is significantly higher than most process gases ($\gamma = 1.1$). The source of air could be a manhole on the suction vessel (fitted with a guard to avoid human ingestion) and exhaust via a removed NRV. The pressure in the suction vessel must be monitored as well as the discharge temperature. Noise profiles should also be checked as well as the sealing arrangements. The flowrate is usually limited by *stonewall*[10] or sonic flow in part of the machinery.

Many process fluids have a lower specific gravity than water so it may (or may not) be possible to test pumps on water. It may not be possible to test run all pumps due to the low density of the process fluids and the specific seal arrangements. Where rotating equipment is run on non-process fluids (air instead of gas, water instead of hydrocarbons), carful checks on metallurgy, density difference and heat removal must be considered. For obvious reasons cryogenic pumps should not be tested with water. It may be possible to check the performance of heat exchangers using water, but this could result in contamination. If water is used, subsequent drying might be necessary. Power and other utilities must be available by this point.

15.6.1 Housekeeping

Housekeeping around the site must be maintained to a high standard with packaging and construction debris removed before it accumulates to a dangerous level. Eating should be in controlled areas with all food waste promptly removed to avoid insect or rodent infestations. Waste containers should be available around the site to avoid drink bottles and other debris being left on steelwork or behind equipment. Cigarette ends and alcoholic drink containers are signs of a serious failure of site discipline.

All piping and vessels should be cleaned before being "boxed up ready for use". Specific procedures will be necessary for some types of plant. For example, in the case of low-temperature systems, after drying, it is good practice to leave cleaned or dried equipment filled with inert gas at a slight positive pressure.

15.6.2 Cleaning

Cleaning activities must be carefully controlled and appropriate. The amount of effort required should not be under-estimated as significant amounts of debris will need to be removed. Piping over 0.75 m diameter can be entered provided appropriate safeguards are in place (for PPE, see Section 3.5, and for a confined space entry permit, see Section 16.6) and, if necessary, mill scale can be brushed out. Smaller bore piping can be water jetted in straight runs. Thereafter it is a case of deciding if water flush out or air blowing is better. If water is used, water quality is important, and the comments made earlier regarding hydraulic testing also apply here. Air blowing is very noisy and creates flying debris. Adjacent areas will need to be barricade off; these could be quite large as a stone or welding rod can travel significant distances. Nearby work should be suspended whilst air blowing takes place and even workers at a distance may require ear protection. Jet reaction forces on small-bore piping must be considered when cleaning by air blowing.

High-pressure steam mains are a special case as loosened mill scale or other dirt may damage steam turbine blades. Steam can be used to clean the lines, but it will be necessary to cycle the temperature in order to loosen the scale. This will require a number of noisy steam "blows".

Lubricating and seal oil systems can be water washed and then circulated with oil. Once again temperature cycling will be beneficial in loosening scale. Bearings and seals should be inspected for debris following oil circulation. Circulation times may be extensive, and it could be weeks before inline filters in the system are clean. Beware of "dams" created by uncentred gaskets in flanged piping where fine debris may lodge and not be trapped in the filters—creating damage at a later date.

The cleaning schedule must be developed with care to avoid re-contamination of cleaned piping. Like many aspects of construction, a method statement for cleaning should be developed. The statement should define the starting point and the line number should be identified followed by the subsequent lines to be cleaned. Different lines might require different cleaning methods and these should all be recorded on the P&ID. The sequence should be "clean to dirty" and "small to large". Planning the cleaning sequence is a complex time-consuming task and is often iterative. Due to the fragmented nature of cleaning, it may not be perfect but attention to detail is essential. The method statement will probably require revision as it is almost inevitable that the last piece of the system will involve re-contamination of a cleaned section.

A set of master P&IDs should be used to log those lines which have been notionally cleaned and those which require to be cleaned. The cleaned lines should be highlighted and show the initials of the person responsible for the cleaning.

15.6.3 Drying

A similar approach to cleaning can be applied to drying low-temperature systems. A start point and line number should be defined and a method statement produced indicating the sequence in which lines are dried. A dew point of −50°C is adequate for many systems, but much lower figures will be needed for plants handling very low-temperature materials such as liquid hydrogen or LNG. Hot air can be used for drying, and swabbing with methanol is an alternative solution but a safe route for the disposal of used methanol will be necessary. P&IDs should be marked up when satisfactory tests results are obtained with the location of the test point and the test results noted. After drying it is useful to maintain a small flow of dry air to detect any water pockets that may have been missed by routine testing at sample points,

As with cleaning, drying may take several iterations before a satisfactory dew point is reached.

15.6.4 Inerting

Inerting will follow a similar approach to drying and cleaning. It is essential that the method statement provides a defined flow path from inerted to not fully inerted piping—it is easy to contaminate a purged (inerted) line.

15.6.5 Final Preparation

The punch list will again require to be reviewed to ensure that all critical items have been completed before commissioning starts.

Procedures vary but there is often a second set of certificates required to indicate that each system or subsystem is ready for handover. Piping subsystems which are ready for commissioning should be isolated by blinds (and tagged) to separate them from systems still under construction or otherwise at a pre-commissioning stage. Similarly, energised electrical equipment must be clearly labelled.

15.7 COMMISSIONING AND HANDOVER

An essential part of the commissioning programme activity is the preparation of operating instructions and the training of operators. Operators will normally be selected well in advance of commissioning and have months of training covering both methods of working and the specific operating instructions for the plant (see Section 18.5). In many cases the plant contractor will have responsibility for startup and employ a specialist commissioning team, experienced on similar plant. The eventual plant owner's operators will observe and, often, assist. The operating instructions will be subject to review and possibly amendment, as operators are trained.

After mechanical completion and pre-commissioning work are complete and no critical punch list items remain, the plant is *ready for startup.* Any maintenance work or late construction work undertaken at this stage must be performed using an operational PTW. A final safety check, a *pre-startup safety review* (PSSR) using a prepared checklist tailored to the individual plant should be used. Detailed records of all commissioning activities must be maintained. These should be reviewed in conjunction with the PSSR to ensure nothing has been missed. At this point the piping and vessels will be filled with air, nitrogen, water or other non-process materials. These fluids must be displaced. Systems being filled with flammable materials must be purged with an inert gas prior to the entry of process materials to avoid a flammable atmosphere being created. High point vents and low point drains are often in use at this point and must be closed and capped after use. Inert materials may create issues with condensing in low-temperature systems with frequent purges of inert materials to a safe location being required. Inert gases may accumulate in liquid systems such as condensers and require venting.

Prior to commissioning, process vessels will have no liquid level and process fluids will need to be introduced. In some cases this may involve defeating process trip systems which would normally cause a plant shut down. Some modern control systems allow a trip to be bypassed for a limited period of time before the trip is re-established. Whichever method is used, bypasses should be controlled by a comprehensive procedure (see Section 16.7).

There is sometimes no clear boundary between pre-commissioning and commissioning. For example, commissioning a refrigerated storage tank will involve slowly chilling down the tank to operating temperature with process fluids. The commissioning process will be slow and methodical with extensive note-taking, so that any incidents can be investigated. Where operating conditions are significantly above ambient temperatures or pressures these should be increased slowly to avoid damage. The steps in the commissioning process should be defined in detail and followed precisely.

The ideal startup will involve everything working perfectly at the first attempt. In practice this does not happen, although a well-performed HAZOP should identify most of the issues that can occur during commissioning. The commissioning process may need to be halted numerous times before successful sustained operation is reached. Issues such as leaks, incorrectly set process instrumentation and contamination are common. For a large plant commissioned, section by section, the process can last several months and involve some minor, and possibly major, modifications which must be performed in line with a MOC procedure (see Section 16.9).

During the commissioning stage routine maintenance will be necessary. This is generally carried out by the contractor, possibly using spare parts supplied as part of the contract (generally the construction contract includes two years consumable spares). Often maintenance is performed without the involvement of the eventual plant owner and there have been occasions when the owner has taken over maintenance activities to find the expected stock of spare parts has been severely depleted and limited records of pre-handover maintenance exist,

The commissioning process generally concludes with a *proving run* with the plant running for a specified time (often 72 hours) at the maximum production rate. However, proving runs may be significantly more complex if the plant is making numerous different products or uses alternative feedstocks.

15.8 WHAT CAN GO WRONG?

The commissioning of process plant is, arguably, the most hazardous time in the whole plant life cycle. Numerous incidents, often errors of omission or deviation from the startup procedure have occurred. Many issues will be related to process errors and equipment behaving in unexpected ways—which may tempt the commissioning team to bypass process trips.

During the commissioning of a methanol plant problems with the flange on one of the manholes on a pressure vessel occurred. The commissioning team could not get the flange to seal adequately and the system had to be depressurised and the gasket changed on three occasions. A tight seal could still not be obtained and as the team had run out of new gaskets, they decided to reuse one of the old gaskets. In their hurry to return the plant to service the gasket was not inserted correctly and leak of high-pressure gas resulted. The leak ignited and a significant jet flame resulted. An ESD then took place. Although the plant was not significantly damaged, commissioning was delayed whilst extensive metallurgical checks of equipment exposed to high levels of thermal radiation took place.

Although not common, incidents of sabotage have occurred. At the end of the project construction personnel will be demobilised and, sometimes, sabotage equipment either in the hope of extending their employment or just out of malice.

A packer was found in a heavy-duty strainer on the inlet to a 30 MW steam turbine (Figure 15.4). This line had previously been thermally cycled and blown down six times with all debris cleared. The packer had not been displaced, it was the result of industrial sabotage when a construction team was laid off.

FIGURE 15.4 Industrial Packer (mounted on plaque) found in the Strainer of a steam turbine

REFERENCES

1 Davies V J & Tomasin K, *Construction Safety Handbook* (2nd edition), Thomas Telford, 1996
2 Scott D, Process Industry Construction Risk Management—A View from the Insurance World Doug Scott, *Loss Prevention Bulletin 245*, IChemE, October 2015, www.icheme.org/media/2134/lpb245_17.pdf
3 UK Health and Safety Executive, *Avoiding Danger from Overhead Power Lines (GS6)*, HSE, 2013, www.hse.gov.uk/pubns/gs6.htm
4 False brinelling is bearing damage caused by fretting (with or without corrosion). Visually, it looks like brinelling which is the permanent deformation of a hard surface
5 British Compressed Gases Association (BCGA), *The Storage of Gas Cylinders* (CP44) BCPA, 2022, https://bcga.co.uk/publications/cp44-the-storage-of-gas-cylinders-2022/
6 Anon, Blast kills worker at Yangshan, *Shanghai Daily*, 7 February 2009, www.china.org.cn/china/news/2009-02/07/content_17240000.htm
7 Horslet D M C & Parkinson J S, *Process Plant Commissioning*, ICHEME, 1990
8 Anon, Blast Kills Worker at Yangshan, *Shanghai Daily*, 7 February 2009, www.china.org.cn/china/news/2009-02/07/content_17240000.htm
9 A golden weld is a welded joint that does not undergo pressure tests. As an alternative such welds go through extensive non-destructive testing (NDT) to ensure they are defect-free
10 Stonewalling or choking of a centrifugal compressor occurs when the compressor is operating at low discharge pressure and very high flowrates. These conditions represent the highest flowrates the compressor can achieve. Decreasing the outlet pressure will not increase compressor throughput

16 Hazard Study 6—Operation

16.1 INTRODUCTION

This chapter outlines safe working practices during the operational phase of the plant life cycle.

The Operational Phase is ruled by procedures which have been well proven over many years of operation. Details will vary between organisations, but the basic principles remain the same. Critical aspects are covered in this chapter, and Chapters 18 to 22 provide greater detail on the activities of the various plant functions.

The operational phase of a project should be longer than all the other phases of the project. A design life of 25 years is common and in some cases the process is still operational after 50 years or more. There are a few sites which were developed over 100 years ago but little, if any, of the original equipment is still in use. In a few cases, plants have been shut down soon after commissioning due to changes in process economics or the technology becoming obsolete. There are also cases where war or terrorism has resulted in plant shut downs. In some cases, the plant has been shut down and evacuated so quickly that mothballing was not possible (see Section 17.4). If the plant is later to be restarted significant physical inspection and testing will be needed to ensure asset integrity. Assessments will be necessary to ensure the physical integrity of the plant and equipment: to ensure, amongst other considerations, that corrosion is not excessive, PSVs are correctly calibrated, electrical and instrument systems are undamaged. Numerous, less obvious studies are also required, for example: to check that the spare part inventory is accurate, and components have not been damaged. It will also be necessary to ensure operating instructions and maintenance procedures are available and up to date.

Even for a plant operating economically over an extended period, a variety of safety studies will be needed, often repeated over the lifetime of a plant as regulations change and modifications are made to the plant (see Section 1.6 and Chapters 5 and 6). Over its lifetime a plant will undergo numerous changes. Equipment may be replaced, piping and control configurations changed and spare parts may be difficult to source. As equipment ages the maintenance and inspection schedules may need to be changed. Inspection frequencies may be increased due to increasing damage from corrosion, fatigue, and temperature creep. Alternatively, improved inspection techniques may allow an increase in inspection frequency (see Chapter 20). Maintenance techniques may also change and increasingly *risk based maintenance* practices are being followed. For plants operating beyond their design life, the concept of "routine maintenance" must be reviewed critically. Corrosion allowances built into the design may be used up, creep may be a problem with hot systems and equipment may become worn out or obsolete. A specific example is the loss of *interference* on bearings which might require a new shaft to be installed. See Chapter 19.

16.2 THE LINK BETWEEN DESIGN AND OPERATIONS

The operation phase of the project lifecycle is illustrated in **Figure 16.1**. There is a complex interface between the Designer and the Operator. They have different and complex perspectives which can create problems. Design engineers often think of the process as operating in steady state and may not consider in detail the following dynamic situations:

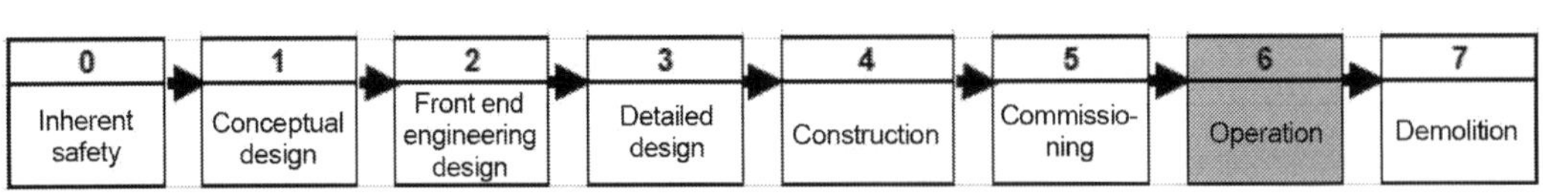

FIGURE 16.1 Project Lifecycle Operation

DOI: 10.1201/9781003510680-19

1 Startup
2 Shut down
3 Failure of a pump or compressor during operation
4 Remaining in a safe and stable "holding" state whilst repairs or investigations are undertaken

The design must handle these dynamic states and also have adequate buffer storage to prevent minor upsets in one place requiring the whole plant to be shut down. Many of these issues should be identified in the HAZOP. Last-minute modifications and retrofitting additional facilities, such as buffer storage, are seldom easy and not cheap.

Once production starts, the operations team are likely to look for changes to the "design" operating instructions, for example to facilitate startup and shut down or to hold the plant in a stable condition whilst a minor upset is corrected, particularly during startup. Changes to the mode of operation not anticipated by the design team will require physical modifications to the plant.

Some common issues which are not necessarily recognised in the design process are as follows:

1 Standards of isolation
2 Diagnostics for trouble shooting
3 Special features for startup and shut down
4 Safe hold points where a process is safe and stable during startup or following a shut down.

These may not always be identified during the FEED and detailed design HAZOPs. In some cases, these features have been given limited attention by contractors—additional valves and piping increase costs and a contractor—especially in a fixed-price contract—will want limit costs as far as possible. The importance of experienced operations and maintenance personnel working for (or on behalf of) the plant owner cannot be overemphasised.

16.3 SAFETY STUDIES

Requirements for safety studies will vary with a number of factors such as location, age and type of plant. Specific requirements will apply to many industries such as food manufacturing, pharmaceuticals and explosives. The most commonly encountered studies are summarised in Table 16.1.

TABLE 16.1
Common Safety Studies Required during Plant Operations

Regulatory Requirements (Process Safety)	Many industrialised countries have specific safety regulations, such as the Seveso Directive in the European Union or OSHA 1910.119 in the USA (see Section 1.6). Many companies, particularly international companies, operating in less developed countries will adopt the same regulatory approach as required in the EU. These often utilise techniques such as HAZOP, What if?, HAZID and QRA (see Chapters 5 and 6). A common requirement is a safety case approved by the regulator. This will identify operating and maintenance requirements for the plant with any deviations requiring a risk assessment and formal approval by senior management. To ensure that safety is not degraded over time it is common practice in highly regulated regions of the world particularly for high-hazard sites that key process safety studies be repeated at intervals (often five yearly). Ideally, the studies should be started "from scratch" with up-to-date P&IDs and a blank sheet of paper. There is often resistance to this as it is time-consuming. However, any other approach is likely to be less rigorous especially at the interfaces between the parts that are new and those where the previous study is accepted without question.

(Continued)

TABLE 16.1 *(Continued)*
Common Safety Studies Required during Plant Operations

Regulatory Requirements (Occupational Safety)	Most countries will have a series of regulations covering mechanical equipment which will require regular auditing and inspection (see Chapter 20). Common examples include the following: 1 Inspection of pressure vessels (as a minimum boilers and air receivers generally require periodic inspection). The regulations for other pressure vessels differ widely. In some countries the requirements for inspection are based on *pressurised systems* (which includes pipework) rather than individual pressure vessels. 2 Pressure Safety Valves are generally inspected at fixed intervals. 3 Cranes and lifting equipment. Cranes (including lifting beams) require periodic inspection and load testing. More frequent inspection of lifting slings, ropes, shackles and other components is generally required. 4 Sampling of liquid and gaseous effluents (including fugitive emissions), see Chapter 4. 5 Medical examinations of staff will often occur both before original employment and, in many cases, at intervals during their working life on-site (see Chapter 3): • Medical tests on staff handling specific chemicals (such as lead, dusts, radiation or biological materials) • Hearing damage assessments • Physical fitness tests for some roles such as firefighting • Eye tests • In specific industries it may be necessary to carry out colour blindness tests The exact requirements for inspection of equipment and examinations of the workforce will be industry and country depended.
Modifications of Change (MOC)	Over the life of the plant there are likely to be numerous changes ranging from small additions through to major plant revamps which will probably involve an external engineering contractor. Changes to process operations and increasingly to management systems are also subject to review. All changes should be covered by an MOC procedure (see Section 16.9). Changes should be assessed using the same, or better, methods used in the original design with results fully documented. At major hazard sites, the safety case will need updating.
Job Safety Analysis (JSA) and Task Analysis (UK)	Maintenance and construction work may create unexpected hazards at the site. It is increasingly common for a formal review of hazards to be undertaken before work is approved. However, there can be a tendency for the JSAs to become "rubber stamping" or "tick box" exercises. For routine work a generic JSA may be acceptable but for more complex tasks a step-by-step analysis such as "task analysis" (or even a HAZOP) will be necessary. The plant manager decided it would be easier to remove a PSV using a crane instead of the lifting davits available (and specified in the manual handling study). No assessment was made of the forces around the crane outriggers as the "load on hook" was acceptable. The cost of the repairs was significantly more than the potential cost saving by use of a crane. The "load on hook" was well within the limits but the toppling moments had been ignored.

Crane Tilt (reproduced courtesy of the Kletz Estate)

TABLE 16.1 *(Continued)*
Common Safety Studies Required during Plant Operations

Risk Assessments	There will be numerous situations where a risk assessment will be necessary to determine if some proposed change will increase or reduce the safety (or environmental impact) of the plant. If the latter, should any form of risk mitigation be applied? A common example is the bypassing of a critical instrument system for maintenance whilst the plant is operational. Risk assessments are covered in Chapter 6 and the procedure for trip bypassing in Section 16.7.
Safety Audits	Safety audits of various types will be conducted over the lifetime of the plant. There are like to cover a wide range of issues: 1 Item-specific audits for example to check on the safety of handrails, tripping hazards, completion of PTW forms and many other issues. 2 Housekeeping to check for good levels of cleanliness and tidiness. Good housekeeping is a factor in developing a positive safety culture. 3 Regular audits to ensure that critical isolation valves are in the correct position 4 Firefighting equipment audits to ensure that all firefighting equipment are available and operational 5 Checks to ensure that standard operation procedures (SOP) are being correctly applied, for example the TW, MOC and critical trip bypass procedures. Audits are discussed in Section 21.7.
Safety Observation Programmes	Most companies have had a reporting system for unsafe acts, but many have been extended; to identify all safety issues (positive and negative). Safety observation programmes[1] form parts of behavioural approach to safety with the intention of developing a deeper safety culture.[2] Analysis of safety observations can often identify systemic problems which can then be analysed, and solutions implemented. Safety observation programmes are covered in Section 21.3.
Third-Party Audits	Third parties are often involved in audits or surveys of a site: 1 Banks and other financial institutions—to ensure that there are no situations which are likely to affect loans 2 Insurance companies—to determine the "quality" of the risk, its insurability and likely financial loss "if the worst happens". 3 Auditors—conducting benchmarking studies for compliance with international standards, for example ISO 9000, ISO 14000 or ISO 45000. 4 Customers—many customers will want to check on the production process and quality assurance programmes in place. This is particularly important where products have potential to cause significant harm if they are not within specification. Examples are aviation fuel, foodstuffs, and pharmaceuticals. 5 Audits by corporate head office personnel or, in the case of joint ventures, the joint venture partners. Reports of human rights abuses in the process industries are few, but the requirements for inspection by non-governmental organisations (NGOs) in the future cannot be ruled out.
Competency Assessments	Assessment of staff to ensure they are adequately trained to fulfil the requirements of their job. Assessments can also be used to identify personnel with the right skills and experience for promotion.

16.4 MANAGEMENT STRUCTURE

There are numerous variations in the organisation of a process facility. Management structure will depend on factors such as size and complexity of the site (such as, single or multiple plants, extent of storage and handling). A typical structure is shown in Figure 16.2.

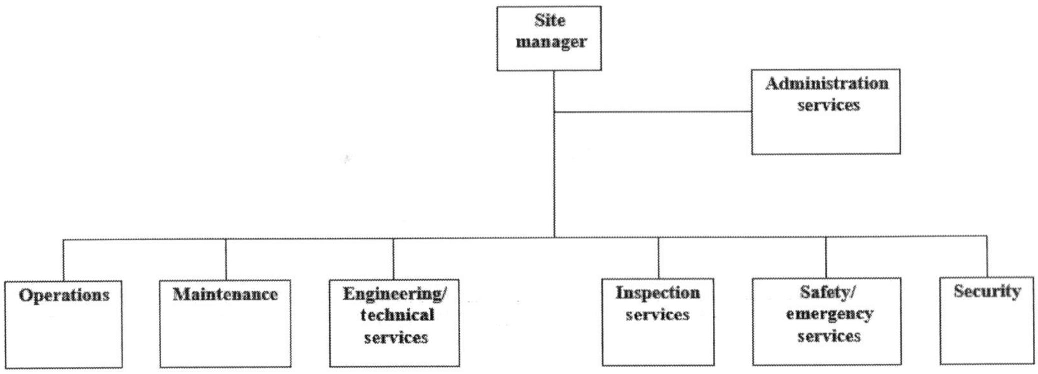

FIGURE 16.2 Simplified Site Management

When establishing the organisational structure, a number of points need to be considered:

1 Span of Command — Large numbers of departments or subgroups all reporting to the same manager can result in problems with communication and decision-making. Ideally there should be no more than five or six groups reporting to one manager.

2 Complexity — Overly detailed organisations can result in difficultly in communication with information needing to go up one "leg" of the management structure and then down another leg to get to the recipient

3 Conflicts of Interest — The most common conflict of interest is the safety function reporting directly to the operations manager. This can allow safety concerns to be overridden by production pressures, possibly with tragic consequences.

An equally important potential conflict between the inspection (now increasingly called *asset integrity*) function and the operations (and sometimes maintenance) function, where there may be pressure for scheduled inspections to be delayed for production reasons. This often occurs with storage tank inspections where tanks may be out of service for weeks, if not months.

4 Contractors — Non-core activities are often subcontracted to third parties. Cleaning, catering and gardening are common examples. Increasing tasks such as painting, scaffolding, PSV calibration and some inspection activities are also performed by third parties. Operations and maintenance (O&M) contracts with all day-to-day activities being undertaken by a third party are also becoming common in some industries. Often the O&M contractor will operate several similar plants and have a good understanding of the likely issues. If staff shortages occur, the contractor can bring in experienced staff from similar plants on a temporary basis. There is the drawback the O&M contractor will know more about the plant than the owners.

Contractors may also be used to fill gaps in the organisational chart where the company has been unable to recruit staff.

Whilst, in principle, there is nothing wrong with the use of contractors, problems can occur:

1 Frequent changes of personnel with the consequence that those working at the site are unfamiliar with its geography, processes and procedures.
2 Contract staff may have a poorer work ethic and safety culture than the company's own employees.
3 Specialist knowledge will be in the hands of contractors rather than staff. If the contractors leave the site—often following a bidding process for future work—that knowledge can be lost.
4 Records may be held offsite and not available to site staff.

Whilst it should be obvious, it still needs to be frequently repeated that safety is the responsibility of all site personnel, including contractors and visitors. The safety specialists cannot "magic" safe working conditions but have to work as advisors and auditors. There are cases—particularly in construction—where safety personnel are ignored (or even threatened) when they try and ensure safe working practices on an unwilling management.

When restarting a hydrotreater after maintenance, an explosion occurred in a hydrogen compressor. The investigation discovered that the hydrogen piping had not been purged with nitrogen prior to filling the system with hydrogen. Not surprisingly, a flammable atmosphere occurred in the system, and a source of ignition in the compressor caused the explosion.

Part of the company procedure required the safety officer to ensure that purging had taken place, and when asked why he had not done so, his response was that the shift supervisor had threatened to vandalise his car if he did not agree to omit the purging step and allow the startup to proceed without delay.

Operations, maintenance, inspection (asset integrity), safety (and emergency response) are covered in Chapters 18 to 22, but the other functions in the management structure are briefly discussed later.

16.4.1 Engineering

16.4.1.1 Organisation

Large organisations may have a small site engineering team and access to larger resources at a corporate engineering department that can be consulted when required. The site team will generally handle small projects with larger projects being handled by the corporate engineering function, possibly with the aid of an engineering contractor. Smaller organisations are likely to use external contractors for all but the smallest tasks.

16.4.1.2 Process Technology

Many processes in widespread use are licenced from the technology developer, who will be involved in the plant design. After construction, it is common practice for the licensor to provide ongoing technical support over the lifetime of the plant. This may include updates to the technology, supply of new or specific (confidential formulation) catalysts or similar materials plus notification of incidents that have happened on similar plants. In some cases, user groups have been established from the owners of plants using licensed technology.

Some equipment suppliers operate similar systems with users able to obtain information on equipment failures and technical updates.

16.4.2 Security

Most sites may will have an individual responsible for security, for small locations this might form only part of a job description. Day-to-day security activities are very often subcontracted to a dedicated security company.

Entry to the site is now almost always controlled for both staff and visitors. Road barriers are generally in place to control movement of vehicles. Pedestrian access is often via turnstiles. Airport-style security is uncommon but does exist in some places. The identity of visitors is generally checked, and bag searches are common. In politically unstable areas substantial road blocks, chicanes and ramps may be in place to stop or slow down approaching vehicles.

Security around and within a site varies significantly between companies. In some cases, the main concern may be theft or trespass, but terrorism is an increasing worldwide concern.

> An old but often repeated story concerns theft from a chloralkali plant using mercury cell technology. At one shift change, the security guard watched the outgoing shift leaving the site. Some staff were walking, but others had bikes. One man dropped his bike and had difficulty pulling it up upright. The security went to help and discovered the bike frame had been modified so that it could be filled with mercury.

The first requirement is to provide a secure perimeter. This is generally provided by a chain link fence topped with barbed wire, but concrete or masonry walls are common (these can inhibit dispersion of gas leaks). Electrified fences are sometimes installed, but this is uncommon and illegal in some countries.

Frequently there is an outer parameter around the entire site with an inner perimeter into the *hazardous* areas—process, storage and utility areas. Access into the outer and inner areas is now commonly controlled and increasingly *radio frequency identification* (RFID) cards are used to identify the location of personnel. An older alternative is the use of *T cards* (the name is based on the shape of the card) which are kept in racks at the point of entry to the site. Whilst partly for security purposes, these can also be used to identify personnel in the area if an emergency arises. An alternative, but less desirable, option is for staff and visitors to sign a *visitors' book* when entering or leaving.

The hazardous area may be further sub-divided, a common example is the separation of truck loading facilities from the rest of the site. This will prevent drivers and other unauthorised visitors entering production and storage areas. Similar arrangements are in place at marine facilities. Often these areas are also separated from process areas.

A range of measures are available for both detection of intruders and control of access, including intruder alarms of various types and CCTV. Patrols around and within the parameter of the site frequently take place but controls need to be placed on security guards to prevent them entering hazardous areas—particularly if they are carrying firearms.

No security system is perfect and determined intruders may succeed in entering the site. However good security precautions—usually a mixture of technology and human activities—will reduce the potential. Problems are more likely in remote parts of the site where fewer people work, especially at night.

Although intruders are unwanted visitors, the site owner still has a duty of care to ensure their welfare and sometimes accidents can occur.

> A body was discovered during the cleaning of a toluene tank. Investigations determined that it belonged to a local youth who had disappeared some years previously. It is assumed that he broke into the site, climbed the tank to inhale toluene vapour, lost consciousness and fell into the tank where he drowned[3].

16.5 STANDARD OPERATING PROCEDURES

All plants will have a range of SOPs (sometimes called *standing instructions*) to define and regulate activities, such as the conduct of maintenance work, accident reporting and other activities. These are different to operating instructions which describe steps to be taken to start up, shut down or operate the plant under normal conditions. Many SOPs whilst still necessary to ensure activities are performed consistently are relatively trivial. However, some are vital to safe operation. The most critical are

1 permit to work (PTW) (see Section 16.6)
2 critical trip isolation (see Section 16.7)
3 isolation for maintenance (see Section 16.8)
4 management of change (MOC) (see Section 16.9)
5 emergency response plan (ERP) (see Section 22.3)

16.6 PERMIT TO WORK

Major accidents in the process industries often occur during startup and shut down (see Chapter 18) or during maintenance activities (see Chapter 19). For this reason, PTW procedures are employed. PTW systems are believed to have been developed in the early 20th century by the United States Navy as an accident reduction tool. They are now almost universally applied in numerous industries. There is no universally correct PTW system, factors such as the technology, hazards present and culture of the workforce vary too much for a "one size fits all" approach. The procedure must be practical but relatively comprehensive. Its main purpose is to ensure safe methods of working are employed and all personnel involved—operators and maintenance or construction personnel—are aware of the precautions. Some procedures have been designed to be "fool proof", these are generally so complex and restrictive that they are often bypassed—the exact opposite of what is intended. The key aspects of a PTW system are discussed later and more detailed guidance is available from various sources. A UK HSE publication, *Guidance on Permit-to-Work Systems: A Guide for the Petroleum, Chemical and Allied Industries*,[4] is particularly useful.

A PTW can be considered a form of contract between operational staff and maintenance or project personnel who will be working on part of the plant. For them to work safely, the equipment must be made safe and isolated from operating plant and other sources of energy, such as process materials, electrical power or utility steam. It is not to be confused with a work order (WO which describes the work to be performed; see Section 19.3).

A wide variety of work is undertaken on process plant and the PTW system will usually have a number of permits for different activities. The basic types are:

1 Cold work (work that does not involve an ignition source)
2 Hot work (work that does provide an ignition source)
3 Confined space entry (work within a partially or totally enclosed space)

Hot work is sometimes split into work with a possible ignition hazard (such as grinding) and work involving an open flame with two different permits for the different types of activity. Whilst hot work might intuitively be considered as the most hazardous kind of activity, cold work in the form of "line breaks" (opening a process system to atmosphere) has produced some of the worst accidents in the history of the process industries. Whilst good planning and procedures should prevent breaking into a line that has not been depressurised and made safe, it happens depressingly often.

A variety of additional permits are commonly found to cover specific activities; the most common are shown in Table 16.2.

Inter-plant Clearance certificates are found on large sites where activities in one part of the plant could affect other areas. A typical example is work on a service line (air, nitrogen or steam) where isolation can affect other users of the service. In essence, this is similar to SIMOPS discussed in Chapter 15.

Maintenance work often involves mechanical or electrical isolation (see Section 16.8) or bypassing of a critical trip system (see Section 16.7).

The basic steps involved in controlling maintenance and project work in an operational plant are shown in Table 16.3.

TABLE 16.2
Common Types of Permit

Type	Purpose	Example	Comments
Excavation	To prevent excavation work damaging underground piping or utilities	Excavating a trench with a process area	Approval will be required from the site civil engineer and specific safety precautions may be required. For example, the use of hand tools rather than mechanical excavation.
Line Break	Opening a section of line which may contain pressure or hazardous liquid or gas	Removing a flanged section of line	
Radiography	To control work involving radioactive sources.	Weld inspection or maintenance of instrumentation which contains radioactive sources	The safe operation of radioactive sources is a specialised activity outside the scope of this book.
Working at Height	To ensure safe working practices are followed	Working on a ladder or scaffolding erection	

TABLE 16.3
Steps in the PTW process

	Activity	Comments
1	Planning	Generally, there will be a meeting the day prior to the maintenance activity taking place. This will involve both operations and maintenance personnel and discuss what activity is to take place, who will do, ensure that spare parts are available and that there are no conflicts (safety or otherwise) between the operational activities and the proposed tasks. In large sites (for both construction and operation) there will be dedicated *permit coordinators* who oversee all aspects of the PTW system and keep track of open and overdue permits.
2	Define the task to be completed	This should be a short but detailed description including the tag numbers of all the equipment involved. If the scope of work changes, a new permit will need to be issued. For example, a simple inspection might identify the need for welding work.
3	Define the method of working	For example, tools to be used; PPE required; site preparation; isolation requirements and evacuation procedures; the need for local gas detectors, firefighting equipment, safety harnesses and stand-by person, often called a *fire watcher*. Responsibility for task supervision also needs to be confirmed.

(*Continued*)

TABLE 16.3 *(Continued)*
Steps in the PTW process

	Activity	Comments
4	Identify hazards and specify safety precautions required	Many activities will be routine and have standard methods of working but more unusual activities may create specific hazards which need to be addressed. It is common practice to perform a risk assessment for each activity (commonly known as a Job Safety Analysis or JSA). These are often generic for common tasks but for complex activities a more detailed approach is required (such as task analysis). As stated earlier, there is a danger that generic JSAs will become tick box exercises and eventually something significant will be missed. Actions to be taken following an emergency (caused by the task or at some other part of the site) must also be considered. An example of the detail required during the JSA for breaking a joint is given below. • Can the integrity of isolation be proved by "leak-off" through vent or drain? • Can slip-plates be installed safely? • Is there a double block and bleed system which allows safe disposal of "trapped fluids"? • Is a stand-by person necessary in case of leakage? • Should there be a fire extinguisher on hand if the fluids are flammable? • Should the joints be "broken" away from the fitter when breaking the joint (to avoid being sprayed inadvertently)?
5	Ensure that all precautions are in place	Common precautions are generally defined on the PTW form but these must, as a minimum, be supplemented by the JSA. Again, there is a danger of a tick box approach. Often valves will need to be closed (mechanical isolation) and electrical equipment isolated (electrical isolation). Draining, purging and sweetening of equipment may be necessary and other safety precautions required. When working at height scaffolding must be installed and inspected before use. A site inspection must be conducted before work starts. This should include checks to ensure that isolations are correctly installed and there is no trapped pressure in piping and pressure vessels. Ideally both operations personnel and the "work performer" should conduct the inspection together.
6	Perform the work activity	If the proper precautions are in place, this will not generally create any issues unless some unexpected hazard is identified during the work activity. Where this happens, the task should be suspended in a safe condition and a new permit prepared.
7	Ensure that the jobsite is clean and tidy and left in a safe condition	After work is complete the jobsite should be left in a clean state and checked by operations personnel before any isolations are removed.
8	Remove any safety precautions that are in place	A site inspection to ensure completeness will be required.
9	Confirm that the work is complete, safety precautions are removed, and the plant is ready for operation.	At this stage the PTW should be signed off to show that the work is complete. A different procedure will be necessary if the work is suspended in an incomplete state.

A simplified cold work PTW is shown in Figure 16.3.

PLANT	**AREA**	**DATE**	**REF. NO.**
1. EQUIPMENT		TITLE PLANT/EQUIPMENT NUMBER	
2. TASK 2.1 EXACT DESCRIPTION OF TASK 2.2 LIMITATIONS OF SCOPE OF TASK 2.3 TOOLS PERMITTED			
3. PERIOD OF VALIDITY 3.1 (TIME-DATE) UNTIL (TIME DATE) 3.2 MAY BE RENEWED Y/N?			
4. HAZARD IDENTIFICATION (SPECIAL ENGINEERING FEATURES) 4.1 MECHANICAL REVIEW 4.2 PROCESS REVIEW 4.3 ELECTRICAL INSTRUMENT REVIEW 4.4 CHEMICAL HAZARDS REVIEW			
5. EQUIPMENT PREPARATION 5.1 MODE OF ISOLATION (INDICATE TAG NUMBERS) 5.2 DRAINING/PURGING 5.3 POTENTIAL HAZARDOUS FLUIDS IN THE EQUIPMENT			
6.1 TRAPPED PRESSURE 6.2 SERVICES 6.3 OTHER PROCESS EQUIPMENT IN THE AREA 6.4 OTHER SOURCES OF HAZARD 6.5 OTHER PERMITS IN THE AREA			
7. AREA PREPARATION 7.1 DRAINS 7.2 ROPING OFF 7.3 SCAFFOLD 7.4 OTHER ACCESS 7.5 OTHER			
8. PERSONAL PROTECTIVE EQUIPMENT TO BE USED 8.1 PERSONAL PROTECTION 8.2 SITE PROTECTION 8.3 GAS DETECTORS			
9. SUPERVISION 9.1 SITE TO BE INSPECTED EVERY ...HOURS BY....... 9.2 PERMANENT STAND BY PERSON REQUIRED?			
10. WHEN MUST WORK BE SUSPENDED?			
11. WHEN MUST THE SITE BE EVACUATED?			
12. ISOLATION CERTIFICATE		PROCESS: NO. INSPECTED BY ..... ELECTRICAL NO INSPECTED BY ..... NUCLEONIC: NO INSPECTED BY	
13. OTHER CERTIFICATE/PERMITS		13.1 HOT WORK 13.2 ENTRY PERMIT 13.3 CLEANLINESS CERTIFICATE 13.4 SCAFFOLD CERTIFICATE	
HAND OVER 1. I agree the site has been inspected and is in accordance with this certificate. 2. I agree that the site is safe so far as is reasonably practicable. 3. *I agree that the work will be carried out exactly as specified.* Issuing Authority (process) Performing Authority (engineering discipline) (Time .... Date .....) (Time ..... Date HAND BACKI 1. I agree that the site is safe so far as is reasonably practicable. 2. *I agree that the work has been carried out exactly as specified* Performing Authority (engineering discipline) Issuing Authority (process) (Time .... Date .....) (Time ..... Date...)			
LOCATION OF COPIES OF PTW Top copy to be held in the work site. Second copy to be held in the main control centre Third copy to be held on the local control centre Fourth copy to be held by issuing authority			

FIGURE 16.3 Simplified Cold Work Permit.

As previously stated, there is no perfect PTW system, but there are a number of points which will help to produce a system which is workable:

1	Permit types	As shown in Table 16.2, there a several permits in common use for different types of activities. Frequently activities cover work for which more than one type of permit is required. For example, welding in a confined space or at height. Where this is necessary, all permits and other documents must be cross-referenced to ensure they are all correctly signed off before and after the completion of work.
2	Isolation Requirements	Piping arrangements for isolation are covered in Section 14.5. Control of isolation requirements is covered in Section 16.8.
3.	Validity	PTW forms should have a limited validity; this varies from company to company but usually requires site conditions to be checked on a daily basis. PTWs for more hazardous work often have limited validity, often either 24 hours or a single shift. Permits for less hazardous activities may be valid for a longer period, often a week, but with revalidation every day or every shift. It is common for work to be started but not completed before the associated permit expires. "Open" permits must be registered and the site left in a safe condition. A new permit will be required for completion of work.
4	Signatures	All PTW forms will require, as a minimum, an operations and maintenance signature before the start of work and once work is complete. This is not a bureaucratic exercise but demonstrates the transfer of equipment "ownership" to maintenance after the first set of signatures and return of ownership to operations after the second set. The signatures demonstrate that approval for the work to take place and define who is "in charge" at any given time. The requirement for additional signatures varies greatly with individual PTW systems and the hazards involved. For example, some systems require a signature from the operations supervisor, plus an additional signature from a field operator to confirm that the isolation is in place and site is suitable for work to start. Permits for more hazardous work may require additional signatures, for example, to show that a gas test has been completed before hot work starts and repeated at intervals whilst the permit is valid. For specialist activities additional approvals may be required, for example, a civil engineer for excavation work or a radiological protection officer for work involving a radioactive source. Some permits require a safety officer to approve every permit, but this is not considered good practice—operations and maintenance staff will generally have a more intimate knowledge of the work activity. The involvement of the safety officer in all cases can lead to a less rigorous risk assessment, with the safety officer being the "fall guy" if something goes wrong. The *permit receiver* will often be a maintenance supervisor who may be remote from the actual activity. Those actually performing the work should sign the permit, but where a team is involved, every member of the team should have been briefed on what to do and how to do it safely as well as emergency actions.
5.	Risk Assessment	The assessment should involve as many people as possible who will be involved in the task—particularly for complicated activities. A copy of the assessment should be displayed with the PTW form.
6.	Training	The days of on-the-job (OTJ) training for PTW systems have long since vanished, and training for the control of maintenance will form part of any operator or maintenance technician's training. For staff who will sign to authorise work or accept a permit to start work, additional training is usually provided with staff certified as being competent. Certification should be reviewed at two- or three-yearly intervals. Contractors who are only occasional visitors to the site should have basic PTW training as part of their safety induction (see Section 21.6) and should be closely supervised by site staff.
7.	Number of Copies	It used to be standard practice for PTW forms to be available in pre-printed pads with two or three duplicates of each permit. One copy was to be kept in the control room and another (together with copies of the JSA, isolation arrangements, certificates and any other necessary documentation) to be available at the job site. A copy in the control room would allow control panel operators to identify work underway in the plant, and the site copy would be available for checking by operations and safety personnel to ensure that the work is being performed as required.

	The use of computerised PTW systems is growing, and there is a danger that the need to comply with the format of the computer input can distract from analysing the hazards of the maintenance task. Operators can usually check for active permits from a computer screen in the control room. But a hard copy of the PTW and supporting documentation is required at the job site. Each PTW must have a unique serial number to avoid confusion. Where two or more forms are required for a single activity, the serial numbers of each form must be recorded on the others. All PTW and isolation certificates (see Section 16.8) should be cross-referenced on the other forms.
8. Filing	PTW forms should be available for a period of at least two to three months to allow auditing of completed permits and for use as evidence in the event of an incident.
9. Communication	Whilst signatures on the PTW forms show approval for the activity to take place, they are equally important as a means of communication. Various methods are in use to ensure all staff are aware of the work being undertaken and the safety precautions required are in use. These include: 1 The provision of a copy of all documentation at the job site 2 A site plot plan in the control room showing the location and type of work underway through the site (this can be using either a marker board or screen based) 3 A *Tool Box Talk* (TBT) prior to the start of work to ensure all staff are aware of the nature of the work, how it is to be carried out and actions to be taken in an emergency This is not an "either/or" list. Many companies will use all the methods listed.
10. Auditing	To ensure that the PTW system is working correctly, regular audits should be undertaken. These should identify incorrect use of the system and allow corrective action to be taken. Further information on audits is given in Section 21.7. A checklist to assist with auditing is shown in Table 16.4.

Procedures, like steelwork can "corrode" over time and regular auditing of the PTW system is necessary to ensure that "short cuts" are not being taken. The UK HSE guidance includes a useful checklist. A simplified version is shown in Table 16.4.

TABLE 16.4
Checklist for Permit to Work Auditing

1	Is the scope of work clearly specified?
2	Are necessary risk assessments available for review?
3	Are identified hazards listed on the permit-to-work?
4	Are appropriate precautions listed on the permit-to-work (including clearly specified isolations)?
5	Is the operational time limit of the permit clear? (Are extensions properly authorised?)
6	Are certificates completed properly and appended to the permit?
7	Are other area or system activities cross-referenced correctly via the permit?
8	Are copies of permits, certificates and attachments legible?
9	Are signatures and initials traceable and legible?
10	Are copies of permits and certificates posted at correct locations?
11	Are attachments, drawings and so on held at the correct locations?
12	Are users briefed on the permit-to-work, and have they acknowledged understanding of requirements?
13	Do people know what to do in the event of emergency?
14	Are isolations appropriate for the task clearly specified on the permit or isolation certificate and correctly implemented?
15	Are common isolations cross-referenced?
16	Are the right people aware of isolated equipment?
17	Is the area authority aware of the work?
18	Is the work carried out in conformance with the permit?
19	Are control measures and personal protective equipment appropriate for the task?
20	Are tools and equipment suitable and in good condition?
21	Are housekeeping standards satisfactory?

Source: Guidance on permit-to-work systems (HSG 250), HSE, 2005

The potential complexity of work activities is shown by real life examples in Table 16.5.

The incidents mentioned earlier, typical of failures during maintenance, are tragic but, mercifully, involved only a few individuals. One of the worst oil industry accidents, Piper Alpha (offshore

TABLE 16.5
Maintenance Case Histories

Case 1	A vessel had become contaminated with a granular sludge of heavy oil and organic solids required cleaning.
Risks	Release of fumes, poor access (which could inhibit escape), possible ignition and flash fire.
Work Activities	The vessel had been fully blinded, A man-hole was opened and water put on top of the sludge. This was then heated by a live steam sparge pipe for three hours. Gas tests were carried out—oxygen 20.8% $^{v}/_{v}$, no flammables were detected. Under these conditions, it might be considered that a safe working environment existed but this would be an INCORRECT assumption as hydrocarbon fumes could be released as the sludge was disturbed. The cleaning was performed using phosphor bronze (notionally non-sparking) tools, Gas tests were repeated every five minutes to determine if flammable gases had been released during cleaning. Cleaners wore PVC overalls, gloves and non-sparking boots plus masks to protect against airborne dust generated during cleaning. A fire watcher stood outside of the manhole with fire extinguishers. This was an uncomfortable way of working but the cleaners were protected against dusts and a flash fire.
Case 2	A vessel containing partially polymerised Butadiene required to be cleaned and rubbery solids removed.
Risks	Release of fumes, poor access (which could inhibit escape), possible ignition and flash fire.
Work Activities	Gas tests were carried out—Oxygen 20.8% v/v, no flammables were detected. Cleaning proceeded smoothly until there was a small internal flash fire, probably due to steel/steel impact or friction. One person was seriously injured, and a second suffered fatal burns. The incident reports made no mention of a fire watcher who may have been able to respond to the incident.
Case 3	A crude storage tank had a deep deposit of sludge. The solids had to be removed to allow an inspection of the floor of the tank (a known corrosion issue).
Risks	Release of fumes, poor access (which could inhibit escape), possible ignition and flash fire.
Work Activities	Removal of the sludge was undertaken manually. The tank was slip plated and "steamed out". Gas tests were no flammables and 20.8% oxygen v/v. There was a warning sign to indicate that no sources of ignition were to be taken into the tank. Whilst a firewatcher was posted at the entrance, they do not appear to have been diligent in their duties. A flash fire occurred in the tank. One person died of burns and a second was seriously injured. It appeared that one of the teams took matches into the tank so he could smoke whilst working.
Case 4	A diesel pump on a side stream of a crude distillation column was to be taken out of service for inspection due to high bearing noise.
Risks	Release of vapours/fluids, risk of burns from ignition of hot gas oil.
Work Activities	A PTW was issued for the inspection and the bearing was found to be seriously damaged. The maintenance team who discovered the damage to the bearing considered the PTW issued for inspection would also cover repairs. The first action was to remove the pump to a workshop for maintenance. Removal of the pump resulted in a leak of hot diesel oil above the auto-ignition which ignited. The ESDVs did not close. The resulting fire resulted in fatalities. It is obvious that non-intrusive inspection activities are totally different from pump removal which requires electrical and mechanical isolations. A new JSA should have been conducted and a new PTW issued before pump removal started. An obvious but fatal error.

UK) was also caused by confusion over maintenance activities. Damage was in the US$ billions and 167 lives were loss.

On 6 July 1988, PTW forms for the day shift were issued. These included work on one of two condensate pumps. Two permits were issued for work on pump A—one for a pump overhaul and another for the removal of the pump's PSV for calibration.

During the day, pump A was electrically and mechanically isolated and the PSV removed. but containment was not broken. However, the PSV was removed and a blind installed—but only hand tightened. Calibration and reinstallation of the PSV could not be completed before the shift change and the blind flange remained in place. The PTW was reportedly updated to show that the work was not complete and pump A was not available for use. However, this information was not passed to the night shift during the shift handover.

Later in the shift, pump B stopped and could not be restarted, probably due to a hydrate blockage in the downstream pipework. A search through the PTW forms found a permit for the pump overhaul but not the PTW for the PSV calibration. As only isolation of pump A had taken place with no line breaks or dismantling of the pump the shift team removed the isolations and restarted the pump. They were unaware that the PSV, located some distance away and not easily visible from the pump had been removed and not replaced.

When pump A was restarted, hydrocarbons leaked from the loose flange. The noise was clearly audible and the gas alarms were activated and an explosion occurred shortly afterwards, followed by a series of events leading to destruction of the platform and the deaths of 165 men.

A number of other factors contributed to the disaster, including a change from automatic to manual operation of the fire pumps for the protection of divers. However, the initiating events were confusion over the status of equipment undergoing maintenance.[5]

Many of the worst accidents in the process industries are as a consequence of failures to plan, control or communicate maintenance activities and the importance of the Permit to Work system cannot be over emphasised.

16.7 CRITICAL TRIP BYPASS

Process (and hardwired trips) are installed around process plant to ensure a safe shut down if parameters move out of the operating envelope. However, it is sometimes necessary to bypass them—particularly for plant startup and for maintenance. Active trips can inhibit startup—a low level trip in a vessel, for example, will cause a shut down but after maintenance there will be no level in the vessel. The trip must therefore be bypassed until a liquid level has been established. It is often good practice to have an operator in close proximity to an ESD button during this activity.

Trips are often bypassed for testing, to ensure that they are not in a failed state (see Section 19.6), and to allow operation to continue if a trip is defective and cannot be easily repaired.

Previous generations of instrument technicians had a relatively casual attitude to bypassing critical shut down systems. When most trips were analogue based, it was common practice for maintenance personnel to have a "jumper" (a short length of wire with bare ends which was connected up and downstream of the shut down logic). There was limited control of bypasses which were meant to be in place for short periods of time but could sometimes be forgotten. Operations managers were sometimes complicit in trip bypassing in an attempt to prevent spurious plant trips and a belief that the trip could be quickly reconnected if necessary.

Current practice is much more disciplined and it is accepted practice to have a procedure in place to manage bypassing. Critical trip bypass procedures operate in a similar way to PTW systems.

Where a bypass is necessary, a risk assessment will be necessary followed by authorisation of the bypass.

As with PTW, there is "no one size fits all" solution but a number of criteria are necessary in the design of a critical trip bypass procedure:

1 A written procedure, including the use of approval forms, must be in place.
2 The bypass must be authorised at a senior level and escalated to even higher levels of management if it remains in place for an extended period—usually more than 48 to 72 hours.
3 The reason for the bypass must be clearly explained and a risk assessment to determine if mitigation methods need to be put in place (for example, more frequent field readings of instrumentation or the temporary change to alarm settings on control loops.
4 Any isolated instrumentation must be communicated to operations personnel. This is often achieved by the use of a critical trip bypass register available for inspection in the control room.
5 Once the trip is re-established the approval form must be signed back in the same way as PTW form.

The status of any isolation must be checked regularly (ideally weekly) and reviewed to determine if any action can be taken to speed it's return to service or if additional mitigations are needed. If production pressures require that a trip is bypassed for an extended period, it is common practice to review the bypass and mitigations under the MOC procedure (see Section 16.9).

Modern DCS sometimes includes built-in bypasses for startup with a timer which allows a trip to be isolated for a fixed period of time. Although not universal practice, it is considered desirable that even these timed bypasses are covered by the critical trip bypass procedure.

Many bypass procedures are paper based with the signature of the authorising manager being displayed. The written forms are supplemented by a *trip bypass register* which records all bypassed trips, when and why the bypass was made, when the trip will be re-established and when it was actually re-established. The critical trip bypass procedure can easily be automated but a key feature is to ensure that the length and necessity for bypassing can be frequently reviewed. On modern plants, bypasses trips and alarms are frequently recorded in the DCS historian.

Whilst the discussion earlier relates to process trips, they are equally relevant to the bypassing of other safety systems such as vibration and lube system monitoring major equipment and fire and gas detection systems (see Section 22.5).

16.8 ISOLATION PRACTICES

Fifty years ago, it would have been standard practice to simply close isolation valves around equipment which required maintenance and install blinds (or remove sections of pipe) for safety-critical activities such as vessel entry. Current practice is for isolation valves to be locked and tagged. The various piping isolation arrangements are discussed in Section 14.5.

Electrical equipment should always be positively isolated. Modern electrical switchgear can be locked but there is still a significant amount of older equipment in use where locks cannot be installed. Where this is the case, isolation is achieved by removing fuses in low-voltage switchgear and "racking out"[6] the contacts for higher-voltage equipment.

A procedure to control the isolation of equipment is required, generally in the form of mechanical and electrical isolation certificates. These form part of the PTW system and the PTW should not be approved until the isolation and isolation certificates are complete. These certificates should have unique serial numbers and be crossed-referenced with the relevant PTW forms.

Isolation certificates are relatively straightforward to complete and will require signatures to confirm that the isolation is made before any work activity is performed and when removed after work is complete but before the PTW is signed off. Locks for mechanical isolation will normally be installed by an operator and locks for electrical isolation by an electrician. It is common practice for additional locks to be installed by technicians working on equipment to ensure that the isolation cannot be accidently removed. Individual keys are either kept in a "lock box", which is itself locked with the key kept in the operations supervisor's office. An alternative arrangement is the *multi-lock* where a number of individual padlocks can be placed in a frame which slips over the valve handle or other item to be locked. The last lock to be removed should be the first one to be installed. In addition to locks, isolations should have a label (generally numbered) naming the person performing the isolation and the date. This information should be recorded on the isolation certificate.

Increasingly isolation procedures are being formalised into a lockout-tagout (LOTO) procedure.[7] This was first developed by the US Occupational Safety and Health Administration (OSHA) and formalised into an OSHA standard 1910.147.[8] The principles of LOTO have been expanded into lockout tagout tryout[9] (LOTOTO) to ensure that an attempt to start the equipment after the isolation has been made. Whilst this would be considered good practice, historically, it was not always done.

16.9 MANAGEMENT OF CHANGE

As soon as a plant is operational, sometimes before, there are likely to be requests for modifications (staff working for a former major UK chemical company, ICI, frequently joked that the initials actually stood for I've Changed It"). Any change must be fully assessed before implementation. The economics and practicality are obvious factors to review but the potential safety impact must also be considered. As with other procedures, communication and coordination are as important as the checks to ensure that the inherent safety built into the plant is not diminished. Many modifications have been started and not fully completed, leaving P&IDs, operating procedures, and maintenance practices out of date.

> The classic example of an uncontrolled change occurred at Flixborough in the UK on 1 June 1974. An explosion equivalent to between 15 and 45 tonnes of TNT occurred following a release of cyclohexane. A key part of the process was a train of six reactors in series connected by flexible metal bellows (to allow for thermal expansion). One reactor suffered cracking and the plant was shut down and the reactor removed. To allow operation to continue a spool piece was installed between the inlet and outlet flanges of the cracked reactor. The spool piece was supported on scaffolding, with no calculations or engineering drawings being produced. The modified system was not hydrotested and subsequently failed in service. The released cyclohexane resulted in a major VCE which killed 18 plant personnel and injured 36 employees. 35 offsite injuries also occurred. Two hundred houses were damaged[10].

A comprehensive MOC procedure is therefore a necessity. Originally MOC procedures covered only physical changes to the plant with any replacement that was not a *like-for-like-replacement* of a component being considered a change. More recently the scope of MOC has expanded to cover changes to operating conditions (instrument set point changes, increases in temperature, flowrate or other parameters) and organisational change, such as changes to organisational structure and job descriptions

The essential steps in an MOC are as follows:

1 Identify the need for a modification.
2 Develop a basis for the change (including cost and benefit).
3 Review the proposed change for practicality, operability and safety. This will often require some form of "filter" to determine if any of the proposed changes impact safety at the site. A typical example is given in Figure 16.3. Note that safety whilst the change is being implemented needs to be considered as well as safety following the change.

4 Approve, reject, or revise the change (if revised steps 2 and 3 must be repeated).
5 Install the change.
6 Provide appropriate training to all staff that will be affected.
7 Ensure that all documentation and procedures are updated, and out-dated documentation removed and destroyed (except an archive copy).

At an early stage in the modification the hazard potential must be assessed using some form of analysis or checklist. A simple, generic example is given in Figure 16.4; this will need modification for individual locations but demonstrates the basic principle. One feature of a well-designed checklist is that a hazard might be identified by more than one of the guidewords. For example, an increase in process temperature should be identified under process conditions and design pressure (engineering hardware and design). Figure 16.5 can then be used to perform a simple risk assessment. Again the format will need to be tailored for each location.

Engineering standards change overtime, so the date/revision number of the standard used should always be specified. When modifications to a plant, particularly an older plant, are made it is prudent to check for changes between the original and current version of the relevant standards to ensure there are no incompatibilities. Some companies have gone a stage further on older plant and ensured that all equipment—much of it original is compatible with current standards.

Plant	**Modification No.**	**Modification Title**	
Underline those factors which might be affected			
Process Conditions	**Process Safety**	**Engineering Hardware & Design**	**Occupational Safety**
Temperature	Trip Systems (ESD)	P&ID	Handrails
Pressure	alarm settings	Plant Layout	Means of escape
Flow	Pressure Relief	Design Pressure	Protection against Hot/Cold
	ESDVs	Design Temperature	PPE
Corrosion	ROIVs	Materials of Construction:	**Environment**
Level	Sources of Ignition	Control Systems	Open and Closed Drain Systems
Composition	Changes to ATEX	Foundations	Vent Systems
Reaction Conditions		Structural Supports	Liquid and Gas Discharges
Toxic / Pyrophoric Flammable Materials	Isolation	Access (including stairs and ladders)	Effluent Treatment
Safety Systems	**Work Activities**	Lightning/Earthing Protection	Noise
Fire and Gas	Operations Activities (including startup/shutdown)	Piping	Light
Fire Water	Maintenance Activities	Instrumentation	**Organisation**
Fixed protection	Inspection Activities	Radioactivity/Radiation	Reporting Routes
Mobile Fire fighting	Civil Work	Corrosion	Job descriptions
Passive Protection	Overhead work	Erosion	Competencies
	Temporary staging	Stress Corrosion	
	Temporary equipment	Jointing	
	Portable Equipment (Electrical safety)	Cabling	
		Seals	
		Ladders/Stairs/access	
		Loads on structures	

FIGURE 16.4 Minor Modification Safety Assessment.

Plant	Modification No.		Modification Title	
Does the proposal affect any category below?	Y/N	Action	Signed	Date
Relief and Blowdown				
1. Introduce or alter potential causes of over/under pressure (or raising or lowering temperature) in all or part of the system?				
2. Introduce the risk of a vacuum in all or part of the system?				
3. Negate in any way the protective systems installed to prevent over/under precure?				
Area Classification (ATEX)				
1. Introduce or alter the location of potential leaks of flammable materials (gaseous, dust or liquid)				
2. Affect the chemical composition or physical properties of the process material? Consider by-products.				
3. Introduce new or modified electrical equipment?				
Safety Equipment (PPE)				
1. Negate existing safety equipment				
2. Require additional safety equipment (escape sets or other)?				
3. Require stand-by man (rescue or fireman)?				
4. Are all means of escape or reuse fully satisfied and unimpaired?				
Structures				
1. Are new or temporary structures designed against a recognised code and fit for purpose"? See also checklist 1				
2. Are structures liable to physical overload?				
3. Are new structures liable to external corrosion?				
Metallurgy				
1. Does any change affect the risk of embrittlement, stress corrosion cracking, fatigue, corrosion or erosion?				
Operation and Design				
1. Alter or introduce new hardware?				
2. Require consideration of new codes and specifications				
3. Affect up or downstream equipment?				
4. Affect access for personnel and equipment as well as safe working and laydown?				
5. Are temporary/permanent structures fit for purpose?				
6. Require upgraded equipment inspection routines?				
7. Affect ESD/alarm systems or require additional trip/alarms?				
8. Affect process stability or controllability?				
9. Affect or require new maintenance or operating procedures?				
10. Affect means of, or control of, effluent disposal				
11. Alter noise profiles?				

Signed	Date	Production Mangar	Signed	Date	Assessor

FIGURE 16.5 Simple Management of Change Risk Assessment

16.10 WHAT CAN GO WRONG?

A multitude of incidents, some minor, some humorous and some more serious will occur over the operational life of a process plant. Most plants will not suffer a catastrophic incident during their life cycle; sadly a few will. Many incidents will be repeats of similar accident from the recent or distant past where lessons, once learnt have been forgotten. Some incidents can be confidently predicted such as a failure of the PTW system, a minor leak (or vibration) which was not reported for investigation or poor communication between operators. Some simple—but largely avoidable incidents are briefly listed in Table 16.6.

TABLE 16.6
Typical Operational Incidents

Incident	Cause after investigation
Leakage of hydrocarbons from a refinery additives tank.	Poor supervision of operators had led to bad practice in the opening and closing of valves. The operator should have opened the suction and discharge valves before starting a transfer pump and then closed the valves after transfer had finished. However, the operators had got into habit of leaving the valves open to make the operation simpler. On one occasion, the operator mistook the drain valve for the pump suction valve and opened it before moving onto the next tank, leading to a US$ 1 million loss of chemical additive. The drain and the pump valves were not easily confused, and the operator appears to have been paying insufficient attention. The operator was disciplined but, surprisingly, there was no criticism of the supervisor who should have kept a closer eye on operations. To prevent a reoccurrence, the refinery management installed an expensive instrument system which raised an alarm if the tank level dropped too quickly. Whilst better training and supervision were obviously required, a cheaper solution would be to lock drain valves in the closed position.
Repeated high vibration on a compressor	Repeated high vibration on a compressor occurred. The uppercasing had been removed on several occasions and the compressor rotated, but no problem was found. Eventually it was decided to lift the rotor out of the casing, and a nut was found at the bottom of the casing. This was not an isolated incident and it is common practice to restrict access around machinery when the cover is removed to prevent items from being dropped, unobserved, into the casing. Components from the rotor or stator of a machine must be carefully stored and checked to ensure none are lost.
A turbine "threw" a blade.	The process had been poorly analysed and during Startup the upstream knockout drum became flooded due to unusual dynamics which were not identified during design. Liquids entered the vapour line and entered a power recovery turbine casing which caused impingement damage to the blades and eventually led to fatigue failure at the root.  Damaged Turbine Blade showing grey marks on the non-active side, indicating impact by liquids
A turbine "threw" a blade.	In another case, an exhaust line drained *towards* the power recovery turbine. Eventually condensed water filled the casing and the turbine threw a "shoe". The designer did not recognise the need for the piping to slope away from the turbine. Once the vector analysis is carried out for gas and liquid, the damage becomes obvious.  Damaged turbine blade with a silver colouring on the rear (exhaust side) which indicates liquid impact during start up

(*Continued*)

TABLE 16.6 *(Continued)*
Typical Operational Incidents

Incident	Cause after investigation
A small explosion occurred in a compressor lubricating oil tank.	The oil return line had been changed from "drowned" to splash filling. The oil was contaminated with hydrocarbons due to a minor internal leak, resulting in an explosion caused by static ignition.
Low-pressure explosion in a seal oil returns tank after a number of abortive startup attempts.	The investigation showed that inerting gas system was only sized for diffusion and not repeated startups—when oil would be pumped out of the tank and the level reduced, causing air "breathing" into the tank.
Reverse flow in a slurry pump	A slurry pump needed to be removed for maintenance. The first step was the isolation of the suction and discharge valves and the disconnection of the pump from the motor. Although isolation valves on a slurry pump were closed when the coupling was disconnected the pump began to rotate. 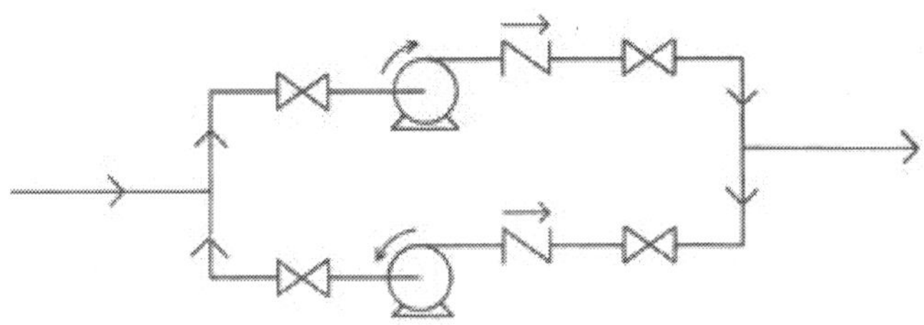 Reverse Rotation of Pump (reproduced courtesy of the Kletz Estate) The pump was on a slurry duty. The NRV leaked—a common problem—and the gate valves installed for isolation did not fully seal. Alternative valve types such as diaphragm, ball or plug valves would be preferable on a slurry duty.
Failure of PTW system	An instrument technician was asked to remove an actuator on a plug valve. The plug was released, and process fluids escaped. 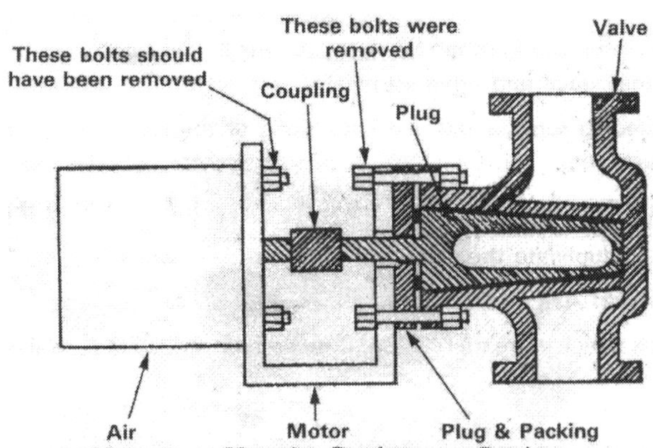 Valve Actuator (reproduced courtesy of the Kletz Estate) The technician did not realise that there were two sets of bolts—one was safe and the other dangerous. The PTW was imprecise, and a full risk assessment was not performed.

(*Continued*)

TABLE 16.6 *(Continued)*
Typical Operational Incidents

Incident	Cause after investigation
Damage to the vent of a steam PSV.	The piping designer had not considered jet reaction from the discharge—also note the bend in the tip of the pipe.  Effect of Reaction Forces on PSV Tailpipe (reproduced courtesy of the Kletz Estate)
Failure of a vessel due to vacuum.	A large vessel located 10 m above ground level with a design pressure of 4 bar g was used to contact two fluids in a continuous extraction process. The vessel operated full of liquid with the two fluids flowing counter current. During maintenance the fluids in the vessel were displaced from the bottom with clean water. The operator then opened the drain located at ground level to drain the water. Initially he was a little surprised that the flow was very low until he heard the vessel groan and crumple inwards. The operator had not been told how to drain the vessel and as every schoolchild knows a column 10 m high, when drained, will have a full vacuum at the top.
Blocked discharge on a pressure safety valve	A PSV had a vertical "tail pipe", to avoid any steam burns to bystanders should it lift. A 5 mm drain hole was drilled in the pipe so that condensate and rainwater would not build up in the tail pipe. If not drained the head of water in the tail pipe would impose a back pressure on the PSV, increasing the discharge pressure, and causing corrosion which could cause the valve to "stick". This is a common arrangement on PSVs discharging to atmosphere. One day there was a violent eruption of hot condensed steam from the tail pipe! 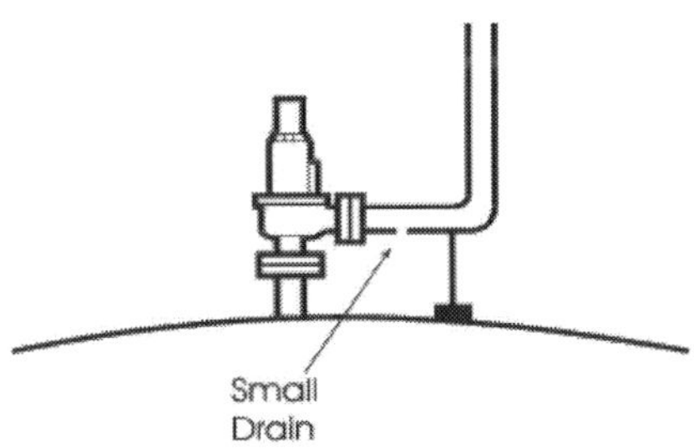 PSV Drain Line There had been no regular inspection of the system and, over time, the drain hole had become clogged with dust and debris. Problems of this type are common on tail pipes discharging to atmosphere; tail pipes and drain holes have been blocked with anything from mud to birds' nests and beehives. A further problem can occur where water constantly drips onto the metallic surface (causing corrosion), and frequently a short tail line is installed to drain water away. However, the longer the drain line, the greater the potential for blockage. Finally, note the "duck foot" installed close to the vertical section of the pipe in Figure 16.9 to counter jet reaction forces when the PSV operates.

Some of the events described earlier are the consequence of poor design and construction and, if identified and corrected during design, the problem would not have occurred. This reinforces the importance of getting it right first time. Plant operation comes with an almost continuous series of unexpected and sometimes unexplainable events. Whilst good procedures and training should reduce the number of events and improve the response the unexpected will often happen. A few years ago, the standard way of recording process upsets was to note them in the shift logbook. More recently, most companies have established an incident reporting system as well as the use of the logbook, to identify upsets and ensure they are investigated. During commissioning, start-up and operations lessons will be learned. These should be recorded so that the mistakes are avoided in future or so that problems can be addressed by improved design.

The price of safety is eternal vigilance, a questioning attitude and never taking things for granted.

REFERENCES

1 Irish Health and Safety Authority, *Behaviour Based Safety Guide*, ISA, www.hsa.ie/eng/publications_and_forms/publications/safety_and_health_management/behaviour_based_safety_guide.pdf
2 Cheung C N, Burch G & Bridges W G, Process Safety Culture—Making This a Reality, *Process Improvement Institute, 10th Global Congress on Process Safety*, March/April 2014, www.process-improvement-institute.com/_downloads/Process_Safety_Culture_-_Making_This_Real-website.pdf
3 Anon, Boy Died in Solvent Vat, *Independent*, 13 December 1994, www.independent.co.uk/news/uk/boy-died-in-solvent-vat-1389373.html
4 UK Health and Safety Executive, *Guidance on Permit-to-Work Systems: A Guide for the Petroleum, Chemical and Allied Industries (HSG 250)*, 2005, www.hse.gov.uk/pubns/books/hsg250.htm
5 Lord Cullen (chair), *The Public Enquiry into the Piper Alpha Disaster*, HMSO, November 1990, www.hse.gov.uk/offshore/piper-alpha-disaster-public-inquiry.htm
6 "Racking out" is the act of connecting, or disconnecting, electrical switchgear from bus bar
7 US Occupational Safety and Health Administration (OSHA), *Control of Hazardous Energy (Lockout/Tagout)*, OSHA, www.osha.gov/control-hazardous-energy
8 Legal Information Institute, *29 CFR § 1910.119—Process Safety Management of Highly Hazardous Chemicals*, www.law.cornell.edu/cfr/text/29/1910.119#:~:text=%C2%A7%201910.119%20Process%20safety%20management,toxic%2C%20fire%20or%20explosion%20hazards
9 Human Focus, *LOTOTO—Why You Need to Upgrade Your LOTO System to LOTOTO*, April 2023, https://humanfocus.co.uk/blog/what-is-lototo-why-its-essential/
10 Turney R, Flixborough: Lessons Which Are Still Relevant Today, *Loss Prevention Bulletin* (No. 237), IChemE, June 2014, www.icheme.org/media/2229/lpb237_p21.pdf

17 Hazard Study 7—Closure, Mothballing and Demolition

17.1 INTRODUCTION

At some point the plant will be surplus to requirements, the final stage in the project lifecycle, see Figure 17.1. This may be due to age, obsolescent, economic factors or some other reason. Some plants have operated, with modification and updating for much longer than their design life. But others have been shut down soon after commissioning due to technical or economic issues.

Shut downs may be temporary, for example due to market conditions or, unfortunately, due to political unrest—particularly in developing countries. The final plant shut down is a sad and traumatic event for the workforce. More than one plant operator has cried after the final shut down, and there may be understandable concerns regarding future employment.

For recently constructed plant, the design process should have considered shut down of the plant and its ultimate demolition. However, this will not be the situation for most process plants operating today.

Both temporary and final shut downs require detailed planning which may start years in advance. The legacy of plant operations can be significant and many older facilities will have been in operation before modern standards of record-keeping and environmental protection became common. Older nuclear power stations, which produced radioactive materials, require complex long-term demolition plans (with huge costs). The extent of contamination from materials such as asbestos may be unknown as will contamination within the process system from impurities, by-products and other sources. Information may be limited with drawings out of date, details of materials of construction unknown and the current condition of the plant uncertain. Ground contamination of soil may exist in older industries such as coal gas works and metal processing.

For a temporary shut down, the plant will need to be preserved (often called mothballing) and returned to use at a later date. Whatever the shut down, permanent or for temporary *mothballing* the plant must be put into a safe condition free from flammable or toxic materials and with no trapped pressure. Note that mothballing can apply to both operational plant and plant under construction. In the latter case, work is often stopped suddenly and the site shuts down—resulting in a greatly increased workload when/if restarted. Lack of preservation will result in corrosion which in some cases may be mild but complex electronic equipment, if unprotected, will need replacing. If the equipment becomes obsolete in the interim redesign may be necessary. Records also need preservation so that quality assurance records are preserved and inventories of material and spare parts are maintained. Unprotected structural steel may be so corroded after a number of years that the original plant cannot be returned to service, lifting beams and walkways may also be unsafe.

Lack of information may be a problem in older plants with drawings either being unavailable or out of date. The plant layout can easily be verified by site inspection but out of date P&IDs create many problems, particularly with regard to cleaning and decontamination.

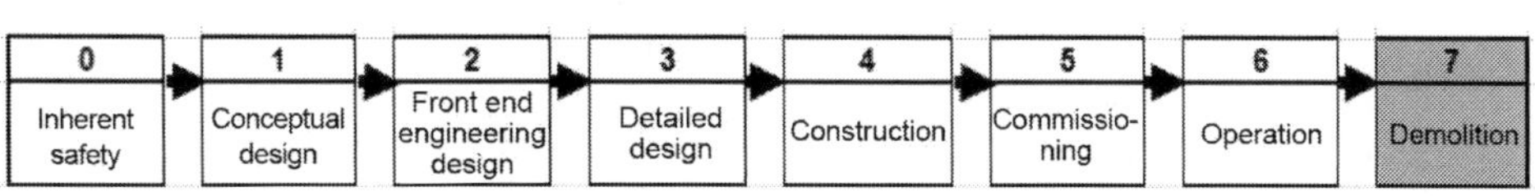

FIGURE 17.1 Project Life Cycle Closure to Demolition

DOI: 10.1201/9781003510680-20

Most guides to decommissioning, preservation and mothballing consider each activity under three phases:

1 Pre-project considerations
2 Project planning
3 Execution

A useful introduction to the subject is provided by Briggs, Buck and Smith[1]

17.2 SAFETY STUDIES

Demolition is possibly the most dangerous phase of the plant life cycle in terms of both human safety and environmental effects. For a variety of legal and financial reasons, demolition may not take place immediately after plant operations cease. It is not unusual for parts of a site to remain in operation whilst the remainder is shut down—a number of small refineries have long been closed but the associated tank farms remain in use.

The final shut down and preparation for demolition must recognise that the demolition team may have little knowledge of the possible hazards at the site Each case will be unique and in high hazard cases a safety case approved by the national regulator may be required.

For modern plant, final shut down of the plant and restitution of the site should have been considered at the appropriate project stages:

1 Conceptual Design — The final demolition of the plant is unlikely to affect the choice of process route but may affect fundamental areas of design. Some concepts which affect earlier stages of operation might also be relevant. For example, skid-mount units are often installed in remote locations as they can be fabricated at a vendors' yard and easily shipped to site for installation and hook-up. The same process in reverse will also assist dismantling.

2 Front End Engineering Design (FEED) — During this phase the design team can influence the decommissioning, demolition and site restitution. Some countries already have legislation requiring this—a practice that is likely to spread further. Relevant factors include:

1 Contamination of equipment (e.g. thermal insulation or concrete) and the disposal of such materials. Could an alternative material be safely disposed of more easily?
2 The potential for spillages to contaminate the ground through loss of localised containment such as bunding and spill trays. Is the containment adequate and fit for purpose?
3 The choice of welded or flanged connections. Flanged connections will be easier to dismantle but are more likely to leak.
4 The need for specialised equipment such as pump and compressor seals
5 Potential loss of containment or spillage during dismantling
6 Stagnant sections in piping and equipment which are difficult to drain or vent—these may also result in localised corrosion (see Chapter 20).
7 Possible future needs for enhancements/de-bottlenecking. Will these affect the initial design?
8 Special features only required at the end of plant life such as isolation valves, access for dismantling and so on.
9 Does the layout have good access for cranes and other vehicles, and can tall structures be safely removed without impacting surrounding areas?

Many of these aspects will be relevant to earlier stages in the plant life cycle, particularly during construction, but will probably have faded from memory by the time the plant is shut down.

3 Detailed design — Many of the details of the FEED will be carried over to the detailed design stage but there is a danger that the detailed design team will either attempt to make the FEED design more "efficient "(generally more economic) or "cut and paste" the FEED design without adequate review.

Two examples from the large tank fire at Buncefield in the UK in 2005[2] illustrate the necessity for detailed thought.

1 The design intent was to contain spillages within bunds. However, the possibility of spillage, followed by a fire was not considered. The sealant used in the annulus between the concrete bund walls and the piping penetrations into the bund was specified as an elastomer, which was damaged in the fire.
2 The overflow in the roof of the critical tank also served as a vent. The designer had not considered the impact of the overflow on the generation of a large vapour cloud—which eventually exploded.

It is difficult to detail every feature that may need attention. It is a case of considering the final phase of the plant life cycle and asking: "How can this be considered without detriment to the operational safety and environmental impact?". It may well be the case that considering the end of plant life will generate ideas which enhance the safety and reduce the environmental impact at earlier stages.

4 Construction and Commissioning

In principle, these phases should not impact the ultimate decommissioning and disposal of the plant. However, poor practices at this time can impact demolition and may be poorly recorded: For example:

1 The construction contractor may have disposed of materials on-site rather than remove them. In extreme cases, this could include hazardous materials or chemicals.
2 Pits, temporary piping or cabling may have been left at the end of construction.
3 Scaffolding may have been left in place in areas which are hard to access following the installation of piping or equipment.

> A long-established production site had a good inspection and preservation policy. All piping had a regular visual inspection and was painted. The plant decided to check the P&IDs were up to date and walk-throughs around the plant started. Confusion occurred in a short pipe trench between two roadways where a section of line, which disappeared into the tunnels under the roadways could not be identified on drawings.
>
> Eventually a confined space entry permit was issued to allow access into the cramped tunnels under the roadways where the "pipe" was discovered to be a scaffolding pole which could only be removed by cutting. The pole had been in place since construction many years before when the contractors presumably decided it would be too difficult to remove.
>
> Before the "pipe" was identified as a scaffolding pole, it had been regularly painted and inspected.

4 Operation

The importance of maintaining plant records up to date has been mentioned previously (see Section 1.7). Significant changes to design, for example the replacement of equipment items with lighter or heavier items may affect lifting plans. Changes to steelwork design during the operational period could also require significant changes to the demolition plan.

Spillages should be recorded but historically this may not always have been the case. Many materials may leach into the soil and if water soluble, may enter water courses and possibly drinking water supplies. Hydrocarbons are generally not water soluble and have a density lower than water. Often spills will disappear into the ground only to reappear when the water table rises after heavy rainfall. It is therefore important that the loss of containment records are as accurate and up to date as possible. Contaminated soil my require removal for specialised cleaning/decontamination.

5 Demolition

A risk assessment should form part of any demolition project and much of the rest of this chapter is devoted to safe systems of work during decommissioning and demolition.

17.3 DECOMMISSIONING

Regardless of whether the plant will be shut down permanently or for a temporarily, but extended, period detailed planning and preparation is necessary. Even if a decision to demolish has been taken, it may not be implemented immediately. Owners might be tempted to delay in the hope that prices for scrap materials improve or changes in legislation might reduce costs. In some cases, plants have been mothballed in the hope of selling to new owners—sometimes in a separate location which will require the plant to be carefully dismantled for re-erection at a new location, which maybe thousands of miles away.

If there are plans to restart the plant, a detailed preservation and maintenance programme needs to be developed (often known as *mothballing*) and implemented (see Section 17.4). If the plant will not be restarted but is not to be immediately demolished operational and maintenance activities will still be necessary but they will be relatively simple by comparison with operational plant. In both cases, staffing levels are likely to be lower than those necessary when the plant was operational. Security will be a major concern to ensure that intruders do not injure themselves and thefts do not take place.

On one plant a thief took a cask of mercury. He was seen by a guard who chased him. The thief held onto the flask and eventually became so exhausted from the weight in the cask that he had to stop and was caught.

The level of detail and analysis required will vary depending on the complexity of the process but there is no situation where the decommissioning, mothballing or demolition process should be considered so simple that risk assessment is unnecessary.

The decommissioning process can be considered in several stages.

17.3.1 Investigation

The condition of the plant at the time of shut down must be thoroughly understood. As stated previously, for modern plant, decommissioning may have been considered as part of design but for the majority of facilities this is unlikely to be the case.

A good level of documentation is desirable and on modern plant and well-managed facilities this should be available. Drawings should represent the current condition of the plant, with detailed information to identify areas of corrosion, hazardous materials (both process chemicals and those used in construction). After reviewing documentation detailed site investigation will be necessary to confirm the condition of equipment and the actions necessary for safe decommissioning.

On sites which have been partially shut down information may be available on the operational assets but limited date on assets which have been shut down for an extended period. In these situations, maintenance will be limited and the condition of corroded structural members may be unclear, the extent of decontamination unrecorded and extent of hazardous materials, such as asbestos, unknown. Small *hydro-skimming* oil refineries prove a good example of this with the process units being long abandoned but the tank farm remaining in use as a fuel terminal.

Even if a mothball/demolition decision has not been taken at the point of decommissioning the site must be left in a safe condition with procedures in place to minimise future safety and environmental issues. Temporary buildings may be necessary to store critical equipment/spare parts and temporary containment for liquid spillages may be necessary.

17.3.2 Planning

The planning process needs to consider the following stages:

1 Planning
2 Shut down
3 Removal of all hazardous fluids (these will mostly be process fluids but in some instances, utility fluids could be hazardous)
4 Cleaning, decontamination and sweetening

5 Service isolation (power, steam, nitrogen, instrument air and possibly other utilities).
6 Services that might be required during demolition, such as fire water and lighting.
7 Documentation for the mothballing/demolition team
8 Preparation for next use of the plant (either mothballing or demolition)
9 Duty of care for staff, visitors and trespassers

Each stage is considered later to allow the continuum and the interactions of the process to be followed. Each phase will require a risk assessment and may become part of a Safety Case preparation.

17.3.2.1 Planning

Planning the final shut down is potentially more complex than any other shut down. The shut down plan should be a detailed step-by-step description of the task, defining at each stage the objective and how it is to be achieved. The plan must be detailed and consider possible setbacks such as the possibility of recontamination of cleaned equipment, loss of utility systems or unavailability of equipment/personnel for the defined plan of action. The use of flow diagrams to illustrate the sequence of actions is useful. As the shut down plan is dependent on-site-specific factors and process chemistry detailed guidance cannot be given here.

The sequence of the shut down may well be different from normal. The means of storage and disposal of process fluids may also be different as they will not be returned to the plant. There may be differences in cleaning/decontamination procedures and sequencing by comparison with a normal maintenance turnaround.

17.3.2.2 Shut Down

The final shut down will be unlike any previously experienced! It will have a different objective from that of a routine "turnaround". Human emotions can play a part with management and staff at a decommissioned plant having low morale. Many experienced staff may have moved and operating personnel may be young, inexperienced and, crucially, have little knowledge of the plant's history. Other staff may be nearing retirement or expecting redundancy when the plant shuts down and therefore may be demotivated. The equipment may be suffering from a "terminal decline in maintenance" where it has reached the end of its life and maintenance is, at best, "make-do-and-mend" (or as an accountant would say, "sweating the assets"),

17.3.2.3 Removal of All Hazardous Fluids

During decommissioning process and hazardous or polluting utility streams will need to be removed from the piping systems and vessels (see Table 17.1).

The following must be considered:

1 Are materials safety data sheets MSDS available for all products, by-products and chemical used onsite (including chemicals used for the cleaning/decontamination process)?
2 How can fluids be collected?
3 What condition and what level of purity will the fluids have?
4 How may the fluids be treated or cleaned for removal from the site?
5 Where and how can they be stored?
6 Where can they be transported to for reuse or disposal?
7 Are there any transport regulations governing transport or storage?
8 How can the condition be verified (for example laboratory analysis)?
9 What records and recording systems need to be maintained?

TABLE 17.1
Materials Encountered during Decommissioning

Material	Comments
Catalysts[3]	Catalysts must be deactivated. In some cases, contamination—such as fine carbon particles—can make them subject to spontaneous combustion. Catalysts may have significant residual value but require special storage conditions. They may also require deactivation.
Molecular Sieve/ Desiccants	These may have some residual value.
Raw Materials and Products	Removal of the bulk of raw materials and products from storage vessels should have been considered as part of the decommissioning process. However, some residual material is likely to remain in piping systems, process vessels and smaller items of equipment. The small quantities and level of purity might make them hard to re-sell. However, they must be made safe or removed for disposal. Process piping and equipment must be free of contamination for resale or reuse. Some of the inventory may be acidic and the storage may be carbon steel. In situ cleaning may be necessary. A specialist contractor may be necessary for this work. If eventual resale of equipment is planned, a study of potential freezing damage in water systems should take place; this is particularly relevant where significant equipment damage could occur.
Chemical Intermediates	These will have the same problems as raw materials and products and there may be little demand apart from other plants using the same process. As with raw materials and products, intermediates must be removed so that piping and equipment is free from contamination.
Utilities	In most cases the utilities such as water, air and nitrogen will provide no problems for disposal but should still be risk assessed. Some more specialist utilities will require greater attention, for example hot oil systems. Some industries have specialist chemicals used in utility systems or use common chemicals in an unusual manner. Some companies handling toxic chemicals use air or nitrogen to transfer materials. "Toxic air" was used to transfer liquid chlorine from one pressure vessel to another in a chloralkali plant. A non-return valve failed on the toxic air line, resulting in backflow into a reservoir of compressed air, which was considered non-hazardous. If the PSV had operated or the vessel had been "blow down" to remove water chlorine would have been released.
Lube and Transformer Oil	This may need specialist disposal if the equipment is to be scrapped but for mothballing or resale is likely to remain in place. A particular problem may occur with PCBs. Whilst these are no longer used in new plant, some other facilities dating from the 1970s/80s may still contain PCBs in transformers or switchgear.
Chemical Additives	A wide variety of chemical additives may have been used for corrosion control, catalysts, chemical additives, water treatment and so on. Unused containers of chemicals may be transferred to operating sites but part-used containers may need specialist treatment and disposal. Some chemicals, particular catalysts, may require storage under special conditions such as refrigeration or in an inert atmosphere.

17.3.2.4 Cleaning, Decontamination and Sweetening

Removal of the process fluids and replacement by "safe" fluids may involve a rather tortuous sequence. For example, the process fluid might be removed by steam stripping. This in turn may require the displacement of steam by an inert gas, usually nitrogen (to avoid a vacuum), until the lack of residual process fluids can be verified. Then the nitrogen may then be displaced by air.

Whilst volatile flammable fluids are usually removed by steam stripping, heavier, higher boiling point, materials will still be left. These may not be flammable at room conditions but may be ignited by heat from flame cutting or other sources.

An oil tank at Dudgeons Wharf[4] London, UK, had been taken out of service and was being demolished. A small fire occurred, the fire brigade was called but the fire was quickly controlled by demolition workers. When firefighters arrived at the site they decided to cool the tank by feeding water into the tank from the roof. The firefighters had no idea what the tank had contained, they asked for hole to be cut through an inspection plate to allow access. Whilst this was in progress an explosion occurred killing a demolition worker plus five fighters and injuring a further five. The explosion is thought to have been caused by a spark from the cutting operation igniting flammable vapours in the tank.

The lack of knowledge of the tank contents was a factor in the establishment of the HAZCHEM code (see Section 3.7).

In extreme cases it may be necessary to fill the equipment with inert foam (such as nitrogen foam) before flame cutting. Under some conditions it may be possible to carry out a slow controlled burn of materials to reduce oxygen concentration.

Toxic materials are often water-soluble and can be washed out by circulation of hot water. Water washing of vapour lines needs to be considered with care as they may not be designed for hydraulic loads. This is an example where consideration during design could assist decommissioning. A safe disposal route for contaminated water and contaminated steam condensate must be established. Acid and alkali materials can be removed by circulating hot water with standard safety precautions.

Spontaneously flammable materials can be reduced under inert conditions but in some cases, if the products of combustion are non-toxic, they can be allowed to burn in a controlled manner.

The cleaning method chosen and the quality of cleaning should consider the ultimate means of size reduction and disposal. The cleaning method and any test results should be recorded on a marked-up P & ID which is individually colour coded showing the various steps. For example, the process may be steamed out, followed by a nitrogen sweep to avoid a vacuum. The cleaning/decontamination process may require a number of marked-up and signed-off P& IDs:

1 Working level at shut down
2 First level of cleaning
3 Second level of cleaning
4 Possible residual materials and hazards

As with many other aspects of engineering, there is no-one-size-fits-all solutions. The methods outlined earlier will be applicable, perhaps with some modifications, in many instances but there will some cases where very specialised techniques will be necessary.

17.3.2.5 Utility Isolation

Frequently, utilities are required during the decommissioning process and may be the last systems to be shut down and isolated. In other cases, particularly for hazardous utility systems shut down at the start of the decommissioning process will be desirable. Further information is given in Table 17.2.

Positive isolation by blinding or removal of a piping spool should be made and not reliance on an isolation valve. It is also good practice to install warning signs on live equipment—this is particularly important for electrical equipment.

Whilst use may be made of plant utilities, it may be more economic and safer to use small portable units to provide utilities. In some cases, services not required during operation may be needed such as diesel fuel for vehicles.

TABLE 17.2
Decommissioning of Utility Systems

Electrical Power	Once production has ceased there is unlikely to be a requirement for high-voltage electrical supplies, but low-voltage supplies are likely to be necessary for lighting and power during decommissioning. Low-voltage power could be supplied from the sites existing distribution system, or alternatively portable generators could be used. The use of existing supplies will involve less disruption but may require high-voltage power to be maintained and require long sections of cable to be "live" in areas which are otherwise shut down. The use of generators will create noise and require fuel storage. Generators might also involve the use of temporary cables which could be vulnerable to damage. The choice should be determined on the basis of risk assessment. Once decommissioning is complete, it may be desirable to completely isolate the supply if the plant is to be demolished. However, some systems may need to be operational or mothballed. For example: anti-condensation heaters in large motors, periodic running of electrical drives and lighting.
Steam	The steam system may be needed for cleaning of pipework and equipment, either during decommissioning or at a later date if localised residues are found. In cold locations steam heating might be necessary to prevent freeze damage to equipment. Again, a risk assessment is necessary and the risk of scolding on hot surfaces a should be minimised. If steam will be required at any point prior to demolition, the steam system will need to be maintained in an operable condition with all statutory inspections kept up to date. Condensate should be drained from the system. Depending on potential steam demand, a temporary boiler might be a cheaper alternative to using the large, fixed boilers required for production.
Compressed Air	Where compressed air is supplied by in plant systems the system will generally be shut down. Instrument and service air are generally isolated at the plant battery limit on multi plant sites.
Inert Gas	Inert gas is likely to be required during cleanout/purging of piping and pressure vessels and may be required for pipework and vessel preservation if the plant is to be mothballed. If the inert gas system is to be maintained after shut down, the potential for asphyxiation should be considered. Alternatively gas cylinders or a liquefied supply could be used and tied into the existing piping system.
Water	The plant may have a number of different water systems, some of which may be required during or after decommissioning. Raw Water Raw water abstracted from rivers or wells may be required for cleaning and washdown. In some cases, treatment may be necessary, for example for cleaning stainless materials where a high chloride content in the water could result in stress corrosion cracking (SCC). Cooling water With the plant shut down, cooling water will not be needed for process cooling but may be required for services, particularly steam and power generation. Where not required, it may be prudent to isolated and drain cooling water systems to prevent freezing. Water treatment chemicals can be hazardous and their security must be considered. Potable water Potable water will be required until work in the plant has ceased and may later be required as part of amenities for a recommissioning or demolition team. Fire Water Fire water systems should remain operable until the plant is demolished. Detailed planning and risk assessment will reduce the possibility of fire but the risk cannot be eliminated. Some hazards will exist for all types of water and should be considered: 1 In cold temperatures water filled systems may freeze and crack piping 2 Cooling water on pump bearings may freeze, destroying the equipment and reducing the resale value to zero. 3 Stagnant pockets of water could result in enhanced corrosion.

(*Continued*)

TABLE 17.2 ***(Continued)***
Decommissioning of Utility Systems

4 Open stagnant reservoirs of water could become breeding grounds for insects. The possibility of mosquitoes and malaria is a real concern in some parts of the world.
5 Legionella may be present in stagnant water storage areas. This is most likely to be a concern in potable water systems but cannot be discounted elsewhere.
6 Large reservoirs of water (for cooling or fire water use) could represent a drowning hazard.

17.3.2.6 Documentation for the Mothballing/Demolition Team

Recommissioning or demolition might take place months or years after decommissioning. A good level of documentation is therefore desirable but not always available and site inspection to verify conditions is necessary.

Ideally the mothballing/demolition team will know exactly how the plant was shut down, decontaminated/sweetened and exactly what materials were left in the plant (inert or otherwise) plus what might be left in stagnant pipework or adhering to walls of piping or equipment. This is not always the case, documentation should include the following:

1 The shut down procedures with each step highlighted and initialled on a P&ID to show completion.
2 P&IDs with the fluids used at each phase of the shut down, the analyses made and the likely materials that could still exist in the equipment.
3 The toxic or harmful properties of all fluids handled, including process materials, by-products, corrosion inhibitors, chemical additives water treatment chemicals and others.
4 Materials of construction of all equipment. In many processes there will be a wide range of materials with different scrap values. This is particularly the case with low- and high-temperature steels. Rebar in re-enforced concrete also has a scrap value.
5 Potential ground contamination.
6 Location of underground piping and services
7 Any external contamination such as thermal insulation soaked with process fluids
8 Any possible hazardous materials (especially asbestos).

Documentation is covered in Section 1.7. However, there may be reluctance to produce the highest level of documentation for economic reasons.

There might be limited continuity between staff from the operational, shut down and mothballing/demolition stages and in some cases plant ownership might have changed with different companies having very different philosophies.

17.3.2.7 Preparation for Next Use (Either Mothballing or Demolition)

As far as possible all piping connections should be positively isolated, although this may not be possible on some utility systems. Power may still be required, and live electrical equipment should be clearly marked.

Fire detection and protection systems should be maintained in an operable condition.

The temptation to use a disused site for dumping or storage must be prevented—this could result in combustible materials, creating a fire risk, or undocumented chemicals accumulating.

17.3.2.8 Duty of Care for Staff, Visitors and Trespassers

The number of staff and visitors to the site will be limited once decommissioning has taken place. Entry into the site (and for a decommissioned unit within a larger operational complex) should be controlled and site should be surrounded by a fence. Access doors should be locked but emergency exits maintained in an operational condition with exit routes unobstructed.

Even when shut down, the asset owner is still responsibility for the site's safety and security. The safety of visitors to the site needs to be assured as does that of trespassers. In the UK, the site owner will be responsible if a person gains access and then falls through a corroded structure or stairway.

Only if the Plant is surrounded by existing plant may it be possible to leave the plant unattended.

On one large petrochemical complex, one unit was shut down for 25 years before demolition. After this time all the plant personnel had retired or relocated and information on conditions at the plant was limited. Corrosion protection of the plant had lapsed and its condition was uncertain. After such a long period of time with no maintenance, lifting lugs, stairs and walkways were probably severely corroded and dangerous to use. Alternative means of access and lifting was be necessary.

Although the plant may have been left in a safe condition deterioration can occur, particularly if the plant is left for an extended time, steelwork may degrade, metal stair and walkways may have corroded, as could lifting beams. Thefts could also occur if the site is not adequately secured.

The best solution is usually the prompt demolition of the plant.

17.4 PRESERVATION (MOTHBALLING)

Plants which are shut down with the expectation of recommissioning at a future stage will need preservation and maintenance. This will require detailed planning and regular checks to ensure degradation is not occurring. Records should be kept to ensure that the correct checks are performed at the correct intervals on all equipment. There may also be cases where some equipment is to be sold but the rest of the site demolished.

Some delicate equipment may be removed for storage in an environmentally controlled (temperature and humidity) area.

Partially used stocks of materials may be maintained on-site (catalyst, chemical additives and so on) and these will need to be stored in the same way as on an operating plant.

General information on preservation is given later but it would be prudent to check the specific requirements with equipment vendors or preservation specialists.

17.4.1 Piping and Static Equipment

Piping systems and pressure vessels will have been drained and will require to be filled with a dry inert material to prevent internal corrosion. The following steps would commonly be performed:

1. Carry out an internal inspection for debris and corrosion
2. Remove debris where possible
3. Wire brush or sand blast corroded areas and apply a coating of rust inhibiting oil preservative
4. Remove debris
5. Consider the benefits of pressure testing. It will be necessary to carry out a full pressure test as part of the re-commissioning process and if equipment fails a pressure test, this will limit future options. Note the need to use demineralised water with stainless steels.
6. Drain all fluids.
7. Air blow where possible to remove traces of water.

8 If possible, dry the internals indirectly with heated air (combustion products could be corrosive.)

For preservation, systems should be pressured above atmospheric pressure with the pressure monitored to ensure leakage is detected. External examination of thermally insulated material piping and equipment will be necessary to ensure there is no *Corrosion Under Insulation* (CUI; see Section 20.5). Nuts and bolts may be coated with grease to prevent corrosion as well as to facilitating unbolting.

17.4.2 Rotating Equipment

Whilst much of the electrical equipment may be shut down, power will often still be required for ancillary systems such as lighting and anti-condensation heaters in motors. It is common practice to uncouple motors from their associated pump/compressor and run them on a scheduled basis. Rotating equipment may require a change from lubrication to anti-rust oil to prevent corrosion and the internals of reciprocating units will require to be treated with a preserving anti-rust oil. All units must be rotated on a routine basis.

Oil-lubricated systems (journal bearings and gear boxes) should be circulated for a few minutes every few months to ensure a fully "wetted" system. Oil seals should be circulated at the same time. The casing drains should be "blown" using an inert gas at the same time to remove any seal oil leakage. Exposed shafts are often wrapped with protective tape and checked periodically. Seals should be periodically sprayed with dewatering fluids.

Long-shaft centrifugal/axial compressors will be inside some form of protective structure. Where possible, these should be rotated half a turn whilst oil is being circulated. Chalk marks are frequently used as a reference line.

There are numerous types of centrifuge and, as with other equipment, manufacturers' recommendations should be followed. However, the following general rules apply.

1 Remove any filter cloths.
2 Clean, dry with air and store rotating elements in a clean dry environment.
3 Clean all other elements and dry with a jet of air.
4 Treat lubricated features, shafts, coupling and bearings as a motor or pump.
5 Blank all inlets and exists with blanks or tight wrapped heavy polythene covers.
6 Separate any touching parts and store separately.

Fans should be sprayed internally with de-watering fluids and then blanked off at the inlet and exit.

Before restarting rotating equipment, seals and bearings will require inspection and/or replacement. Exposed equipment is sometimes covered with tarpaulins, but care is necessary to avoid condensation under the coverings. The use of silica gel or a similar desiccant may be necessary to ensure electrical and instrument systems remain dry.

17.4.3 Reciprocating Equipment

Cylinders and pistons should be cleaned and cylinders, pistons and valve systems sprayed with oil. Piston rods should be coated with de-watering fluids. These precautions should be repeated periodically.

17.4.4 Cooling Towers

Cooling towers are a special case. Wooden packing can self-ignite from biological action if not correctly treated with biocides. Traditionally these are regularly "wetted" on a routine basis. However, there can be problems during cold weather when freezing may occur. Towers with plastic internals should require no special treatment apart from draining.

17.5 DEMOLITION

In most countries there will be specific regulations covering demolition—either as a specific activity or incorporated within wider safety and environmental regulation. In addition, discussion with local authorities will be necessary on various issues, as discussed later.

Much of the planning for demolition may have been completed before decommissioning but should be reviewed—particularly if the plant was decommissioned before a decision to demolish had been made. Sometimes facts may not be verifiable until decommissioning is complete. Decisions will need to be made prior to the start of decommissioning and these will affect the overall cost of demolition and should ensure that safe working practices are followed.

There must be a detailed plan for the whole demolition process as well as for each piece of equipment. This will detail the order in which equipment is removed. Logically, the process will start at the outside and work inwards (the reverse of construction), but the site may be littered with pieces of equipment that require to be sorted and removed. The sequence may also be complicated by the requirement to remove a particular piece of equipment for reuse/resale. Process equipment will have several piping connections, and these must be severed before the equipment can be removed. It is not acceptable to find that there is a "last minute" connection to break—a further reason why it is important that P&IDs are up to date and thoroughly checked. Should this occur, it must be treated as a potential incident and investigated.

Specific considerations include the following:

1 The order in which items, small or large, are to be removed or dismantled.
  Large pieces of equipment, like compressors, are often delivered early in a project. These may be the last items to be removed.
3 The removal of the wire ropes and the impact of this must be considered to prevent "whip back" when demolishing vent and flare stacks.
4 Wind speed will need to be considered when demolishing tall structures such as columns, vents and flares.
5 Structural steel may be some of the last pieces of equipment to be removed. However, it might be appropriate to remove some sections earlier to facilitate removal of other equipment. If this is done, a check must be made to ensure that the overall integrity of the structure is not affected.

> A poorly planned demolition activity involved a demolition contractor cutting steelwork with a gas cutting equipment. His work was not planned, and he cut away all piping and steelwork within reach. This included the access way to his place of work. He was marooned and had to be rescued: (witnessed by one of the authors – but at a safe distance).

As with construction, there are specific requirements during demolition which need to be considered and are, in some countries mandatory. For example:

1 Vehicle movements should be minimised and avoid roads through operating areas or where there is likely to be significant vehicular or pedestrian traffic. A dedicated site entrance might be desirable.
2 Impact with piping or services crossing or close to the site needs to be controlled. Examples include buried or overhead piping and cabling.

3 Temporary facilities could impact on surrounding areas (noise from air compressors/generators, dust, spillage from temporary fuel facilities, light from welding or illumination from night time work). Demolishing walls and the like can result in significant vibration which can sometimes impact on delicate equipment.
4 Safe storage of vehicles and equipment outside working hours.
5 Storage and separation of materials awaiting transport out of the site.
6 All crane activities should be accompanied by an approved lifting plan.
7 An ERP should be developed for incidents that could occur during demolition work and resources allocated.

17.6 SIZE REDUCTION

As with many other subjects within this book, there is only space for an introduction to a complex topic. Only the broad principals are covered and each case must be treated individually.

Whilst some parts of the plant may be easy to dismantle and remove, most components will require some form of size reduction to reduce the number of loads of material to be removed from the site. The most common means of size reduction for metallic materials are as follows:

1 Unbolting
2 Flame cutting
3 Saw cutting
4 Cutting by hydraulic croppers
5 Concrete breaking by hydraulic jackhammer
6 Explosives

Each is discussed in the following table:

1 Unbolting	The safest means of size reduction will probably be by unbolting flanges. However, in many cases piping will be fully welded and flanges only fitted where necessary for equipment removal/inspection. Where this option is not available methods such as flame cutting and grinding will be needed. Note that equipment that is to be recycled may be at height and a safe means of lowering to the ground will be necessary. On an ill-disciplined site piping or equipment is left to fall with the risk of injury to those below. As a minimum the area should be roped off and "danger overhead work" notices displayed.
2 Flame Cutting	The most common means of size reduction is flame cutting. This carries several potential hazards. For new plants these should be analysed during design but are less likely to have been considered on older units. Common problems include: 1 Equipment may have been fabricated in sections on the ground and then lifted into place during construction. Size reduction may require the equipment to be lifted out of its final location and placed onto the ground. The following must be considered: • How can the equipment be accessed? • Have structural steel and walkways suffered severe corrosion after being exposed to wet weather for an extended period? Attention to their integrity is important to avoid injury due to structural/walkway corrosion. • Is the equipment in a state that it can be accessed and handled by cranes? • Can the original lifting lugs for installation be used or are they damaged? Do they require NDT? • Is the equipment fully decontaminated or is further decontamination required before size reduction? • Could the use of a flame vaporise trapped toxic of flammable materials?

2. Piping may be supported for welding during construction but supports may not be adequate for size reduction.

- What will be the effect if a section of piping is cut and falls?
- Where and how will it land?
- Will it constitute a hazard?
- Is the piping fully decontaminated and is further decontamination required before size reduction?
- Could the use of a flame vaporise trapped toxic of flammable materials?

3 Saw Cutting	This is the second safest means of size reduction, but it is slow. The comments above still apply.
4 Hydraulic Cropping	This process can be treated as saw cutting and has all of the potential problems associated with unsupported equipment. There is a size limitation for its use.
5 Explosive Cutting	Explosive cutting using shaped charges can be a very effective means of size reduction and has been used to demolish equipment/structures damaged by fire where other means of demolition could be hazardous as the strength of the damaged structures is unknown. It is a tool which requires specialised skills but may be a means of reducing structures into more readily handled sizes or for toppling structures so that they can be accessed more readily. The removal of structures by explosive cutting will probably involve "selective weakening" of structural elements and should be the subject of detailed load/stress analysis. The safe/secure transport and storage of explosives also needs to be considered.

Factors to consider in size reduction of other materials are discussed in the following table:

1 Concrete	Concrete, including reinforced concrete) can be "chipped" to size by hydraulic jackhammers. Where appropriate the steel can be cut with a flame cutter. Consideration of contamination that may be in the concrete and how it is handled will be necessary. As stated earlier, steel rebar can be recycled, and concrete reduced and used as aggregate in new construction projects.
2 Masonry	Buildings can be demolished using a jack hammer or, for a small structure by pushing with a backhoe digger. Very large structures such as chimneys or hyperbolic cooling towers may be demolished using explosives.
3 Timber	The amount of timber in a process plant is likely to be limited but there are still cased where wooden walls and floors in buildings may exist, as well as roof trusses. Previous practice has been to burn timber but apart from the fire hazard, timber may have become contaminated and require special treatment.

17.7 RECYCLING OR REUSE OF EQUIPMENT

The potential for equipment to be reused will need to be assessed. This may be known by the asset owner who should have a full listing of the materials of construction. However, the equipment may have deteriorated since decommissioning. Changes, possibly unrecorded, to the original equipment specification may also have occurred.

Equipment may be in good condition but unusable due to obsolescence, this is likely to be particularly true of instrumentation. This may also be true of older equipment constructed to imperial units but in some cases, there may be scope for resale for use in other old plant. Where reuse is considered full inspection of static equipment will be necessary whilst substantial refurbishment of rotating equipment, for example to replace seals and bearings, will often be necessary. Pumps and centrifuges may also be resold but larger vessels such as heat exchangers and distillation columns may be too duty specific and thus be less attractive for resale. Other items for resale may include electrical motors and isolation or control valves. The seller has a duty of care to ensure that equipment is fully decontaminated. This will almost certainly require "local retreatment" to achieve the required cleanliness. Following retreatment, a certificate of cleanliness will be required for the sale. The past maintenance history and any modifications should form part of the sales documentation as are all appropriate drawings and performance data.

Equipment which cannot be reused can often be recycled. Metals can be sold as scrap. The asset owner should have detailed knowledge of material composition from equipment data sheets or other sources. For alloy materials, quick and accurate identification of materials is possible with an alloy analyser. Care is necessary to ensure scrap material is not contaminated, for example mercury and radioactive scale might be present in process piping from the upstream oil and gas industry.

Regardless of reuse or recycling, the following need to be verified and documented:

1. What is an acceptable condition?
2. How can the condition be verified?
3. How can the composition be verified?
4. How is the material labelled for segregation?
5. By whom?
6. How can the materials be segregated into compatible storage areas?

The most common examples of recyclable materials are:

1. Structural steel—which generally has a low scrap value.
2. Stainless steels—probably the most valuable material. There are many grades and analysis will be necessary. Note that in some cases the incorrect grade material may have been installed at the design stage. The value of the chrome and nickel will be variable depending on market conditions.
3. Alloy steels (often used in steam mains) once again, determination of exact grade will need to be determined.
4. Copper, aluminium and other materials. Note that some heat exchangers use cooling water with a high copper content which could confuse material identification.

Materials other than metals may also have a scrap value. For example, "engineering" bricks in older sites may be re-useable.

Equipment to be sold as scrap must, as a minimum, be decontaminated for transport. Due consideration should be given to the possible contaminants that might affect its reuse in steel recycling. This in turn may require that the equipment be further reduced in size not only to achieve decontamination but also to segregate different materials. At the demolition site, equipment should be segregated into stacks of similar materials and labelled clearly. A final inspection before removal from the site is prudent.

There is a good case for having a metallurgist on-site to identify materials (such as grades of stainless steels, copper and aluminium). All recycled materials should have a certificate of cleanliness before removal from the site. This should list:

1. Methods used to clean the item.
2. The results of any analysis carried out on that item.
3. Risks from any residual materials that might be present.

17.8 SAFE DISPOSAL OF NON-RECYCLABLE/REUSABLE MATERIALS

There are likely to significant amounts of materials which are not recyclable and require identification and disposal by specialist contractors. Whilst the site owner will make an initial assessment, it is likely that contractors, as part of the bidding process, will conduct their own site surveys. Some examples are obvious such as asbestos but others may be less obvious, for example chlorofluorocarbons (CFCs) which are now banned under the *Montreal Protocol*[5] but are still sometimes found in air conditioning, refrigeration and gaseous firefighting systems. Process chemicals and intermediates recovered when decontaminating the plant will generally require disposal if they cannot be reprocessed in a similar process unit. Contaminated materials, which cannot be cleaned will also need disposal, for example oil-soaked insulation. Disposal will require detailed assessment and communication between the site owner, demolition contractor and a specialist disposal contractor to ensure safe working practices are in place at all stages. A clear audit trail must also be established.

Typical points to consider include the following:

1. Are MSDS available for all products, by-products and chemical used in any onsite cleaning/decontamination process?
2. What is the condition of the material for disposal?
3. How can the condition be verified?
4. What cleaning/decontamination is necessary?
5. How is cleaning/decontamination to take place?
6. What will be the material condition after cleaning/decontamination?
7. How can the condition be verified?
8. Where is the Registered Disposal Site?
9. How is it to be labelled?
10. By whom?
11. How is material/segregated at each stage of the process?
12. What documentation is required?

Material should be recycled where at all possible. Alternative forms of disposal can incur significant cost. Targets of over 95% recycling can be achieved. The duty of care still lies with the plant owner to ensure that the materials are fully decontaminated and that the disposal site is properly managed. To this end the operator may wish to examine means of clean-up which avoids dumping and allows recycling of some or all of the materials.

17.9 PROCESS CHEMICALS

Process chemicals (and others such as lubricants) should have been removed at the decommissioning stage and any residues documented. However, it is always possible that chemicals will be discovered during decommissioning and a procedure should be in place to identify, safely store and remove any chemicals discovered.

17.10 SITE CLEAN-UP OR RESTITUTION

Environmental awareness has increased massively over the last few decades but the sins of previous owners are often still apparent during demolition. Contaminated or off-spec products might be buried and spillages inadequately cleared—most probably near loading and unloading facilities. Secondary containment may be limited with tank dikes not having membranes installed and limited secondary containment around pumps, hose points and lube/transformer oil systems.

Site restitution is, potentially, one of the most expensive costs after construction and as a result cleared sites are often only used for brown field development. Contamination may not just be from liquids. Metallurgical sites can often be contaminated with heavy metals. Clean-up may require a number of expensive techniques. These include the following:

1. Sinking boreholes and pumping out any leachate for disposal. (This can take years.)
2. Subsurface oxidation of organics by air injection. (This too is relatively slow.)
3. Biological treatment of soils. (This is contaminant specific.)
4. Stripping soils and off-site cleaning. (This is also expensive.)
5. Removal and replacement of soils.

17.11 WHAT CAN GO WRONG?

The potential for minor injuries should be clear. Broken concrete and steelwork cut into sections for size reduction will have sharp edges and PPE will be necessary (see Section 3.5).

Whilst there are numerous reports of casualties and fatalities in demolition work, few relate specifically to process facilities. Many accidents may be similar to construction incidents where cranes, scaffolding and vehicles are involved.

REFERENCES

1 Briggs M, Buck S & Smith M, *Decommissioning, Mothballing and Revamping*, IChemE, 1997

2 UK Health and Safety Executive, *Buncefield Response Reports and Recommendations Arising from the Competent Authority's Response to the Buncefield Incident*, HSE, www.hse.gov.uk/comah/buncefield/index.htm

3 Conservation of Clean Air and Water in Europe (CONCAWE), *Catalyst Handling Procedures to Minimize Exposures |(report No. 95/57)*, CONCAWE, June 1995, www.concawe.eu/wp-content/uploads/2017/01/rpt_95-57-2004-01710-01-e.pdf

4 Anon, *Looking Back at the Dudgeon's Wharf Disaster*, 2019, www.london-fire.gov.uk/news/2016-news/looking-back-at-the-dudgeon-s-wharf-disaster/

5 United Nations Environmental Programme (UNEP), *About Montreal Protocol*, www.unep.org/ozonaction/who-we-are/about-montreal-protocol

Part D

Aspects of Plant Operation

18 Operating Practices

18.1 INTRODUCTION

Almost any actions by operational staff can have safety implications, and indeed, a number of publications are dedicated to this subject. The key safety studies and safety-related procedures are discussed in Chapter 16. This chapter provides a broader overview of the operational activities.

18.2 SITE MANAGEMENT

Overall site management is outlined in Section 16.4. Operations personnel can be broadly split into categories, as shown later. There will often be a few personnel with "dotted line" relationships; these would include trainers, trainees and personnel with particular and possibly time-limited roles such as technical writer updating operating procedures.

18.2.1 Managers

Managers will normally have an academic qualification appropriate to the process. It may be chemistry, chemical engineering or biochemical engineering. The person will have undergone training and assessment plus an annual appraisal as part of their career progression. They will also participate in continuing professional development (CPD). Regardless of background a "new manager" will not be fully competent and should be "shadowed" and mentored by a competent individual until proved "competent". Safety cases often include details of the training and experience of managers.

Young graduates with little, or no, industrial experience often start life under the supervision of a plant manager or supervisor to "learn the ropes".

18.2.2 Shift Supervisors

Shift Supervisors will rise through the ranks and/or have an intermediate-level qualification supplemented with considerable operating experience. Their competence will be assessed by their manager by interview and/or observation.

18.2.3 Panel Operators

Panel operators (operators based in the control room who operate the control panels) are often experienced operators who have had additional training covering the control system.

18.2.4 Operators (Often Called Field Operators)

There has been a tendency for "on-site training" using the trickle-down process. "A" tells "B" what he/she should do. This can lead to poor understanding of operations and the propagation of poor operating practices. A formal training programme which teaches good practice is a better alternative. Once a foundation of competence is established "on the job" experience and specialised training by a supervisor may be acceptable.

18.3 SELECTION AND TRAINING OF OPERATORS

Many process plants operate continuously with shift working of staff being required. Out of normal working hours operations will be managed by a shift manager/supervisor with one or more control

DOI: 10.1201/9781003510680-22

panel operators and field operators who patrol the plant. In the event of serious incident, senior managers will be "on call".

Most companies would ideally like access to a pool of trained operators with a few years' experience on similar plants who can start work immediately. Whilst this may happen occasionally it is unusual. Where trained operators are available, they often live some distance away, possibly in another country and will still need to become familiar with the layout of plant and the procedures of their new employer. Reliance on older operators has other disadvantages, their working life at the plant may be limited before retirement and differences in technology may cause problems. Some operators, for example, who have worked with analogue control systems for many years may find it difficult to use a screen-based DCS. Operator turnover can cause a problem and it is prudent to maintain a pool of additional operators who can be deployed if there is a sudden spate of resignations—there has been more than one occasion when an entire shift team have left to work together at a new location.

To keep a good age distribution of operational staff most companies would regularly recruit trainees to ensure an adequate pool of new and experienced operators. Generally new recruits will have a technical education but will also need good interpersonal skills and a good verbal and written communication ability. Often more recruits than are likely to be required are recruited to allow for trainees who "fail to make the grade" or leave for new roles in other companies.

Initial training is generally classroom or screen based followed by a period of *on-the-job* (OTJ) training as a field operator working with an experienced operator. The trainee's performance is constantly monitored and they should eventually be deemed competent to undertake the role unsupervised. Most plants, have several different operational roles and after the trainee gains competence on one role they will start to learn the remaining roles in the plant. This adds versatility to the shift team and helps in the overall understanding of "how the plant works". Typically, a six-month period is necessary to learn the first operational role, and in most plants an operator would expect to be competent in all roles in approximately two years. After gaining experience in field operations, and, if a suitable vacancy exists, they can move up to become a control panel operator then shift supervisor (team leader in modern parlance).

Most companies now operate a *skills matrix* system which monitors the individuals training and experience to determine their suitability for promotion. The matrix will cover the technical skills necessary to perform their role and also include "soft" skills such as communication and initiative. The use of a skills matrix allows succession planning to replace operators who retire or leave for any reason.

Whilst operators will quickly become adept at day-to-day activities, they may gradually lose knowledge of activities which take place infrequently such as startup, shut down and emergency situations. Refresher training is therefore required.

A specific case for regular and repeated training is the case of ESD. A good practice, followed in many plants, is a weekly emergency simulation based on a likely scenario such as utility failure, equipment failure (such as seal failure) gas leak, fire or explosion. Field operators will place tags to indicate the valves they would open or close in this event and panel operators would list the actions they would take if the incident was "for real". This minimises the chance of operator error in the stress and panic of a real emergency.

18.4 STANDARD OPERATING PROCEDURES

SOPs can be considered as procedures covering general activities such as PTW, and MOC (sometimes called *standing orders*) or specific operational procedures for the operation of process plant and equipment items. The most critical SOPs have already been considered in Chapter 16.

Well-written procedures can make plant operations considerably simpler for all staff and some general rules are given in the following list:

1 Each document should not be more than around 20 pages long. If this is not possible consider splitting the procedure into two or more parts. For example, procedures for PTW might consist of an overall procedure referring to separate procedures for hot, cold and

vessel entry procedures, plus additional SOPs for isolation and specific activities such as excavation and working at height.

2 Each page of each document should have a header and/or footer giving the document reference number, revision number or letter and date. The front page should include a block showing who approved the procedure plus a signature and date. Often multiple signatures are needed, most frequently from the author, the reviewer and the approver.
3 For clarity, the document should be written in sections and subsection with a contents list at the front. Paragraphs should be short and individual sentences should ideally be limited to 10 or 12 words.
4 Diagrams should be used in place of long descriptive text.
5 Documents should list actions in a logical stepwise manor. Often the steps include a checklist which can be printed off and used on-site with each item ticked off as it is completed. This will also provide a record of the activity.

A widely repeated, but unverified (and probably untrue) story, tells of a section in a bomb disposal manual with the following instruction:

1 Cut the green wire
2 Cut the blue wire
3 Warning do not cut either the green or blue wires until the red wire has been cut

Not a very logical way to proceed!

6 Information that could frequency change, such as staff members and contact number should be in either an appendix or a separate document referenced in the SOP. This will minimise the number of minor revisions needed.

Procedures should be regularly reviewed and updated where necessary.

There are numerous guides to the writing of procedures and other technical documents. Cooper[1] provides one of the best, if dated, guides to technical writing. Document control is covered in Section 1.7.

18.5 OPERATING INSTRUCTIONS

Detailed operating instructions should follow the guidelines described in Section 18.4 and should have an adequate level of detail but be as brief and precise as possible. Where equipment items are mentioned, tag numbers must be given. Diagrams and flow charts often provide greater clarity than text.

Operating instructions need to cover startup and shut down situations as well as steady state operations. They should include a description of the system with valves and equipment clearly labelled. It is useful to install checklists as part of the operating instructions which can be reproduced and used to check steps are not omitted during startup/shut down. A simple example is given in Appendix B.

As well as the instructions for startup, shut down and steady state operation, operating instructions include additional information including warnings of the hazards that can occur in operation and how to manage them. The detail required in the instruction should not be underestimated. Their accuracy and practicality should be checked with the shift teams. If properly applied safe working practices are also efficient working practices.

18.6 STARTUP/SHUT DOWN

Contractors operating instructions may be basic and the plant owner will frequently produce new instruction which are more detailed and follow the company's standard format. Operating instructions should be reviewed at intervals as experience of operation increases and modifications take place.

Plant startup will, for many plants, be an unusual feature and will need careful planning. Startup of individual process equipment items and subsystems must be performed in sequence, generally starting with utility systems. Whilst not required for plant operations, fire and gas detection, fire water (and other emergency systems) must be fully operational before startup begins. Procedures often include instructions on the rate at which the startup proceeds. For example, fired heaters are generally heated slowly to avoid stress and fatigue in tubes. Similarly, the rate of cooling of cryogenic equipment needs to be controlled. When restarting the plant after a process trip additional caution may be necessary. For example, depressurising of a cryogenic system can result in cold temperatures in process systems—over rapid heating could result in thermal stress.

On 25 September 1998, an explosion occurred at Longford, Australia following the release of hydrocarbon vapour. Prior to the incident, one of the three interconnected gas plants had been taken offline following a major upset (thought to have been a high liquid level in one of process vessels). The shut down resulted in some equipment reaching low temperatures due to the JT effect. An attempt was made to restart the plant and this included feeding high-temperature oil into a heat exchanger which had been chilled to –48°C which then suffered brittle failure.

Approximately 10 tonnes of hydrocarbon vapours were released and drifted towards fired heaters located approximately 170 m away where they ignited, resulting in a deflagration which "burnt back" to the leak source causing a significant jet fire at the point of rupture. Two people were killed by the fire. The jet fire impinged a number of hydrocarbon-carrying pipes which subsequently ruptured leading to the release of more hydrocarbons which increased the intensity of the fire.

Operational issues were identified as the underlying cause of the event, described as "deficiencies in their initial and subsequent training and the absence of current operating procedures that could guide them in the problems they faced"[2].

Startup is often considered as a number of steps, each being completed before the next takes place. Sometimes the plant may be held at a step for an extended period before continuing the next step. All steps in the startup sequence must be fully documented for later investigation if a problem occurs. Modern plant with DCS control will generally have a digital *historian* which will record key parameters, but manual records are just as important to record leaks, unexpected vibration and so on.

Shut down can be either a controlled shut down or an ESD. ESDs may be manually activated but are often initiated by instrument systems and require little manual intervention. Fully automated instrumented systems are becoming more common as plant automation increases but need rigorous assessment at the design stage or if there is a modification. This is covered in Section 13.9. In some plants, particularly older plant, ESD will involve a greater level of manual effort with field operators opening/closing valves within the plant and panel operators making sure that no critical parameters are exceeded. Whilst the shut down instrumentation may react quickly it may take some time for the plant to reach a safe situation. For example, a compressor may take up to an hour to slowdown and stop rotating after being tripped due to the amount of kinetic energy present. During this time a supply of lube oil (and for older machines seal oil) must be supplied to prevent damage. This is generally achieved by an emergency lube oil pump powered from the emergency power distribution system and/or an elevated tank which feeds into the lube oil system by gravity.

After any trip a thorough investigation must take place before the plant is restarted. Often this will identify the cause of the problem and allow a solution to be determined. Generally, a more detailed investigation will then take place to determine and correct the *root cause*.

Routine shut down will be less stressful to operators than an ESD and will often proceed in stages. If the shut down is to be followed by maintenance a series of activities to allow safe maintenance will be necessary, such as draining of process fluids, and purging to remove hazardous vapours.

18.7 ROUTINE OPERATIONS

Safety of an operating facility is the responsibility of everyone not just operations and maintenance personnel but also management, contractors and visitors to the site. Anything that does not appear "quite right" should be investigated (it probably is "not right") and, if necessary corrected. Senior managers should visit the plant regularly. Not only does this help to keep the workforce on their toes, but it also demonstrates that mangers are interested in the problems that occur in the workplace. Staff at all levels should practice the time-honoured technique of "Look, Listen, Feel!"

Look uses the eyes for something that does not look "right".
Listen to people and to strange process noises.
Feel involves both safely touching equipment (bearings "running hot" and vibration) but also feeling the concerns of site personnel.

The focus of plant operation is to produce adequate quantities of product within specification at the lowest cost. Much operational activity is routine (and boring) ranging from regular sampling of raw materials, intermediates, and products, recording instrument readings and completing production and maintenance records. This mundane work is important and the temptation to skimp must be avoided. Discipline and oversight are required to make sure "short cuts" are not taken and records are complete, correct and not forged, particularly during inclement weather. Even if they do not affect safety, short cuts can result in off-spec product being produced with the consequence of poor customer relationships and, in extreme cases loss of sales contracts. A useful guide to routine operations and troubleshooting plant upsets is provided by Leiberman.[3]

Panel operators will monitor the key parameters and make adjustments to maximise production, minimise waste and effluent generation as well as minimising utility use. Field operators will make regular patrols around the plant and take field recordings, generally every two to four hours. These provide a useful check against control room parameters but also help ensure that field operators regularly visit the site. Traditionally readings have been recorded on paper log sheets but some companies are turning to computer tablets which read a bar code for the instrument with the instrument reading being input manually and then transmitted directly to a server. In both cases, there is a danger than the field operator will write down a reading without considering it's significance.

Frequent sampling of raw materials products and effluent streams is undertaken. In some cases, continuous online analysers are installed but there are still many cases where manual sampling takes place. A problem with any sample is ensuring that it is representative of the vessel or piping contents. Ideally fixed sampling systems should be installed with a "quill" located away from the pipe wall, where the sample is more likely to be representative of the pipe contents (see Section 14.5). This is known as *isokinetic sampling*. Previous practice in some plants has been to release the contents of piping systems to atmosphere until the sample is considered representative and a "sample bomb" then filled. This process creates a discharge to atmosphere which can be hazardous—especially for pressurised hydrocarbon gases.

Discharge of LPG during routine sampling at a refinery LPG storage area resulted in the gas detectors located over 100 m away being activated several times per day. As a consequence, whenever the detection system alarmed, it was assumed that sampling was taking place and no check was made to ensure a more significant gas escape had occurred.

18.8 SHIFT HANDOVER

Good communication is a key factor in safe operation. Operational staff generally work either 8- or 12- hour shifts and, therefore, individuals operating the plant will change two or three times per day. It is essential that the incoming shift team understand the status of the plant, including any defects, when they take over from the outgoing shift team. Standard practice is for the incoming and outgoing teams to have a 15–30-minute meeting before the end of each shift. This will discuss planned throughput for the next shift plus any factors likely to affect safety or production. The meeting should include all operators. Ideally, incoming and outgoing field operators should walk around the plant together with the outgoing operator pointing out any areas of concern.

Most plants will have at least one shift logbook and it is not uncommon for there to be several log books for different plant sections or even for each individual operator. Both incoming and outgoing operators should, after reviewing events during the shift, sign the logbook as confirmation that an adequate briefing has taken place.

Many logbooks are tabular in format to record key parameters such as production rates, raw material and utility usage but there must be space available for comments, such as details of equipment leaks, equipment "running hot" or other defects. There is increasing use of electronic logbooks which can be accessed remotely by senior staff. Whilst this may be advantageous, electronic logs must have the capability of free text entry and safeguards to ensure that entries are not subsequently altered. The possibility that the site network is unavailable will always exist and it is desirable that a paper copy of the log is always available.

REFERENCES

1 Cooper B M, *Writing Technical Reports*, Pelican, 1969
2 Weiss B, Lessons from Longford—20 Years On, *Loss Prevention Bulletin (No. 262)*, IChemE, August 2016, www.icheme.org/media/1973/lpb262_pg04.pdf
3 Lieberman N P, *Troubleshooting Process Operations Second Addition*, Pennwall Books, 1985

19 Maintenance Practices

19.1 INTRODUCTION

Maintenance of equipment has always been required. Roman soldiers would need to sharpen their swords and the axles on horse-drawn wagons would need greasing. The Industrial Revolution increased the amount and complexity of maintenance work required. Originally this would be carried out on an ad hoc basis but now has every increasing levels of sophistication—just like the equipment requiring maintenance.

Good maintenance requires the correct tools, skills, supervision and replacement parts. Whilst improvisation might work for a short period of time, it is not a long-term solution. Maintenance activities should follow the cycle:

1 What can go wrong/what did go wrong?
2 What is its implication?
3 What can be done to correct it?
4 Was it corrected properly?

This coincides with the concept of continuous improvement.

Maintenance activities are one of the major sources of hazards as they frequently involve interacting with the process and equipment. Frequent maintenance increases the potential for hazards and all maintenance activities must be performed using a PTW system (see Section 16.6).

Maintenance procedures must be clearly defined, defining how the repairs should be carried out and what makes or types of parts should be replaced. Each type of spare part will have a unique reference which should prevent incorrect components being used. This does not always work as the reference numbers can be long and the series of letters and numbers may not have a clear meaning to personnel.

For each piece of equipment records must be kept. The reason for maintenance intervention must be recorded together with details of the equipment condition before and after maintenance plus details of the work carried out and any information related to the future condition of the equipment. Maintenance activities, and frequency, are often defined by the equipment supplier and are frequently modified in the light of experience (ideally with input from the manufacturer). Routine maintenance should not be carried out too frequently or too infrequently as this will lead to *wear-in* (infant mortality) or *wear-out* (old age) failures (see Section 6.3). Repair frequency may have to be adjusted for each piece of equipment to optimise the running life without entering "wear out".

Performance of maintenance functions are frequently monitored using *Key Performanc Indicators* (KPIs). These vary with different industries, but common features are the ratio of co tive or breakdown maintenance to planned/predictive maintenance (see Section 19.5), the ar overdue maintenance. Performance against budget is also an important parameter.

19.2 ORGANISATION

The maintenance function in most organisations follows a similar pattern covering:

1 Planning
2 Mechanical maintenance

DOI: 10.1201/9781003510680-23

3 Electrical maintenance
4 Instrument maintenance
5 Civil/structural

Often additional specialists are employed in the maintenance function or the inspection function (see Section 20.2). These could include a metallurgist or a condition monitoring specialist.

Some activities may be contracted out to third parties. For example, external specialists are often employed for DCS and fire detection system maintenance. Simple maintenance activities such as air conditioning system and vehicle maintenance might also be performed by a third-party organisation as might the control of vegetation.

Some specialist activities will report either directly or indirectly to the maintenance function. Typically, these include spare parts procurement and the warehouse. Other functions, such as inspection (see Chapter 20) and reliability engineering, sometimes form part of the maintenance function but more often form part of the separate engineering/technical services team.

The maintenance function will generally be composed of the site owner's staff but there is a trend towards contracting all maintenance activities to a third party as part of an *operation and maintenance* (O&M) contract (see Section 16.4).

19.3 PLANNING AND RECORD-KEEPING

The key component of maintenance is the *work order*. This will define the work to be undertaken and should either specify how the work is to be done or refer to a procedure or manufacturer's maintenance instructions (see Figure 19.1 for an example of what can go wrong if procedures are not properly followed). Method statements are designed to be followed and are not suggestions of how work should be done.

Following maintenance, staff decided to replace the standard studs and nuts on a valve with long studs which stretched across both flanges of the valve due to differential thermal expansion.

The long studs expanded more than the short ones. The tension at the bottom of the valve was therefore lost and leakage occurred, as shown in Figure 19.1.

19.3.1 Case History

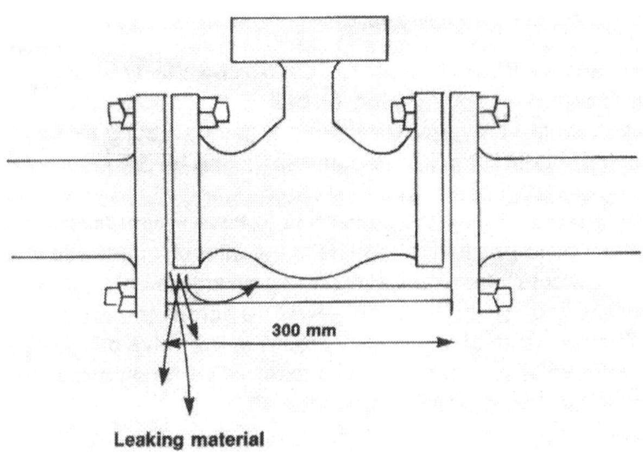

'RE 19.1 Short Flanged Studs Replaced with a Long Stud Allowing Leakage (reproduced courtesy of ·tz Estate)

In simple terms, planned maintenance can be split into three broad categories:

1. Routine maintenance
2. Opportunity maintenance (work which cannot be performed whilst the plant is operation but can take place during short or unexpected shut downs. This often involves overtime or round the clock working.
3. Shut down (often called turnaround) maintenance. This involves a total shut down of the process unit(s), decontamination and entry into equipment. Shut down maintenance would typically take place a year after plant commissioning and then at intervals ranging between two to four years. As experience of the plant increases the periods between shut downs increase. The frequency is usually defined by either the need for a catalyst changes or statuary inspection of pressure vessels. Shut downs are usually planned for the summer months when, in many parts of the world, the days are longest and the weather warmer and dryer. In hot climates, a shut down at a cooler time of the year would be preferable.

Emergency maintenance is likely to be needed from time to time, but good maintenance practices should minimise the number of occasions when equipment failure results in a shut down.

Whilst a few sites may still use paper-based planning, most organisations have some form of computer-based system. These vary significantly from spreadsheets to specialist computer-based systems. The latter are becoming increasingly sophisticated and will link with spare parts inventory modules to ensure that the necessary components are available before work starts. The system will also record any issues with each task as well as reorder components when minimum stock levels are reached.

Most planning systems can also manipulate data to show performance, for example proportion of maintenance activities completed, ratios of different maintenance types and performance against budget. Accurate records will also allow changes to schedules, identification of poorly performing equipment (frequently known as "bad actors") and changes to required spare part inventory levels.

Records should be kept of all maintenance activity, either paper-based or electronic. Details of wear and corrosion should be recorded for static equipment (see Section 20.3) and full details should be kept for rotating/moving equipment (dates of work, condition, work undertaken and spare parts used). There may also be comments or recommendations for future maintenance interventions. Data may also be used for trend analysis or as a tool to find better ways of performing work as well as for "reliability analysis".

19.4 WORKSHOPS

Most sites will have one or more workshops for activities which cannot easily be performed on-site. Usually, the mechanical workshop will be larger than areas for electrical and instrument activities and contain traditional workshop equipment such as lathes and milling and shaping machines. Welding is also conducted in a suitably equipped area of the workshop but gas cylinders should be located outdoors.

Many industries will have specific hygiene requirements for workshops, particularly where equipment operates in a high-purity environment such as the food or pharmaceutical industry. Even more exacting standards may be required in semiconductor and other high-tech industries.

Workshop hygiene and housekeeping is important and regular audits (by non-workshop personnel) should be conducted to ensure that there is no build-up of debris, walkways (and emergency exits) are not blocked and equipment is maintained and operated in a safe manner.

19.5 MAINTENANCE TECHNIQUES

Gone are the days when equipment was vastly overdesigned and so robustly constructed that it would, with limited maintenance, last almost indefinitely—some beam engines, for example, have been in operation for well over a hundred years. However, changing economics, increased efficiency and increasing complexity make regular maintenance essential for reliable operation. Beam engines were slow moving with limited friction of moving parts and over-designed. This does not apply to sophisticated modern equipment which may operate at high temperature, pressure or speed where tolerances are much "tighter".

Maintenance philosophy is developed at the plant design stage and maintenance work will start at or before commissioning. Some maintenance activities are defined by regulation or company procedures; others will be based on equipment vendors' recommendation, company procedures or experience. Often, maintenance practices change and evolve over the life of a plant. Maintenance activities can be divided into the following main categories:

1 Planned	Activities performed at regular intervals, based on either a fixed number of days or number of operating hours. Frequency will be based on manufacturer's recommendations and operating experience. There is increasing use of statistical based methods to determine MTBF to define intervals between maintenance activities. Note that in some cases, particularly gas turbines, maintenance is based on *equivalent operating hours* (EOHs) rather than actual operating hours. EOH takes account of the number of starts, stops and other factors as well as the number of hours in operation.
2 Predictive	Monitoring of equipment to determine its status and the need for maintenance intervention.
3 Corrective	Corrective maintenance can be simply defined as all work that is not planned or predictive, Frequently, planned/predictive work will identify either noncritical repairs that are necessary or the need for repair work to prevent a breakdown. These should form the bulk of corrective maintenance.
4 Breakdown	Breakdown maintenance is self-explanatory. Increasingly breakdowns are considered a failure of the maintenance philosophy. The use of planned and predictive measures should minimise breakdowns.

The terms "corrective" and "breakdown" maintenance are sometimes used interchangeably. Confusing corrective and breakdown activities can seriously under, or over, estimate the efficiency of the maintenance function. As previously noted, the ratio of the different types of maintenance is commonly used as a KPI of the maintenance function.

Predictive and preventive maintenance techniques are increasingly being used with very little equipment being run to failure. One company when asked for examples of breakdown maintenance could only think of one example: light bulbs.

Simple but effective techniques have been superseded by evermore more complex and comprehensive methods. However, sight, sound, and feel of equipment can still be useful indicators to a field operator that a problem is developing. Maintenance techniques are discussed in Table 19.1 but new methods are constantly being developed and older ones refined.

TABLE 19.1
Common Maintenance Techniques

Vibration Monitoring	Vibration uses both the absolute value of a parameter and/or a full frequency analysis of the signal as a diagnostic tool. The parameters are usually displacement, velocity or acceleration. The analysis of the spectrum can be steady state or dynamic (when a machine is started or shut down). Large machines have fixed vibration systems both for monitoring machine health and to warn (and shut down the machine) if vibration levels are out of tolerance. Some manufacturers, particularly turbine manufacturers, now provide remote monitoring of machines. Whilst this allows the manufactures special knowledge to be available in real time, there may be dangers of a cyberattack if suitable precautions are not taken. Smaller machines are monitored using portable equipment with frequency dependent on equipment criticality and health. Modern vibration monitoring equipment has the capability of down loading data to a computer for detailed analysis. Mills[1] has provided a detailed review of vibration monitoring.

(Continued)

TABLE 19.1 ***(Continued)***
Common Maintenance Techniques

Magnetic Plug	Oil flows across a magnetic plug and ferromagnetic particles can be removed and analysed. This provides a first level of wear detection and looks for small debris in an easily accessed form.
Ferrography	The use of graduated magnetic flux[2] to separate metallic debris in oils. Course particles are separated first and smaller ones last. Course particles are caused by break-up and finer ones by attrition of course ones. The ratio is a measure of damage. It is a more sophisticated version of the magnetic plug.
Spectrum Analysis of Oil Pollutants (SOAP)	This involves a detailed analysis of "debris" in oils, and from this an area of wear can be identified. The spectrum analysis of the particles will direct the engineer to the area of wear or damage.
Electrical Signature Analysis (ESA)	Current[3] and voltage data are monitored and analysed to identified distortions in the sine wave which indicate potential issues. The technique is non-intrusive and can be applied as a one time health check or on a regular basis to monitor equipment health.
Thermography	A technique[4] that uses a thermographic camera to identify hot spots on equipment. It is used in a range of applications, including checking for hot spots on electrical connections and damage to insulation and refractory linings.
Dielectric Testing	To determine the strength of insulation in oils in electrical equipment.[5]
Dissolved Gas Analysis (DGA)	To determine the concentration and type of gases in transformer oil. The technique indirectly provides information on the quality of the oil and possible electrical failures which may be caused by oil degradation.
Gamma Ray Scanning	A technique used to monitor flow in columns and identify tray flooding and back-up in downcomers.[6]
Pulse Isotope Injection	A very useful means of non-intrusive flow measurement.[7] If the internal bore of a pipe is known, a sharp pulse radioactive material can be injected into the flow and hence be used to verify the flow-rate. The interpretation of the *centre of mass* of the two radioactive pulses (one at the injection point and another some distance along the pipe) requires interpretation.

TABLE 19.2
Simple Maintenance Techniques

Polytrophic Efficiency Monitoring	Polytrophic efficiency of compressors can be measured by Pressure ratio$^{(n-1/n)}$ = Temperature ratio. Increases in the temperature ratio indicate wear in the compressor, possibly dirt in a centrifugal/axial flow unit or valves in a reciprocating unit
Monitoring of Heat Transfer Coefficients	Reductions in heat transfer will indicate fouling of the heat exchanger or possibly leakage.
Pressure Gradients	Increases in pressure drop along piping will indicate fouling or partial blockage.

These techniques mentioned require specialist equipment and trained personnel. There are also common techniques which can easily be used during plant operation, see Table 19.2.

The dynamics of the shaft in a rotating environment are complex and influenced by the oil viscosity. Although the clearance in a white metal (journal) bearing is less than a fraction of a millimetre, changes in oil temperature will change both the spring stiffness and damping characteristic of the rotating shaft. This is a study well outside this text but is essential information.

The trace in Figure 19.2 shows a signature of a compressor with two shafts rotating at 214 Hz and 321 Hz. An odd frequence=t at 540 Hz is the sum of these and not usual, indicating some dynamic instability. The unit of this design sometimes generated a very high vibration level known as "*trumpeting*". During a startup the vibration level rose over 15 seconds and became unacceptable. The main component of the vibration (95+%) had frequency 214 Hz. The vibration was effectively pure forced frequency. The unit was shut down, the oil temperature increase (so softening the spring and dashpot constants and restarted successfully. A simple solution. The orbit may have been that of a diablo or simple pitching. If it continued, the bearing would be damaged within a few minutes.

The trace in Figure 19.3 is a unique problem where the oil film rotated at less than half of the shaft rotational frequency (as the cage in a rolling contact bearing). The solution, while not elegant was to cut overlapping grooves 2/3 of the bearing width such that the rotation of the film is arrested. Simple but effective. (X1 is the shaft rotational frequence but 0.2 0.4 and 0.6 are subharmonic frequencies). 0.6 is also a natural bending frequency.

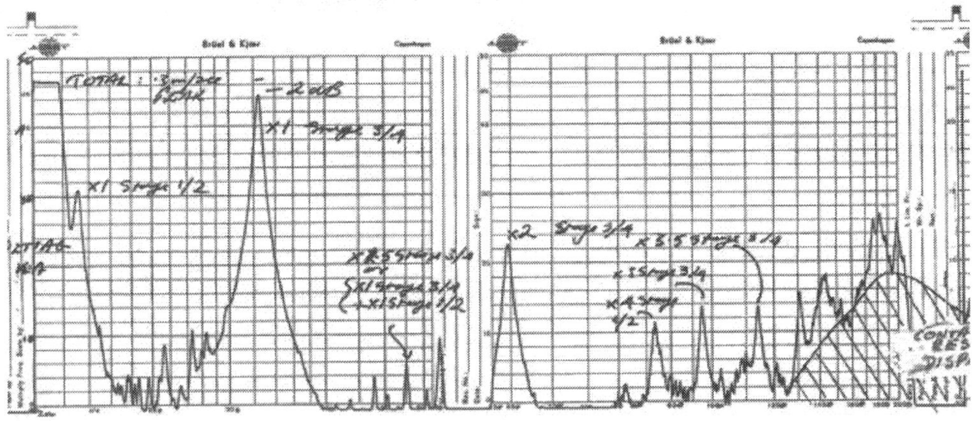

FIGURE 19.2 Vibration Traces

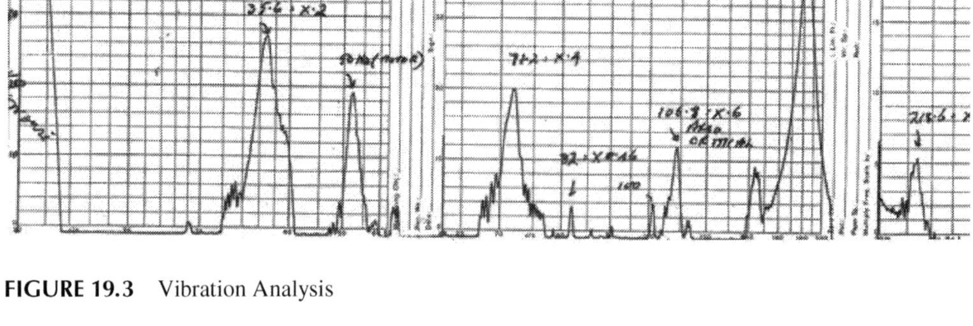

FIGURE 19.3 Vibration Analysis

19.6 CRITICAL SAFETY EQUIPMENT

Some equipment is required for safe operation or shut down of the plant and therefore requires regular attention to ensure that it is in working order. In some cases, inspection and maintenance are a statutory requirement. Plant designers must be aware of the legal requirements for testing critical equipment and incorporate appropriate test facilities in the plant design. All test results must be fully recorded and, if problems identified, they must be investigated and corrected.

Critical safety equipment includes a number of types of equipment:

1 Process trip systems
2 Control systems
3 Pressure Safety Valves (PSVs)
4 Equipment protection trips
5 Non-Return Valves (NRVs)

Each is considered later.

The equipment must be complimented by a safety management system (SMS) to ensure a high level of reliability as a consequence of adequate testing and inspection.

19.6.1 Process Trip Systems

These are now frequently called Safety Instrumented Systems (SIS) and basically consist of a sensing element, some form of switching system to give an output signal to a final element, usually a shut down valve. There is usually an isolation (often called a bypass or override) to facilitate the plant startup and maintenance. A simplified trip system is shown in Figure 19.4.

The reliability of a system is often defined as PFD. The performance of a protective system is assessed on a given test interval and the failure rate of the items of the system (see Section 6.2). The reliability of instrumented shut down systems has increased significantly in the last 20 years or so. This is partly as a consequence of the use of improved design but in large part due to the use of digital rather than pneumatic components in control systems.

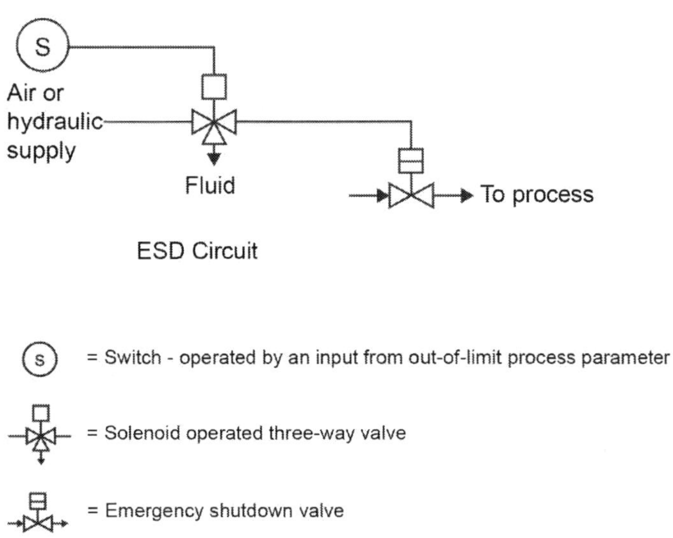

FIGURE 19.4 Simplified Trip System

Instrument function tests must attempt to simulate the real event as accurately as possible. Short cuts will lead to improper testing and, worse, increase the potential to incorrectly reset the trip after testing. Testing must be carried out regularly as part of a scheduled maintenance routine. As in many other areas, the UK HSE has provided useful guidance on trip testing.[8]

Ideally the test should be conducted by simulating the trip condition, making sure that an output signal has been sent to the control device which then processes the input signal to generate an output signal which operates a final element—usually an actuated valve. It has been practice, in some companies, to generate a false signal to the control device and check for an output signal. However, this technique only checks the most reliable part of the system and will not indicate if the sensing device or final element is operational. Comprehensive testing involves simulating the fault condition and ensuring the final element functions as designed.

The SIS test routine must wherever possible simulate the real conditions, without creating unnecessary hazards to the plant or personnel. Where testing indicated failure rates higher than anticipated and does not meet SIL requirements, investigation and corrective action is necessary. Increasing the test frequency is not necessary the answer as more tests mean more time when the loop is out of service and a greater chance that it will not be reset correctly at the end of test.

Complete loop testing may not always be possible. Testing of the ESDV presents a particular problem unless the plant is shut down. Whilst there are ways of testing the ESDV on line these could result in a spurious trip of the plant. The requirement, and means, of online testing of all components should be considered during detailed design and is considered in the following sections.

19.6.1.1 Sensing Element

The method of testing the sensing element depends on the parameter to be measured. Methods used for the most common trip systems are given in Table 19.3.

TABLE 19.3
Common Methods for Testing Sensing Elements

Parameter	Method
High/Low Pressure	High-pressure sensors can be tested by injecting a calibrated pressure signal into the sensor and checking for an output. Low-pressure sensors can be tested by venting the signal through a choke or flow resistance.
High/Low Temperature	High- and low-temperature devices can be tested by injecting a calibrated electromagnetic field (EMF) to the sensor (generally a thermocouple). Moving the set point to generate an output signal will not test the sensor.
High/Low Flow	Low-flow sensors can be tested by isolating the flow signal—this does not involve isolation of the detection device and is fail-safe. High-flow switches can be tested by injecting a calibrated signal into the flow device—this involves isolating the switch with the potential for human error.
High/Low Level	Testing procedure will depend on the type of device installed. For float switches, high level sensors can be tested by isolating the bottom tapping and injecting a compatible fluid—so floating the sensor. This only involves isolating the lower isolation and does not involve isolating the detection device. This can be considered fail-safe—if the lower tapping is left isolated the device will still be in a "tested shut down" condition and the isolation must be opened to drain the sensor and cancel the shut down signal. Low-level switches can be tested by draining the float or detection device; this does not involve isolation of the detection device and is fail-safe.

19.6.1.2 Final Element

Detecting input and output signals from the controller is relatively easy. The biggest problem is ensuring that the final element, usually an ESDV, is operational. Fully closing the valve is not practical as this will interrupt the process and a common practice is to "bounce" the valve (send a signal to close the valve to ensure it moves off its seat but then return it to the open position immediately). Proprietary systems have also been developed which clamp the valve stem as it moves towards the closed position and prevent full closure. Whilst these procedures will determine that the valve will move, they will not ensure that scale or debris in the valve will prevent a "tight shut-off" when the valve is fully closed.

The only way of ensuring a tight shut-off, other than during a shut down, is by installing two ESD valves in parallel and closing each in turn to ensure a tight closure. This arrangement is installed in a few cases, particularly on large steam turbines but the increased complexity increases the possibility of the system being incorrectly reset after the test is complete.

19.6.1.3 Frequency of Testing

A detailed procedure for trip testing should be established and followed. Testing will require the trip system to be taken out of service for a short period of time using a critical trip bypass procedure (see Section 16.7).

The reliability of trip systems affects the overall plant risk assessment. It is therefore important that they are tested regularly and fully in line with the test procedure. The probability of the trip systems failing on demand, or the FDT is assessed based on a fixed interval between inspections. if the interval is irregular, trip performance may not reach target. FDT for a simplex trip is

Weekly	0.0109
4 weekly	0.0172

If the test is carried out every four weeks the FDT is 0.0172 but if performed irregularly between two and eight weeks but averaging four weeks, it averages 0.0182—the error is small but real.

If tests indicate that required level of reliability is not being reached, more frequent testing or modifications to the trip system may be necessary. Incomplete and irregular testing will also degrade the reliability of the protective system.

The performance of all elements—the response time and the stroking time of the shut down valve—should be recorded, analysed, and a report issued. Performance (or probability of fail to danger) should be analysed against the required performance.

It may be difficult to test manually operated (often called "hard-wired") ESD systems and these are generally checked during shut downs. They are often activated for a planned shut down to check reliability.

19.6.2 Control Systems

Distributed control systems have periodic upgrades and checks are often performed to ensure changes are not slowly being introduced. There is also increasing concern over the vulnerability of control systems to cyberattack and periodic checks would be a useful defence. Increasingly access to control systems is being restricted by the removal of USB ports and the regular change of passwords.

19.6.3 Pressure Safety Valves

The design and specification of pressure relief systems are discussed in Section 13.10 and PSVs/BDs in Section 14.8. Whilst these are generally robust devices, they require regular checking to ensure they are operational and correctly calibrated. In some cases, statutory regulations indicate the frequency of testing (particularly for steam and air) but generally intervals of two to four years are common. A few companies have moved to a *Risk Based Inspection* (RBI) basis (see Section 20.1), where the frequency is based on previous inspection history and the consequences of failure. There may be reluctance to take valves out of service—particularly if an online spare is unavailable or the PSV is inaccessible as this may involve a plant shut down. Calibration and maintenance of PSVs and other overpressure protection devices should be considered during design.

Normal practice would be to take the valve out of service and conduct a "pop test" to check that the valve lifts at the correct pressure. In some cases, the valve will then be directly returned to service but usually it will be stripped to check for internal damage and then rebuilt. The valve will require a second pop test and a leak tightness test. Testing methods are discussed in API 575 RP Inspection of Pressure—Relieving Devices.[9] In a few cases, valves cannot easily be removed for calibration and proprietary methods for in situ inspection of have been developed.

Where PSVs discharge to atmosphere the tail pipe needs to be checked to ensure it is not blocked. Similarly drain holes at low points in the discharge line must also be checked for blockage. It is also not uncommon to find the drain hole blocked and the entire tail pipe full of water. Any blockage of the tail pipe will affect valve performance.

19.6.4 Equipment Protection Trips

Over-speed protection of turbines can be achieved by electronic or mechanical devices often with some form of voting logic. Modern devices use a digital counter. Older systems used a centrifugal bolt and impact on a shut down device.

Other equipment trips such as low lube oil pressure of high temperature can be tested in the same way as process trips discussed earlier. Vibration probes can be checked by the use of portable vibration monitoring equipment and comparison of the outputs of the fixed and portable systems.

19.6.5 Non-Return Valves

NRVs (also known as check valves) are not generally safety critical items but in some cases, they have an important safety function. A common example is on the discharge of a centrifugal pump—without the NRV, the pump might rotate in the wrong direction with serious consequences, particularly for an electrically driven pump. Another example is on the outlet of a fired heater. In the event of a tube rupture the normal exit pressure will be higher than the pressure at the point of rupture and backflow will occur. If the process fluid is flammable, additional fuel will be added to the fire.

NRVs, have relatively low reliability and can only be conveniently examined during plant shut downs.

19.7 SPARE PARTS

Spare parts can be considered as follows:

1 Consumable spares—to replace components which become damaged in service. These will generally be machinery parts but will also include bulk items which may be damaged or needed in association with other components, for example, cabling, nuts, bolts and gaskets. Spare parts can suffer from *ageing in storage* due to environmental factors such as temperature, humidity and low-level vibration. Some consumable items, particularly instrument and electronic components require storage in climate-controlled conditions.

2 Critical or *Insurance* Spares which will only be required in the event of a major machine failure where the lead time for a replacement would have serious economic impact. The designation of "critical" will vary from plant to plant but would generally include rotors for major rotating equipment such as compressors, turbines, and extruders. Other common

examples include spare transformers, reciprocating compressors (or crankshafts, cylinder liners and valves). In some cases, the spare rotor may be exchanged with the installed rotor during a major overhaul and then sent to the manufacturer for refurbishment. This may mean that a spare unit is not immediately available.

For a new plant, the contractor will normally provide two years of consumable spares and a record should be kept of usage with replacement spares being order where necessary. The level of spares will depend on various factors such as the anticipated number required for maintenance and the lead time for replacement. Lead time for consumables may be short where the supplier is relatively close to the consumer but can be extended significantly if transport to remote locations is required. Delays in clearing customs often causes delays.

Issues can occur when ordering spare parts. Many companies operate procurement functions independently of the maintenance personnel. Procurement personnel may not have technical training and are focused on "value for money". They may deviate from specifications either through lack of knowledge or because cost savings appear significant for "something similar to the required specification".

One set of roller bearings on a hot-duty pump suddenly require more frequent replacement. After investigation, it was determined that the procurement department had changed the supplier and the tolerance (clearance) in the new bearings was only just within specification causing lubrication to be impaired. The next set of bearings were purchased from the original supplier and the problem disappeared.

As plants are modified and new equipment installed, the size and variety of the spares holding may increase substantially. On older plant this might include spare parts for equipment which is no longer operating. Periodic review of the spares inventory is desirable to weed out items which are no longer required. One example where larger-than-normal spare holding may be justified is for equipment which can still be used economically but where components will no longer be manufactured by the supplier. Generally, the supplier will warn that a "last run" of production for the components is about to take place.

Storage of spares requires consideration. Components must be easily accessible and identified. This is easily achieved by providing weatherproof storage with climate-controlled areas where necessary. Note that spare parts may have a limited shelf life, particularly if stored incorrectly. Chemicals and oils should not be stored in the same area as spare parts, spill protection should be provided and MSDS should be available nearby. Chemicals which could react, such as acids and alkalis should not be stored adjacent to each other.

Large rotating components need careful storage to ensure the shafts do not *hog* or *sag*. This sometimes involves storing rotors vertically or in specially cradles. The use of an anti-rust oil coating is often necessary. Some components are protected against corrosion by storage in a nitrogen environment.

REFERENCES

1 Mills S R W, *Vibration Monitoring and Analysis Handbook (INST397)*, The British Institute of Non-Destructive Testing, 2010
2 Van Rensselar J & Dalley R, Fundamentals of Ferrography, *Plant Engineering*, June 2020, www.plantengineering.com/articles/fundamentals-of-ferrography/
3 Burstein N, Electrical Signature Analysis (ESA), *EE Times*, September 2001, www.ee.co.za/wp-content/uploads/legacy/wilec.pdf

4 Bagavathiappan S, Lahiri BB, Saravanan T, Philip J & Jayakumar T, Infrared Thermography for Condition Monitoring—A Review, *Infrared Physics & Technology*, 60, Elsevier, September 2013, www.sciencedirect.com/science/article/abs/pii/S1350449513000327
5 Tommasini D, *Dielectric Insulation and High-Voltage Issues*, European Council for Nuclear Research (CERN), https://cds.cern.ch/record/1342701/files/335.pdf
6 International Atomic Energy Authority (IAEA), *Industrial Applications of Sealed Radioactive Sources (IAEA-TECDOC-1925)*, IAEA, 2020, www-pub.iaea.org/MTCD/Publications/PDF/TE-1925web.pdf
7 International Atomic Energy Authority (IAEA), *Radiotracer Applications in Industry—A Guidebook (Technical Reports Series 423)*, IAEA, June 2004, www-pub.iaea.org/MTCD/Publications/PDF/TRS423_web.pdf
8 UK Health and Safety Executive, *Proof Testing of Safety Instrumented Systems in the Onshore Chemical/Specialist Industry*, www.hse.gov.uk/foi/internalops/og/og-00054.htm
9 American Petroleum Institute (API), *Inspection of Pressure-Relieving Devices (API RP 576)*, API, latest edition

20 Inspection

20.1 INTRODUCTION

A process plant should start life in pristine condition (although this is not always the case) but degradation will begin from the start of construction. Inspection programmes are therefore essential to ensure that equipment does not deteriorate or degrade to the extent that it may fail dangerously causing damage or loses integrity with the loss of process materials or energy. Inspection activities date back to the railway age where explosions of locomotive boilers occurred with depressing regularity.[1] From these early beginnings in the 19th-century inspection techniques have become more complex and sophisticated. An early example of inspection activity on railways were the "wheel tappers" who checked that wheels were not cracked or damaged.

Inspection work was originally conducted on a time-based frequency but over recent decades there has been a move to *risk-based methods*[2,3] with inspection frequency being varied depending on the results of previous inspections and the criticality of the equipment. The risk-based approach requires identification of likely corrosion mechanisms and the consequences of leakage. Resources can then be concentrated to the areas of greatest risk.

Equipment is increasingly being operated beyond it's design life (see Section 23.3), and critical assessment of its integrity will be necessary. Corrosion allowances built into the design may be used up, creep may be a problem with hot systems and fatigue with equipment subject to repeated thermal cycling or vibration. Enhanced levels of inspection may be necessary.

20.2 ORGANISATION

Inspection (sometimes called asset integrity) is the plant function responsible for monitoring and recording the condition of the physical assets. Most of the activity is associated with static equipment such as piping, pressure vessels and storage tanks but in some organisations includes monitoring of rotating equipment. Within the process industries, inspection activity is usually controlled independently from maintenance. This may not be true in other industries, in the power generation industry for example, there is generally no separate inspection function. The mindset of inspection and maintenance staff is significantly different. The former are concerned that the physical condition of the assets is satisfactory whilst maintenance personnel are more focused on ensuring that production is not interrupted. Within the process industries, the inspection personnel generally report to the engineering/technical services function. One of the inspection function's greatest problems is getting equipment released from service—operations teams prefer to keep equipment available even if not in use to allow for potential process upsets and breakdowns. Ensuring that equipment is inspected to schedule is a common problem and it is not unusual to find equipment, especially atmospheric storage tanks overdue for inspection, sometimes by periods of several years. The extent of overdue inspection is a useful KPI.

For a large complex a team of 10–12 may be employed; for simple locations a single individual may be adequate. It has been traditional practice for inspection personnel to be company employees but with third-party personnel suppling additional labour at times of peak demand, for example during maintenance shut downs. Specialist contractors are also used for inspection activities which will only be required on an occasional basis.

In some cases, the entire inspection function may be contracted to a third-party specialist company with possibly in-house personnel to coordinate activities. This raises the danger that a change in contractor could result in knowledge of the plant residing with the original inspection company

DOI: 10.1201/9781003510680-24

being lost, and the new contractor may have to build up a level of knowledge from scratch. There have been occasions, fortunately not widespread, where inspectors take a blinkered approach—they have a remit to inspect a section of pipe, which they do competently, but ignore the severe corrosion on the piping supports or the fact that the insulation is damaged. Inspection personnel must be encouraged to identify potential corrosion incidents even where they are not part of the specific task in hand. It is good practice for inspectors to undertake frequent walkthroughs of the plant and process operators should be encouraged to report, rather than accept, any corrosion concerns they have.

Whilst one or two members of the inspection department might have limited technical knowledge, being responsible for record-keeping and other administration, most inspection personnel will require a level of knowledge and experience on one or more of the key inspection methodologies shown in Table 20.1. Inspection personnel are trained and certified in the various common inspection technique by a number of internationally recognised bodies such as Certification Scheme for Weldment Inspection Personnel (CSWIP)[4] and the American Welding Society (AWS).[5] These organisations provide training and certification not only for inspectors but also for welders. Welding, inspection procedures and inspection requirements for many applications are often defined in company or international standards, for example, API 563 Above Ground Storage Tank Inspector to API 570 Piping Inspector.

Regardless of who performs the inspection work a good level of supervision will be necessary to ensure that the inspections are performed correctly.

Following repairs to a refinery crude column, radiographs were taken at all the locations where welding work had taken place. The refinery owner's inspection manager noticed that several of the X-ray films were almost identical. The weld inspector had found that several of the welds requiring inspection were not easily accessible and therefore decided to radiograph the same weld several times and pass these off as coming from the locations with poor accessibility. This situation is well known in the inspection world and is known as *ringing*.

Inspection activities range well beyond welding inspection, and the remit of CSWIP, now renamed as the Certification Scheme for Personnel, covers a range of other inspection activities.

20.3 PLANNING AND RECORD-KEEPING

Until recently, inspections were conducted with maximum fixed intervals between them. In some cases, maximum intervals are enshrined in legislation. Sometimes a *derogation* would be issued to allow the interval between inspections to be extended but this should only be allowed under strict conditions. As stated earlier, a *risk-based approach*[6] has been developed for piping and pressure vessels. This approach requires an assessment of likely corrosion mechanisms and locations together with the consequence of a failure at these locations. This has resulted in significant changes to inspection philosophy with more attention to small-bore piping and stagnant sections of pipework. Inspection frequency is then based on the consequences of failure and the extent of corrosion measured in previous inspections. Where corrosion becomes more rapid, the frequency of inspection increases and/or action is taken, such as pipe replacement—possibly with a different material.

Typical practice used, and often still is, to measure wall thickness of vessels and piping, collate the measurement results in a report and file them—often without reference to the results of previous inspection. A similar methodology was adopted with PSV inspections.

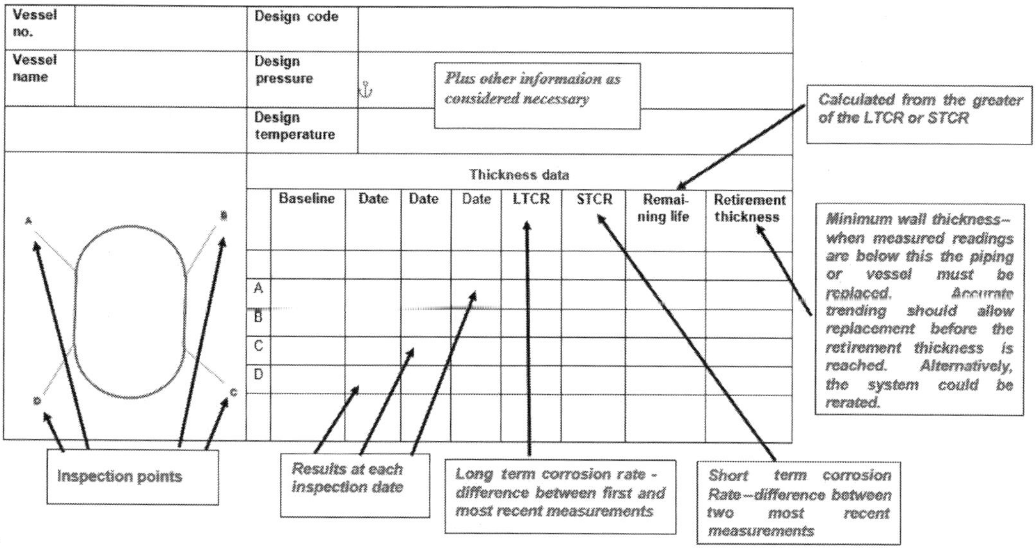

FIGURE 20.1 Example Wall Thickness Record Sheet

This practice does not make use of the data for predictive purposes and makes trending of corrosion rates difficult. More recently, the approach has been to summarise the results of all previous inspections in a table to allow corrosion rates to be calculated and the remaining life of equipment to be determined. A spreadsheet is generally used which can automatically calculate corrosion rates and remaining life. There are also a number of propitiatory programmes with more sophisticated capabilities. A typical inspection record sheet is shown in Figure 20.1.

Whilst piping specifications will give a nominal pipe wall thickness, there will be some variation in thickness and it is standard practice to measure "baseline" wall thickness at specified inspection points before a new plant enters operation. Corrosion rates after the first post commissioning thickness measurements can then be determined. Records should be carefully maintained and duplicated in at least two separate locations—more than one site has suffered problems from lost or damaged records.

Inspection records were stored in a temporary wooden construction building with the authorised smoking point directly outside. A fire (probably caused by a discarded cigarette) resulted in the complete loss of the building and weld inspection data.

Nowadays most inspection records are stored electronically with a backup. It is important that the back-up data is stored at a separate location to prevent a single incident, such as fire or flood, destroying all copies of the data.

20.4 COMMONLY INSPECTED EQUIPMENT

The necessity for inspecting PSVs (see Section 19.6) and pressure vessels is obvious but the need for inspection of other items may be less clear: Items commonly included in inspection programmes are listed in Table 20.1.

TABLE 20.1
Commonly Inspected Equipment

Item	Common Techniques	Frequency	Comments
Pipelines	Intelligent pigging	Variable	Intelligent pigging is the preferred option for pipelines and most pipelines above 6-inch diameter are now designed with this in mind. Older pipelines may have bends which are too sharp for intelligent pigs to navigate and alternative methods may be necessary. For small-bore pipelines, such as oilfield wellhead flow lines, pressure testing is often used. A number of non-intrusive methods are also available, each with particular strengths and weaknesses.
	Non-intrusive methods	Variable	A variety of non-intrusive methods are available but will require specialist contractors and equipment: 1 CIS (close interval survey) 2 DCVG (direct current voltage gradient) 3 ACVG (alternating current voltage gradient)
Pressure Vessels	Visual inspection, ultrasonic inspection and pressure testing. In some cases, more specialised methods are used (see Section 20.6).	Internal inspection is generally at intervals up to five years. The result of previous inspections will also affect inspection frequency	Many countries have statutory requirements for the inspection of pressure vessels (and associated PSVs)
Piping	Ultrasonic inspection is the most commonly used inspection method for the measurement of wall thickness in piping systems.	Variable	It is increasingly common for piping and pressure vessels to be considered as a single *pressurised system* and inspected in its entirety.[7] In some countries, this is a legal requirement
Boilers and Fired Heaters	Ultrasonic inspection, replica testing	Variable. Many countries have statutory inspection intervals for boilers.	Boilers and fired heaters are pressure vessels that operate at high temperatures and the metallurgy of the tubes exposed to flame will be adversely affected over time, particularly *high-temperature creep*, from damage (caused by direct flame impingement on tubes) or *fatigue* (due to repeated startup and shut down). Welds on an "A frame" fired heater between the vertical section and the 45-degree section were damaged due to repeated heating and cooling. This was not foreseen at design and the joins had to be re-enforced.
Storage Vessels	Visual inspection, acoustic emission and ultrasonic thickness tests		Due to the size of storage tanks specialist tools have been developed to allow inspection work to proceed quickly thus reducing the time the tank is out of service. However, draining, and decontaminating a large tank can still take a significant time. A specific issue affecting floating roof storage tanks is the build-up of deposits on the tank floor resulting in uneven surface. The feet supporting the floating roof will not all "land" simultaneously and this can cause damage to the floating roof.

TABLE 20.1 *(Continued)*
Commonly Inspected Equipment

Item	Common Techniques	Frequency	Comments
		Internal inspection is variable depending on duty. Crude oil tanks are usually inspected at <10 yearly intervals. Tanks on "dirty duty" may require more frequent inspection to detect corrosion under deposits and to clean out the vessel. Tanks on notionally clean duty may be less frequently inspected. The inspection frequency for pressurised storage may be defined by statutory regulations storage. In any event inspection intervals seldom exceeds 10 years and may be less. Refrigerated storage tanks are generally not internally inspected unless problems occur. Regular external inspections of tanks take place.	A storage sphere for ammonia was inspected after six years in operation and around 10,000 cracks in welds was identified. A critical analysis of the weld population and the startup mode allowed a model of the crack growth to be developed such that a "go no go" rule for size and age could be defined. The sphere remained in service for a further 15 years.
PSVs	Calibration (pop) test and leak test	Variable but 2–4 yearly calibration is common. Inspection frequencies may be defined by regulation. This will require the use of redundant PSVs with complex interlocking mechanisms. Inspection intervals should be no longer than the inspection interval of the associated equipment.	A specialist test bench is generally used for this work but there are techniques which can be employed where valves need to be tested in situ. Boiler PSVs can be *floated* in situ. PSVs are often calibrated by maintenance personnel but with scheduling and record-keeping being the responsibility of inspection personnel—who also generally witness tests. During a safety audit of a LPG sphere the auditors had difficulty reaching the PSVs due to a swarm of bees. Subsequent investigation found that the bees had colonised the discharge line of the PSV which would have impaired its operation.
Steelwork, Buildings, Marine and other Facilities	Visual inspection (possibly supplemented by specialist techniques)	Variable	Steel work particularly at joints and bolting is subject to corrosion and may require special attention and/or rust treatment.

(*Continued*)

TABLE 20.1 *(Continued)*
Commonly Inspected Equipment

Item	Common Techniques	Frequency	Comments
Welds	Radiography	Immediately after welding. The proportion of welds which require inspection will be defined in the welding code used and will depend on the criticality of the piping.	Welders need to be *qualified* before critical welds are produced. Stringent safety precautions are necessary to prevent exposure to radiation.[8]
Lifting Equipment	Visual inspection and load tests.	Variable, frequency may be specified by regulations	Different levels of inspection may be required, with the use of checklists for daily inspections and more detailed inspections at longer intervals, including a load test. For critical activities, testing requirements may be specified in the *lifting plan*.
Lifting Ancillaries (Slings, Wire Ropes, Chain Blocks and Other Items	Visual inspection/load test	6 monthly	A colour-coding system is often used so that equipment that is overdue for inspection can be readily identified. The current colour is often painted on to a section of wire rope or a coloured ferrule is used on wire ropes. Posters showing the appropriate colour should be displayed around the site.
Hoses	Visual and pressure test	6–12 monthly	This generally applies to critical hoses such as marine unloading hoses or those used for load/unloading road tankers. A colour coding system is often used to ensure hoses overdue for inspection are not used. Inspection sometimes includes checks on electrical connectivity.
Positive Material Identification (PMI)	Alloy analyser	When material arrives on-site and, ideally, before use.	Required to ensure that components to the correct specification are installed. In theory documentation from the supplier (the mill certificate) should ensure that material to the correct specification is installed but there have been numerous failures and additional checks are usually required for stainless and alloy materials.
Corrosion Protection			
1 Sacrificial anodes	Visual inspection	Annually	
2 Impressed Current	Transformer readings	Annually	
3 Coatings	*Holiday* testing	At installation or when a coating failure is detected.	A "holiday" is the term given to a hole in the coating
4 Corrosion Inhibitor and Coupons	Chemical analysis, regular checks to ensure injection systems are operational and injection tanks are not empty	Variable	

20.5 CORROSION ISSUES

Fontana[9] provides a useful introduction to corrosion mechanisms but new examples and variations continuously come to light. By referring to standard works suitable materials of construction for process fluids can generally be identified but should be verified by reference to actual operating experience. Detailed design of the plant can also create unexpected problems with, for example, erosion/corrosion on the outside of bends where the flow-rate is faster than on the inside or corrosion in stagnant sections of pipework. Piping configuration, can increase turbulence, resulting in unexpected corrosion behaviour.

Whilst most attention is devoted to process plant, corrosion can also occur in utility systems and should not be ignored. External corrosion can affect any exposed metal including the outside of piping and structural steel. Degradation of other materials ranging from concrete to plastic can also occur but is beyond the scope of an introductory test. The inspection function should have sufficient technical knowledge, as well as familiarity with the plant to identify and control the various corrosion mechanisms that are likely to occur. However, there are often a few surprises, hopefully identified, at an early stage, but sometimes missed.

On 16 April 2006, a serious explosion and fire occurred at the ConocoPhilip's refinery at Killingholme in the UK. The explosion followed leakage from a failed "elbow" on pipework on a deethaniser column. The rupture occurred downstream of a water injection point. Investigations concluded that, when installed, the pipework complied with piping standards but had subsequently been subject to corrosion and erosion to the point where it failed. There had been a number of changes to the system over the years which were subject to MOC procedures. However, the post-incident investigation considered that there was not an effective system for the inspection of pipework for the piping system that failed[10].

Examples are given in Table 20.2.

Many problems are caused by external corrosion as illustrated by the examples in Table 20.3.

TABLE 20.2
Common Integrity Issues

Example	Comments
Dissimilar Materials	Direct connection between dissimilar materials can result in galvanic action. This is a well-known phenomenon but sometimes occurs when ad-hoc repairs or unauthorised modifications allow incorrect piping materials or bolting to be used. A simple but frequent cause is the use of a stainless-steel plug in a carbon steel vent or drain line. In the unusual situation where the design calls for dissimilar materials to be used, special precautions may be necessary, such as the use of insulated joints (including the bolts on flanges) may be necessary.
Corrosion Under Insulation (CUI)	Damaged insulation covering can allow water to penetrate under the insulation and settle in contact with a metallic surface, causing external corrosion.[11] CUI can be a major problem with hot fluids (which dry the insulation) or cryogenics which create ice in the insulation by condensing water vapour from the atmosphere.
Rusting Flange Nuts/Studs	It is common to finds that regular wall thickness measurements are taken of pipe wall thickness but the condition of exposed studs and nuts is overlooked. Severe stud or nut corrosion will weaken the flange connection which will eventually fail. This problem is often found in conjunction with *short bolting* (where the stud does not fully penetrate the nut).
Fretting	Repeated expansion and contraction of pipe supported on steelwork will result in wear on the external surface of the pipe. The effect is similar to repeated rubbing with sandpaper.

(*Continued*)

TABLE 20.2 *(Continued)*
Common Integrity Issues

Example	Comments
Waste-water Systems	In theory, wastewater should be clean but is often contaminated with corrosive chemicals. Oilfield waste-water systems contaminated with sulphur compounds are a common example. Washing of chemical spills into drains can also cause problems.
Stagnant Piping Sections	Corrosion may take place at an interface between gas and liquid where there is a possibility of concentration gradients leading to a galvanic cell This may also occur at an interface in a vessel where the levels do not change with time.
Corrosion under Deposits	There may be a concentration gradient under the deposits leading to galvanic action.
Small-Bore Piping	Small-bore piping, particularly on drain lines, will often have stagnant liquid above a valve which may result in accelerated corrosion. If not adequately supported, small-bore piping may also suffer fatigue damage from repeated vibration. This is a particular problem with high-pressure reciprocating compressors.
Crevice Corrosion	This occurs in narrow, confined spaces between metal surfaces or between a metal and a non-metal. It is caused by the depletion of oxygen and the buildup of corrosive substances within the crevice. Crevice corrosion can be a problem on exposed screw threads.
Bellows	Bellows are still sometimes encountered as a method of controlling expansion in pipework. Bellows can be subject to cracking but a more common problem is distortion. Bellows are designed to expand and contract axially; if the piping connections at each end are not correctly aligned when the bellows are installed, they will be subject to lateral stresses for which it was not designed. This fact is not always recognised by pipe fitters who often think a bellows can be used as a handy means of correcting misalignment.

TABLE 20.3
Examples of External Corrosion

Corrosion of Underground Pipe	A cross-country pipeline ran above ground and then penetrated the ground surface at 45 degrees before a second 45-degree bend returned the pipeline to the horizontal below ground. The pipeline then ran across a riverbed. The pipeline had an anticorrosion coating which had become detached from the pipe surface and had become damaged at the ground surface. Over time water ingress occurred between the outside of the pipeline and the inside of the coating. Water accumulated at the lower 45-degree bend and corrosion occurred until the wall had thinned to the point where it could not withstand the internal pipeline pressure and ruptured. A significant crude oil spill occurred causing widespread pollution.
Collapse of LPG Sphere	A sphere collapsed during a hydrotest. Investigations showed that water ingress has occurred between the sphere leg and its covering of PFP. Usually a waterproof cover is installed above the PFP to prevent this but, in this case, was not affective. Severe corrosion of the leg occurred and the sphere collapsed as it filled with water[12].
Corrosion of the Base Wall Weld on a Heated (and Insulated Tank) Storage Tank	A heavy fuel oil tank was surrounded by a layer of insulation with an aluminium covering. The insulation was penetrated by supports connecting the stairway to the tank which were not sealed. Over a period of time water ingress through the penetrations occurred, with water draining down to the base of the tank where it was trapped on the base/wall weld. As the tank was heated accelerated corrosion occurred at this weld. The tank ruptured when it was filled to the full working height—when maximum pressure was exerted on the on the base of the tank.
Corrosion of LPG Line	A serious gas leak and explosion at Maryhill Glasgow on 11 May 2004 was the result of a leak of LPG which accumulated in the basement of a multi-storey building and exploded. The building collapsed. There were 9 fatalities and 45 further casualties. Post-accident investigation showed that severe corrosion and cracking of pipework had occurred in the line running from an LPG storage vessel to an oven used in the production process. There were issues associated with the design and installation of the pipework which exacerbated the problem. The inspection history of the pipework was also uncertain[13].

20.6 INSPECTION TECHNIQUES

The number of inspection techniques is constantly expanding with both the development of new techniques and improvements to existing ones. The most common techniques are outlined in Table 20.4. There are numerous sources of information, and these should be consulted to obtain a detail knowledge of the techniques.

TABLE 20.4
Common Inspection Techniques

Technique	Brief Description
Visual Inspection	A common technique to identify obvious defects. To work efficiently, inspectors will need adequate training and experience. This method will often be followed with more extensive inspection if problems are suspected. Visual inspections often goes beyond checking piping and vessel integrity to include other factors such as ensuring earthing connections are correctly installed.
Dye Penetrant Testing	A dye is painted over the surface of the item to be inspected and then carefully rubbed off. Where cracks are present, the, clearly visible, dye will remain in the crack after the surface has been cleaned allowing the crack to be detected.
Hydrostatic Testing	Filling the system with water and slowly increasing the pressure. If the pressure remains stable over a period of time this will indicate to absence of leaks. Note: sometimes pneumatic testing with air is used instead of hydrotesting.[14] This is a much more hazardous procedure due to the vastly higher level of stored energy in the system under test. The dangers of pneumatic testing have been described previously (see Section 15.5).
Leak Testing	During hydrotesting, delicate components, such as control valves and PSVs are replaced with spool pieces to avoid damage. Once the test has been successful, these will be installed, and leak testing takes place.[15] Often the pressure in the system is raised above atmospheric pressure. If the pressure is maintained, there is no issue but if pressure is slowly lost all the flanges and connections in the system will be checked for leakage. In some cases, a *tracer gas* (such as helium) which can easily be detected using an appropriate detector is used, but a soap solution brushed over likely leak sources will also work. Note that some compressor seals have an inbuilt leak-off which must be physically isolated during this test. In some cases, leak detection may be used as an alternative to full pressure testing but this should only be allowable with strict controls and only on low-pressure systems. Stand-alone leak testing is often a sign of failure to identify the need for pressure testing at the design stage.
Radiography (RT)	Radiography is a common method of checking the integrity of welds. The principle is similar to that used for medical X-rays where a radioactive source is exposed to a film beneath the source. After exposure to the source, the film will show any defects or cracks in the weld.
Ultrasonic Inspection (UT)	There are various forms of ultrasonic inspection, the simplest and most common being to measure the wall thickness of pipework. Unless the thickness is dangerously low, a single measurement of wall thickness has limited value. It is standard practice to repeat the wall thickness checks at intervals to check the corrosion rate and the remaining life of the pipework or vessel. To make early use of the wall thickness data, it is common practice to check wall thickness prior to the plant being commissioned. With this *baseline* measurement and the first post commissioning measurement, corrosion trends can be quickly established (see Section 20.3). More sophisticated forms of UT can be used for flaw detection.
Acoustic Emission Testing	A technique based on the generation of waves produced by a sudden redistribution of stress in a material. When a pipe or pressure vessel is subjected to an external stimulus, for example a change in pressure, load, or temperature, this causes the release of energy in the form of stress waves, which can be recorded by sensors on the surface of the equipment. The technique has been widely used in the process industry and a good level of knowledge has been collected.[16]

(*Continued*)

TABLE 20.4 *(Continued)*
Common Inspection Techniques

Technique	Brief Description
Magnetic Flux Leakage (MFL)	A technique used to detect corrosion and pitting in steel structures, piping storage tanks. A powerful magnet is used to magnetise the steelwork. Where metal is missing, for example due to corrosion, the magnetic field "leaks" from the steel. This is measured by a magnetic detector placed between the poles of the magnet to detect the leakage field.
Replica Testing	This is a technique based on the examination of the microstructure of a metallic surface to check for creep. The technique is widely applied to boiler and fired heater tubes.
Positive Material Identification (PMI)	A procedure to ensure materials of construction are to the correct specification. It is most commonly used on alloy or stainless materials. A number of techniques are available with X-ray Fluorescence (XRF) probably being the most common. Other sophisticated techniques such as the use of Laser Induced Breakdown Spectrometer (LIBS) equipment. Some simpler methods such as spark testing or hardness testing (Brinell or Stockwell) are also used. Hardness testing is commonly used to distinguish different types of carbon steel materials.

REFERENCES

1 Hewison C H, *Locomotive Boiler Explosions*, David and Charles, 1983
2 American Petroleum Institute (API), *Risk Based Inspection Risk Based Inspection (API RP 580)*, API, latest edition
3 American Petroleum Institute (API), *Risk-Based Inspection Methodology (API RP 581)*, API latest edition
4 Certification Scheme for Personnel (CSWIP), www.cswip.com/
5 American Welding Society (AWS), www.aws.org/home
6 UK Health and Safety Executive (HSE), *Risk Based Inspection (RBI)—A Risk Based Approach to Planned Plant Inspection (SPC/Technical/General/46)*, HSE, latest edition www.hse.gov.uk/foi/internalops/hid_circs/technical_general/spc_tech_gen_46.htm
7 UK Health and Safety Executive (HSE), *Safety of Pressure Systems Pressure Systems Safety Regulations 2000 Approved Code of Practice and Guidance*, HSE, 2014, www.hse.gov.uk/pubns/priced/l122.pdf
8 International Atomic Energy Authority (IAEA), *Radiological Protection and Safety in Industrial Radiography*, IAEA, January 1999, www-pub.iaea.org/mtcd/publications/pdf/p066_scr.pdf
9 Fontana G F, *Corrosion Engineering* (3rd edition), McGraw Hill, 1988
10 Carter J, Dawson P & Nixon R, *Explosion at the Conoco Humber Refinery—16th April 2001*, IChemE Symposium Series No. 151, IChemE, 2008, www.icheme.org/media/9857/xix-paper-70.pdf
11 UK Health and Safety Executive (HSE), *Corrosion Under Insulation of Plant and Pipework v3 (SPC/TECH/GEN/18)*, HSE, latest edition www.hse.gov.uk/foi/internalops/hid_circs/technical_general/spc_tech_gen_18.htm
12 *What Is Piping*, https://whatispiping.com/hazards-of-pressure-testing/
13 The ICL Enquiry, *The ICL Inquiry Report Explosion at Grovepark Mills, Maryhill, Glasgow 11 May 2004 (HC 838 SG/2009/29)*, The Stationary Office, 2009, https://assets.publishing.service.gov.uk/media/5a75a817ed915d506ee807db/0838.pdf
14 UK Health and Safety Executive, *Safety Requirements for Pressure Testing (GS4)*, HSE, latest edition, www.hse.gov.uk/pubns/gs4.pdf
15 Anon, *ASME B31.3 Leak Testing Requirements Overview*, Piping World, May 2023, www.piping-world.com/asme-b31-3-leak-testing-requirements-overview
16 Canavan H A, *Guidance Notes on the Use of Acoustic Emission Testing in Process Plants* (Third Report of the International Study Group on Hydrocarbon Oxidation), IChemE, 1983

21 Management for Health and Safety

21.1 INTRODUCTION

Safety has become increasingly important over time. In the years following the Second World War, "lip service" was frequently given to safety and the acceptance of a limited number of deaths was considered acceptable. Photographs from the early days of the industrial revolution until around 1970 show limited use of PPE and working conditions in many industries would be considered unacceptable today. Unfortunately, these conditions are still found in some industries in some countries. At this time, a site safety officer may have been required by law and whilst having little authority he (it almost always was a "he") was often held responsible for any incidents that occurred. His responsibilities were largely related to occupational safety (*slips, trips and falls*) with an emphasis on compliance with regulation and recognised safe working practices. A conflict, which still exists, occurred between production and safety functions. Unsafe conditions were frequently tolerated to ensure that production continued. Sometimes they still are.

Since around 1970 the situation has changed. In many countries, good safety practice is not just compliance with government regulations but to reduce risk to ALARP using BAT (see Section 1.5).

Safe operation of a site is the responsibility of everyone at the site including contractors and visitors. The safety function is to advise on and monitor safety and environmental issues and not to be the scapegoat for failings at a more senor level. For this reason, the safety function should be independent of production operations. Nowadays the site safety team has widespread responsibilities including environmental issues and, sometimes, quality assurance. The role of the safety team is widespread including training; record-keeping, incident investigation, monitoring and audits.

21.2 SAFETY CULTURE

The objective of any production plant is to operate profitably by maximising production and minimising costs—be they operational, maintenance or other costs. There appears to be an inbuilt conflict between production to maximise profit and safety which is often perceived to add costs and shackle production. However, Kletz rightly comments"[1] "if you think health and safety is expensive, try an accident". The investigation of safety culture is a relatively recent activity but a number of attempts have been made to develop tools to measure safety culture within organisations operating in the process industries. These are discussed briefly in Section 21.5.

Employees can have negative views of safety and can often see safety requirements as impairments to *getting the job done* and, if necessary, something to be ignored or bypassed. At one time wearing PPE was frowned upon and consider a sign of weakness by the more macho members of the workforce. Fortunately, these attitudes have disappeared. Development of a good safety culture is like growing orchids. A lot of work is needed to nurture the plant which can be destroyed by a short period of neglect or mistreatment.

The development of a positive safety culture is a topic in its own right and requires significant time and effort to achieve. It starts from the top levels of the organisation before working through the more junior levels of management to reach lower-level staff. Staff ranging from the managing director to the humble office cleaner need to be involved. Without top management support—not just lip service—the development of a positive safety culture cannot work.

DOI: 10.1201/9781003510680-25

There is no unique corporate safety culture. One company may believe in "empowering" its employees to take on responsibilities, another may require a detailed audit procedure before taking on responsibilities, and another may request detailed supervision whilst learning a new role with increased responsibilities. Examples of some common cultures are described in the following section.

21.2.1 "No-Blame" Culture

A culture which is perceived as blaming people for poor decisions results in resentful, un-cooperative employees who are unwilling to make decisions or report safety incidents. No-blame culture results in a more open management with employees willing to report untoward events and then discuss the way they can be eliminated. The employee does make decisions and is mentored by his his/her supervision. "No blame" culture also results in easier personnel management.

21.2.2 Domineering Culture

A domineering culture occurs where managers foster the belief that "they know best". Challenges to working practices and management decisions are not encouraged. This approach has a dulling influence on creativity and puts the employee into a subservient position.

A variation of this is the very clearly defined boundaries of authority where only staff above a certain level can make important decisions. This results in frightened or demoralised employees who are incapable or unwilling to challenge the status quo. In an emergency, they are likely to look for a supervisor's approval before attempting an ESD.

21.2.3 Club or Clique Culture

Some companies have a very *cliquey* culture where they employ personnel from certain backgrounds or universities. Employees soon realise that they have to become part of the "club" to advance their careers. If they cannot join the club, they have no option but to leave the company or allow their career to stagnate. This results in a self-perpetuating management style and the management becomes prematurely old or jaded due to the lack of vision and innovation.

21.2.4 Culture of Design/Specification

Some companies write very detailed specifications for a design, based on their own experiences which the contractor must follow. A few go so far as to specify the finest detail ("we know best") but others specify the objective to be achieved and leave the detail to the contractor using international codes/standards. At the other extreme, companies may set the objective and allow a contractor to Engineer, Procure and Construct the plant (EPC).

Some corporate specifications may be a reaction, or over-reaction, to incidents which have occurred in the past (the corporate memory).[2] Other organisations place greater reliance on well-trained personnel and good procedures (the corporate culture). There are no absolute rights or wrongs; each variation is the result of culture and evolution. There is a vast literature on the subject of safety culture but to go further in an introductory text would be to step on the toes of sociologists and psychologists.

21.3 WORKPLACE AND BEHAVIOURAL SAFETY

Most organisations will have a variety of continuing, often occupational, safety-related activities. For example, safety posters on walls, site safety committees and good housekeeping competitions. It is also common practice to provide safety inductions to new employees and refresher training on a regular basis. Depending on the nature of an individual's work additional training may be necessary.

Personnel who use special equipment such as breathing apparatus (BA) will often require training and certification at regular intervals. The best way of boosting safety awareness is for frequent visits around the plant by senior management with an emphasis on safety matters. The visibility of the senior management around the site—not just in the office—will make a positive contribution to safety culture as this demonstrates both ownership and support for safety. More than a five-minute walk-round is necessary, managers should probe in detail where they find areas of concern and listen to comments/concerns from the operations or technical staff.

Some traditional safety practices are likely to have limited effect—walking past the same safety poster day after day will deaden its impact. Safety committees without a strong chairperson can deviate from safety issues to personal grievances and unrelated concerns. Partly for these reasons there has been an increasing trend since around 1990 towards *behaviour-based* safety (BBS) approaches. The approach is based on identifying poor safety performance (even if there has been no incident), investigation the cause/possible consequences and trying to modify worker behaviour. This approach works best during the operational phase of a project and is likely to have a greater impact on occupational rather than process safety. A key feature of the process is the observation and reporting of negative (and in some systems positive) safety actions as part of a *safety observation programme.* For example, not wearing correct PPE, walking under temporary ladders, poor housekeeping and so on. By analysis of the number and type of observations common issues can be identified and management can implement training and changes to working practices to try and prevent future occurrences. An early example of this approach is the Du Pont Safety Training Observation Programme (STOP) but similar programmes have now been developed in many other companies. Such approaches take considerable time and effort to be successful and a half-hearted approach is doomed to failure. The behaviour-based approach is not universally liked, in particular some trade unions view it as an attempt to transfer responsibility for safety from the management to the workforce.[3] This is not how BBS is intended to work—and without the involvement of the workforce, the approach will not succeed. BBS is not a standalone approach but only one weapon in the safety armoury. If properly designed and implemented a behavioural approach can help to move the site towards an improved safety culture.

21.4 SAFETY MANAGEMENT SYSTEMS

Process, or occupational, safety does not happen by accident. Comprehensive procedures need to be developed to ensure that a safe working environment and safe procedures are in place and that environmental concerns are addressed. Regular monitoring is necessary to ensure that standards are maintained and where improvements are identified they are implemented. A typical SMS approach is shown in Figure 21.1.

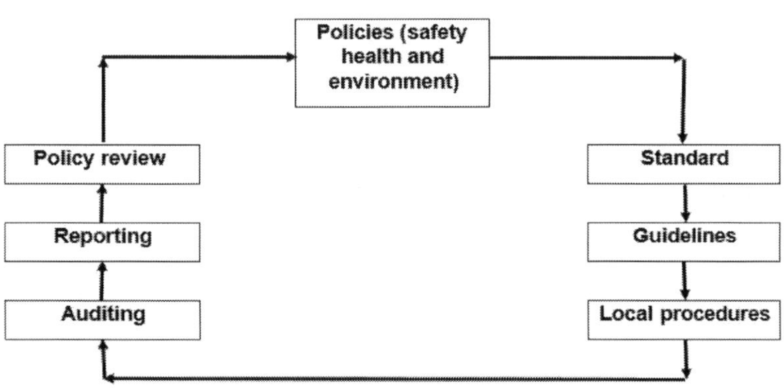

FIGURE 21.1 Schematic View of a Typical Safety Management System

The size and complexity of the SMS will vary with the size of the company and the industry involved. For most large companies, and many smaller ones, paper-based systems have largely been superseded by computerised systems. Some have been developed internally but several preparatory systems are available. As with other computerised systems, there is a danger that management of the system will focus on its minutiae at the expense of understanding the system and making best use of the information it contains. A good introduction to safety management systems has been provided by Wallace[4] and more recently by the UK HSE.[5]

The top level of the SMS will be company policies which define the organisation's attitude to safety, health and environmental issues. The top corporate policy may only be a number of key points but these can then be interpreted appropriately and implemented by lower-level departments. Generic statements are easily found but should be tailored to the individual company. They are widely displayed around company premises but are frequently ignored in day-to-day operations. Unless the policies are followed, they will rapidly fall into disrepute and often allow a cynical attitude to develop amongst the workforce. Compliance with all laws and regulations is a minimum standard and the concept of *continuous improvement* should be adopted.

Much of the SMS will consist of SOPs (see Section 18.4), which define what needs to be achieved and guidelines which illustrate how the guidelines are to be achieved. In a large organisation the SOPs might need amending for individual sites to allow for the specific process hazards involved and, importantly, local cultural factors. Many of the key SOPs, such as permit to work and critical trip bypassing are covered in Chapter 16. Other aspects are covered later (incident reporting (Section 21.8) and auditing (Section 21.7).

Standards and procedures, like steelwork, corrode over time and therefore regular auditing is necessary to ensure there is no degradation in the operation of the SMS. Finally, the policies should be updated at regular intervals.

Most safety management systems will feature a number of key features such as those shown in the following table.

1 Induction Training	All new employees will require basic safety training to cover the hazards they will encounter, safety precautions (including the use of PPE), raising an alarm and escape routes. More detailed training will be required for many staff, depending on their role. Anyone operating or maintaining the plant must, for example, be trained in the PTW system. Appropriate induction training will also be necessary for all visitors and contractors working at the site.
2 Communications	Appropriate methods must be available so that staff, at any level, can make their concerns known to senior management. One example is the use of safety observation programmes (see Section 21.3). Another common method is the site safety committee where safety representatives from all levels of the workforce regularly meet to discuss safety issues. For a safety committee to have any credibility, it must be chaired by a senior manager. Ideally, the site manager—who has the authority to "make things happen".
3 Safety Handbook	Again, this will vary tremendously between organisations but will be the first-place staff will go for safety-related advice. Despite increasing digitalisation, paper copies are still useful. The term "handbook" is important; safety information should be easily portable and the handbook should fit into a pocket. Whilst it will not answer every question it should deal with the most common queries and identify where more detailed information can be found. Intrinsically safe tablets are coming into use within process plant and may be seen as an alternative to a paper handbook but will have shortcomings. For example, screens may not be visible in some conditions and the battery may be exhausted when the handbook is required.

21.5 HOW SAFE IS SAFE?

The acceptability of risk, both within a site and surrounding it, is briefly discussed in Section 1.5; this section covers the slightly different issue of measurement of safety within a particular site and how measurements can be used to guide safety improvements.

Around 1970, safety was generally measured in terms of fatal accident rate (FAR)[6] or lost time incident rate (LTIR).[7] These parameters had some uses, particularly for measuring the relative safety of different industries. When applied within a single industry, they are blunt tools for measurement and group all workers together whereas the level of risk might vary with some workers in the group having a much higher level of risk than others. For example, the chemical industries employ a wide range of people involved in activities ranging from office work (usually well removed from a hazardous process areas) to process operators (in everyday contact with hazardous chemicals at a range of temperatures and pressures). The latter have a much greater level of risk. The situation is not always as clear-cut as the discussion earlier implies. Older installations often had office buildings close to the process areas and there have been cases where office workers have been injured by accidents in the process areas.

> Around 13.20, on 21 September 1992, a jet flame erupted from a vessel being cleaned at Hickson and Welch Limited's site at Castleford UK.[8] The incident was as a result of a runaway reaction causing a deflagration and jet fire which affect both the plant control room (and offices) and a large, more distant, four-storey office block. Two people were killed instantly in the control room/office building and two others subsequently died of burns. A young woman in the more remote office block also died.

FAR was a convenient starting point for measuring and setting targets. More contemporary estimates of risk are sometimes based on *individual risk per annum* (IRPA) which is defined as the probability that a specific or hypothetical individual will be killed due to exposure to the hazards or activities during a year.[9] This approach considers that the hazards that an individual is exposed to over a year and then calculates the total risk to the individual.[10] The calculation will consider the activities the individual performs, the hazards they are exposed to and the duration of the hazards. For example, an offshore worker will be exposed to slips trips and falls; fires/explosion; structural failure/ship collision and transit to from the platform by helicopter (possibly the greatest risk). In some cases, mitigating factors, such as safe refuges or protected evacuation routes, can reduce the overall level of risk.

At a site level, safety is often measured in terms of the number of lost time accidents (LTAs). These are defined as incidents where the employee is off work for a period of time (generally for three days or more). First-aid injuries are also usually recorded. These are the number of incidents per unit time (usually monthly). Other parameters such as the number of releases to atmosphere, property damage and near misses are also often recorded. Trends in these incidents are watched with interest and often seen as indicators of safety performance. Where a site has a good overall safety record a single incident can suddenly transform a good performance into a poor one. This may just be an unlikely "rogue" event. However, often poor safety features are hidden "under the surface" and, until exposed by an accident, remain hidden.

A number of tricks based on these parameters can be used to distort safety performance. For instance, a field operator or maintenance technician who suffers an injury which should be reported as an LTA may be sent on a training course before they have been off work for enough time for the LTA criteria to be triggered. Incidents may go unreported as the staff involved may be concerned that they will be disciplined. A positive safety culture (see Section 21.2) will greatly increase the chance of incidents being reported.

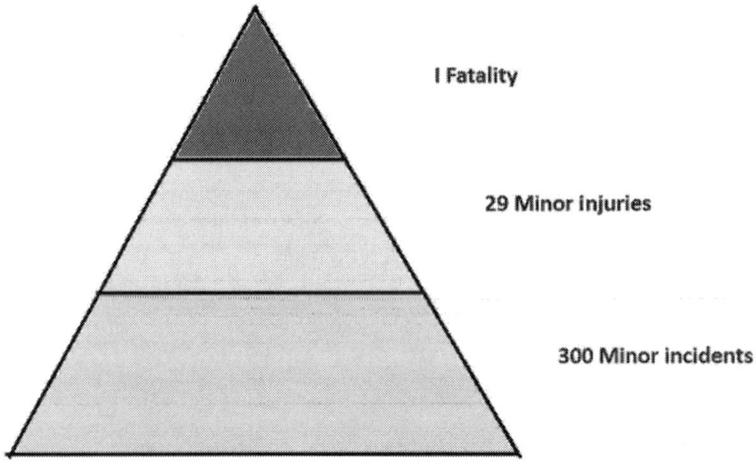

FIGURE 21.2 Heinrich's Triangle

The relationship between minor incidents and fatalities has been the subject of a number of studies. All have established a roughly triangular relationship although the ratio of major injury to minor injury to non-accidents injuries varies. Unoriginally referred to as *Hendrich's*[11] *triangle* after the author of the first study into the subject, the relationship is shown Figure 21.2. The ratio between each layer varies between organisations and may be a feature of culture or working environment.

The recording of significant numbers of near misses and first aid injuries is a warning that a more serious accident may be "just round the corner". The ratio between the different "slices" of the triangle can also provide a rough guide to the level of incident reporting. Safety observations (see Section 21.3) can be considered an additional, lowest, slice of Hendrich's triangle with a much larger population of events.

IRPA and LTIR are *lagging indicators* (showing past performance). Increasingly, organisations are looking to *leading* indicators to give a better indication of safety performance. Road safety provides a simple but illustrative example. If a driver crashes this is obvious—the car is damaged and there may be injury to the driver, passengers, passengers in other cars or pedestrians. A driver's performance may be measured in the number accidents they have had (often few and minor) or by looking at the way they drive. Rapid acceleration and sharp braking suggest a less safe driver than one who accelerates and brakes smoothly. Although there has not yet been an accident, it is not hard to see which driver is more likely to have a future crash.

Marshall, one of the pioneers of process safety was fond of illustrating this point by explaining that "if a driver ran into a group of pedestrians and his brakes and steering were found to be faulty, it would be no excuse for the driver to say he was distracted by a wasp".[12] Vehicle insurers have caught on to this idea and some now offer the potential for reduced premiums for drivers who can demonstrate safe driving practices monitored by an on-board sensor. This approach is still developing within the process industries, but guidance is available from both the UK HSE[13] and the US Center for Chemical Process Safety (CCPS).[14] Webb[15] has also produced a useful introduction to the topic.

Whilst measurement of past performance uses lagging indicators, estimation of future performance involves *leading indicators*. Lagging indicators have traditionally been used to measure occupational safety and leading indicators for process safety. There is, however, no reason why leading indicators cannot be used for any safety situation.

As discussed in Chapter 2, occupational safety differs significantly from process safety. Research[16] has indicated that low personnel accident rates are not an indicator of process safety or give a low probability of a major incident—there are several examples of incidents where sites with good personal accident rates have suffered major incidents. This is the reason why the use of leading indicators is valuable—if used correctly, they give an indication of the direction of travel towards or away from safe operation.

At 13.20 on 23 March 2005, a major explosion occurred at BP's Texas City Refinery. Fifteen fatalities plus 180 other injuries occurred with financial losses over US$1.5 billion. The incident occurred during startup of an isomerisation unit with all the fatalities being near the point of release (a blowdown drum). A review panel was set up specifically to review issues of safety culture across and subsequently issue a report[17] (commonly called the Baker report after a key member of the review team), which was commissioned specifically to review BP's "corporate oversight of safety management systems at its five U.S. refineries and its corporate safety culture".

The review made several recommendations and highlighted the difference between process and occupational safety, emphasising the limited potential for occupational safety activities to significantly improve process safety:

"The literature suggests, and the Panel believes, that the underlying or 'root' causes of the large majority of personal and process accidents are deficiencies in the systems that a facility uses to prevent these two classes of accidents. The literature also suggests, and the Panel believes, that the presence of an effective personal safety management system does not ensure the presence of an effective process safety management system. As discussed elsewhere in this report, BP's personal injury rates were not predictive of process safety performance at BP's five US refineries".

The development of appropriate leading indicators can be difficult, the easier they are to measure, the more trivial, and less useful, they are likely to be. On the other hand, complex indicators will be time-consuming and difficult to calculate or assess. Sometimes excessive effort is expended in the calculation with limited effort on analysis of the results. When developing leading indicators, care must be taken to ensure they are functional and useful and not chosen to ensure a high "score" by the department being monitored. Many leading indicators measure actual performance against planned performance. For example, the actual number of PTW audits against the planned number, or the number of safety audit recommendations completed against the number scheduled. However, this is not always followed up to ensure that the defects have been completely corrected. Whilst this should be done independently of the leading indicator, it is prudent to check that this occurs and the results of the audit are not "parked" without corrective action.

Several organisations have attempted to measure safety, not by performance but by the level of safety culture within an organisation (see Section 21.2). A simple example of the development of safety culture within a process engineering context is provided by Stuart.[18] Lardner, Fleming and Joyner[19] who have proposed an approach for determining the maturity of a safety culture based on an assessment of ten elements:

1 Management commitment and visibility
2 Communication
3 Productivity versus safety
4 Learning organisation
5 Safety resources
6 Participation
7 Shared perceptions about safety
8 Trust
9 Industrial relations and job satisfaction
10 Safety training

An assessment of these elements is used to place the rate the site on a five-stage process as shown in Figure 21.3.

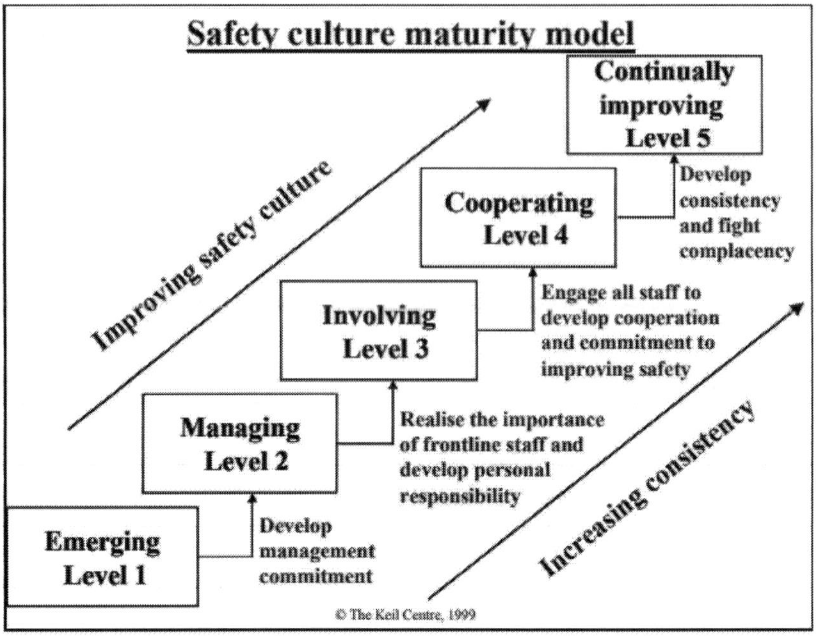

FIGURE 21.3 Safety Culture Maturity Model[20]

In simple terms, the levels are defined as given as follows:[21]

Level 1 Safety is defined in terms of technical and procedural solutions. Safety is not seen as a key business risk with safety primarily being the responsibility of safety personnel. A level of accidents is seen as unavoidable.

Level 2 Safety is seen as a business risk and management time/effort is spent on accident prevention. Accidents are seen as preventable Safety performance is measured in terms of lagging indicators. Management response to accidents is reactive with the use of punishment if accident rates increase. Accident rates are considered average for the industrial sector.

Level 3 Management accepts that a wide range of factors, including management decisions, may result in accidents. The involvement of front line staff in safety-related issues is considered essential to further improvement and the majority of them are willing to work with the management on safety-related issues. The majority of staff accept responsibility for their own health and safety. Accident rates are low and stable.

Level 4 The majority of staff believe that health and safety factors are important from both economic and moral viewpoints. Staff at all levels recognise that there are a wide range of factors which can cause accidents. The contributions of employees at all levels is recognised. Significant effort is put into proactive measures and safety performance is monitored using all available data. Non-work accidents are monitored and a healthy lifestyle promoted.

Level 5 The organisation is continually seeking to improve performance but is not complacent and is aware that an accident may occur at any time. The organisation is not performance driven and has confidence in its safety procedures. All employees consider that health and safety are a critical part of their job and accept that prevention of non-work accidents is important. The organisation makes considerable efforts in promoting health and safety at home. It has had a period of years without a recordable accident or other significant incident.

Based on the result of interviews, group sessions and other activities, the level of a site can be determined and the steps necessary to move sequentially from one level to another can be established. Ambitious attempts to leap over levels should be avoided. Models of safety culture are not fully developed and further work to prove the models is necessary. However they do represent a potentially useful way of determining organisational safety, not only at a site level but also at a corporate level.

21.6 CONTRACTORS AND VISITORS

Most companies now have a list of *preferred bidders* who will have previously been vetted to ensure compliance with required standards. This generally includes checks on the contractor's working practices and safety performance. Where this is not the case, the safety procedures and performance should form part of the bid assessment. A lower price from a contractor with poor safety practices would become expensive if an accident occurred. For larger projects a contractor might have their own safety personnel. This does not avoid the need for the site safety personnel to provide safety oversight as some contractors will avoid safe working practices to cut corners. Contractors' safety personnel often have limited authority and can be overruled or bullied by the contractor's management. There have also been cases where contractor's safety personnel have been sacked for insisting on safe working practices.

A contractor was collecting environmental samples for analysis. He visited various parts of the site, took his sample, climbed back into his van and drove to the next location. Some of the locations were within hazardous areas with hydrocarbon leak potential. The motor was left running whilst the samples were collected. This ignition hazard should have been identified during the Job Safety Analysis or by field operators, possibly during an audit. Clearly the driver had not been instructed on the risks and the management had not used the "look, listen, feel" approach: (witnessed by one of the authors during an audit).

Whilst site staff will be familiar with safety precautions, including actions in an emergency, visitors to the site will not be and specific precautions for their wellbeing are necessary. It is normal practice for personnel who will work at the site for an extended period (not necessarily continuously) to have a safety induction often repeated on annual basis. Even personnel who are familiar with the process industries will need a safety induction as all sites will have different procedures (such as, warning sirens and escape routes).

Even small plants are likely to have a range of non-technical visitors ranging from local dignitaries, salesmen, financial auditors, delivery drivers and so on. Generally, this type of visitor will not need to visit the more hazardous parts of the site, but they do so, a simple safety briefing should be provided, and the visitors must be accompanied throughout their visit.

21.7 SAFETY/ENVIRONMENTAL MONITORING AND AUDITS

Safety monitoring and auditing are two separate but complimentary activities. Monitoring is a less formal approach often consisting of a "walk round" of the plant site (or part of it) looking for things that do not "look right". Typically, a shift supervisor, plant manager or safety representative would monitor the site on a frequent basis. As stated earlier, it is good practice for senior managers and company directors to walk through the site as frequently as possible—ideally without advanced notice. Not only will this keep staff on their toes, but it also shows senior management's commitment to safe working practices.

Safety audits represent a more structured approach, making use of a prepared questionnaire to identify specific aspects to be reviewed by either physical examination or face-to-face meetings. These can be very detailed such as an audit to ensure that all fire extinguishers are correctly located, are not overdue for service and are of the right type (fire extinguisher audits are typically performed monthly) or a more general audit covering a particular area. Similarly detailed audits might cover areas such as the PTW forms, integrity of electrical equipment, missing vent/drain caps, missing padlocks on key isolation valves and a host of other aspects. More widespread audits covering a whole section of the site are also common and should take place on a regular basis. These will be more general in coverage but allow any issue considered relevant to be recorded for investigation.

Staff working in a particular part of the site might be used to seeing a defect daily and the potential danger of a particular *tree* is lost in the overall *wood* of the site. For this reason, audits conducted by people who do not work on the site are valuable. In some companies the site is split into sections with each area subject to an annual audit by staff from other areas. This can be made into a competition with prizes for the winning area.

Less frequently external audits are conducted. These may be undertaken by regulatory authorities, corporate safety departments, certifying authorities for International Standards Organization (ISO) or similar bodies and project lenders/insurers. These audits are generally "high level" activities to ensure that the site complies with relevant standards and to provide a comparison with similar sites.

There are a range of definitions of *audit*; the most useful in the present context are the following:

1 The Institution of Chemical Engineers defines a Safety Audit as
"A critical examination of all, or part, of a total operating system, with relevance to safety."
2 The International Chamber of Commerce (1989) define Environmental Auditing as
"A management tool comprising a systematic documented, periodic and objective examination of how well environmental organisation, management and equipment are performing with the aim of helping to safeguard the environment by:
   1. Facilitating Management Control of environmental practice, and
   2. Assessing compliance with company policies which would include meeting regulating requirements."

There are similarities with financial auditing where an independent examiner verifies that the corporate accounts are correct and recorded in the correct manner. The auditor does not check every penny and entry but samples enough of the entries and checks invoices, receipts and, also, that there is an audit trail to be followed. Any audit must have a defined output, be systematic in its approach, based on some form of sampling, and provide an output which is measurable in some way. An auditor must also be independent from the site or organisation being audited.

Regardless of the type or detail of the audit, an efficient means of recording any issues identified must be established together with a programme to ensure corrective actions take place and the defect is corrected. In some cases, changes can be made quickly but in others a longer timeframe may be necessary (perhaps the next plant shut down). Whilst considerable effort is put into conducting audits efficiently and recording the findings there are frequent examples where corrective actions are not taken and the whole exercise is a waste of time. Not only are defects not corrected but the workforce is given the impression that safe working is not a high priority. Senior management might monitor to ensure that audits take place as scheduled but take no further interest. Therefore, active monitoring of corrective actions is vital and management oversite should ensure that this happens.

> One cash-strapped state oil company carried out detailed safety audits at major facilities on an annual basis. The audits were performed by the corporate safety function and were comprehensive in their description of deficiencies and recommendations. However, many recommendations involved large capital expenditures which were simply not available. The next year, the audit would be repeated and the same recommendations, plus a few new ones, would be made. The audits were considered a success as they were comprehensive and gave useful recommendations but were, effectively, useless as they resulted in no improvements.

There are a number of discrete stages to an audit:

1 Selecting the audit team
2 Setting the terms of reference and scope

3 Planning and preparation
4 Pre-audit sampling/data Gathering
5 Organisation based sampling/data Gathering
6 Forming a judgement
7 Drafting the report
8 Issuing the final report
9 Following up the report

The audit will also have some key objectives:

1 Identification of strengths (and weaknesses) of the local (or corporate) safety/environmental management systems (SMS/EMS)
2 Identification of improvements considered necessary to improve the SMS/EMS
3 Recommendations for improvement of the SMS/EMS

Those must all be measured against a set of references, which may include a whole spectrum of the following parameters:

1 Legislation
2 Corporate policy
3 Corporate standards
4 External standards
5 Best practice

The approach of safety and environmental audits is similar but the causes of non-compliance may subtly vary. For simplicity, these differences are ignored in the generalised description give here. Safety/environmental conflicts are discussed in Section 4.2 and often occur during audits. A common difference between the two types of audit is the hazard-recognition skills of the auditors.

21.7.1 Sampling Data

Where audits involve a large number of similar items, such as PTW forms it is not usually possible to audit every form and it is important that a representative sample is subject detailed analysis. The sample size must reflect both the size of the population and confidence that the sample is representative of the whole. If the total population is two and the sample is only one it would be wrong to conclude that if the one sample was "healthy" the other was also "healthy", but if the population was 1,000 a smaller percentage of the sample would give good indicators of "health".

It is clear that large samples are needed if the auditors wish to have a high level of confidence in their conclusions. In some cases, sample sizes may not be known and a "guestimate" will be needed. For example. a visiting auditor is unlikely to know the exact number of electrical enclosures within a process unit or the number of drain points. With increased digitalisation of data more accurate sample sizes may be predicted.

21.7.2 Identification of Non-Compliance

The audit process requires asks:

1 What?
2 Where?
3 Why?

Audits frequently check physical issues, such as lack of inspection data on fire extinguishers or missing plugs/flanges on drain lines. These are often easily corrected but without changes or enforcement of management systems, the same issues will reoccur. Therefore, the audit must look beyond the obvious to identify the *root cause* of non-conformities.

The "what?" requires an analysis of both *Expectations* and *Required Standards*. The expectations are that the good working practices and management systems that would be expected for a particular activity are in place. The required standards will flow from *Corporate Policies* and *Legal/Regulatory Requirements*. These vary depending on the location and the subject of the audit—an audit to check for safe maintenance activity in an oil refinery would be expected to be much more rigorous than a similar audit in a shop or office block. The identification process requires a large measure of experience, knowing "what?" to look for and "where?" to look for it. Checklists may assist—either standard lists or ones prepared for a specific audit.

Whilst no generalised checklist will be complete or comprehensive for every situation, some typical items examined during an audit of a process plant are given in Table 21.1.

Table 21.2 covers only a few the possible aspects that could be covered in an audit. Numerous checklists are available on the internet but must be tailored to the needs of each industry, company and site. Whilst audits frequently make use of a written checklist, it is useful for an auditor to have his/her own mental checklist. Basing an audit entirely on a checklist runs the risk of it becoming a "tick box" exercise. It is often possible to infer the potential safety implications of an observation, as shown in Table 21.2. Again, these are examples on the myriad situations that might be observed.

TABLE 21.1
Typical Items to be Reviewed During a Process Plant Audit

Task	Reason
Check safety gates (open/closed/stiffness)	Required for entry/exit and during emergency evacuations.
Are PSVs adequately supported against jet reaction?	PSV piping may be bent by jet reaction forces and, if bent, the relief capacity may be restricted
Are PSV tail pipe drains clear?	1 If the PSV lifts someone may be covered with hazardous liquids (hot, toxic, corrosive or flammable). 2 The capacity of the PSV may be compromised.
Are static waste skips on the site?	Supplementary Questions: 1 Do vehicles have to come into a hazardous area to recover the skip? (can the vehicles impact people or equipment) 2 Is it a source of fire/flame? (could there be a source of ignition for flammable materials) 3 What controls are applied for their safe removal?
Are surface drains clear of debris?	If they are blocked, there is a possible risk of the backup due to flooding or fire (due to fire spread up on the un-drained water).
Are all sources of liquid spill or leakage handled effectively and fully contained?	Leaks may enter the water course and hydrocarbons may accumulate in the soil and "float" if the ground becomes wet due to rain or fire water at a future date.
Is combustible waste collected in closed containers?	Waste may blow out of the site into the public area.

TABLE 21.2
Typical Audit Observations and Possible Implications

Observation	Possible Implication
Damage to curbing inside hazardous areas with tyre marks nearby	There may be unauthorised traffic in hazardous areas, creating potential sources of ignition or risk of impact damage.
Tour guide appears to be 'leading' from a specific area of the site	The guided access may indicate deliberate restriction of certain areas, raising concerns that issues or non-compliance could exist in those locations.
Damage to thermal insulation or rust at the bottom of the tank	Possible corrosion under insulation leading to leaks
Roped-off areas with no signs of maintenance	Maintenance may be under-resourced, potentially allowing latent issues to develop.
Oil stains on roads	Oil may enter drains and reach the environment if interceptors are ineffective or overwhelmed.
Open sample points	Vapour release to atmosphere

TABLE 21.3
Audit Observations and Possible Root Causes

	Observation	Possible Root Cause
1	Incomplete PTW forms	1 Poor procedure 2 Poor supervision (and auditing) 3 Limited or no training
2	Liquid spillages	Poor supervision and clean-up by operations personnel
3	Missing machinery guards on rotating equipment	1 Poor operational practices 2 Poor application of the PTW system (see 1 in this table)
4	Litter around site	1 Poor supervision 2 Poor safety culture
5	Cigarette ends around the site	1 Poor supervision 2 Poor safety culture

The "where?" is seeking to identify the source of the non-conformity (sometimes called a *deviation*) which may be a physical or a deviation from expected procedures within the organisation or its management systems. The problem may lie with

1 the organisation
2 personnel
3 other

The "why?" is seeking the root cause of the non-compliance. These are likely to be obvious and have clear indications. The potential list is long and again only a few illustrations can be given (Table 21.3).

21.7.3 The Audit Process

Audits can vary in depth and scope. Simple audits based on checklists and questionnaires require less preparation than comprehensive, detailed audits. The first activity, therefore, is to define the scope and the terms of reference. Often a manager within the organisation will request an audit. In some cases, an audit will be required at fixed intervals. An audit manager will be appointed, and the scope and terms of reference determined. These will generally cover the following:

1 The type of audit to be carried out
2 The objectives of the audit
3 The start date of the audit
4 The delivery date of the audit
5 The person to whom the audit should be delivered (if different from the initiator)

The scope will generally cover the following:

1 The type of audit
2 The areas included
3 The areas excluded
4 Any specific area for detailed attention
5 The benchmark against which the audit should be measured

Detailed planning and preparation will facilitate the audit. Key points to consider are as follows:

1 The list of team members must be finalised and agreed. The team should be chosen according to their skills and the nature of the audit itself.
2 "Domestic arrangements" must be made. For a simple audit, this may consist of ensuring a meeting room is reserved and refreshments are available. A high-level corporate audit will require accommodation for the auditors, site liaison official, office and telephone facilities, filing cabinets/server storage for documents; secretarial support; PPE; security clearances; and safety induction.
3 Unless the team are all from the same site, including the liaison official, they should meet prior to the audit for familiarisation. Following this a list of initial documentation and personnel to be interviewed should be agreed. Any necessary checklists should be developed.
4 An initial timetable for tasks and interviews established.

Following the initial meeting the audit team may need time to examine documents and analyse information prior to the site visit. It is likely that additional information, extra documents or data will be required following this initial review.

The first site meeting must be a "courtesy meeting" with site personnel to discuss the scope and terms of reference, arrangements for the audit and any practicalities. It is important that the audit team start the audit with a positive meeting with site personnel.

> One gas processing company welcomed a young auditor to undertake an environmental audit. The auditor and site guide walked towards the process area and storage tanks when the guide asked to see the lube oil storage drums.
>
> "Don't you want see the process and storage area?" asked the guide.
>
> "No," replied the auditor, "I don't know anything about tanks and equipment. I want to check the drip trays under the lube oil drums."
>
> The auditor was obviously too inexperienced for the task and the site would find it hard to have faith in the findings. The entire audit was a waste of time and effort.

Security passes will need to be issued and safety inductions take place. The audit can now start and the "what?", "why?" and "where?" questions can be asked during interviews with site staff and site visits. Auditing is based partly on experience and definitive rules are hard to apply. However, it is surprising how often a finding is identified by the well-trained eye and ear.

At the end of each day a short audit team meeting should be held to exchange observations and classify details. The meeting need not be highly structured but would generally include the following:

1 Preliminary findings (these may change at a later date)
2 Areas requiring further clarification
3 Gaps in knowledge
4 Areas needing further study

Periodically—not longer than every other day—the audit leader should meet with the site liaison official to provide feedback and clear up misunderstandings—which are surprisingly common. Regular meetings will resolve possible misunderstandings and also prepare the site for the need for a plan of action, if necessary. Regular communication should help prevent resentment of the audit team and the development of poor working relationships. The attitude to audits is beginning to change. Traditionally, auditors have been received with apprehension and suspicion, but with some managers now consider them as a tool for improvements. It is very important that data is recorded and filed. The data becomes the evidence that supports the finding and recommendations. Initially progress may be slow and additional meetings and site visits may be required for the audit team to gain a full understanding.

Ideally, a draft report—or at least a verbal summary—should be available and presented to the site personnel before the audit team departs. Inevitably there will be negative features to report and these should be balanced with positive comments where possible. Feedback should avoid surprises in the final report. However, it is common for information to be provided to the audit team after they have left the site so comments in the final report which have not been discussed on-site may occur. The final report must reflect the observations of those being audited. If the appropriate liaison and supporting information is available, the final report should only be a "polished" version of the draft, possibly including site comments on the draft report.

The audit will, invariably, produce a list of recommendations. These must include sufficient detail that they can be implemented without confusion, together with a timetable and a person responsible for implementation. A mechanism to follow up each recommendation until completion is essential.

Whilst some recommendations will be straightforward, to be credible, they should contain a number of characteristics:

1 The recommendation must reflect the needs and products of that company as well as its corporate structure. For example, a research company which produces a technical product does not necessarily require a complex PTW system, but it would be appropriate to expect a person with specific responsibility for safety during maintenance/construction work to ensure good practice. A large petrochemical complex on the other hand could be expected to have a whole team of safety/environmental specialists and a comprehensive PTW system.
2 The recommendation must be practical. It is no use recommending a large fire water system for a plant located in the middle of a desert (which has happened). Making impractical recommendations or imposing impractical time scales only damage the integrity of the audit.
3 The recommendation must be measurable against a benchmark. This could be corporate standards, accepted good practice or legal requirements.

4 The recommendation must be researched and supported. One person who did not wear head protection or one minor spillage would not be a suitable sample and may be due to one rogue employee. Single minor issues should be highlighted but are not a major finding. However, if there are numerous examples, it will be necessary to search further to find the real source of the problem.
5 The recommendation must address the cause of a problem not its symptom. If numerous fire extinguishers are missing, for example, the recommendation should go beyond recommending they be replaced and determining why they were missing. Was the budget for maintenance/replacement adequate? Was there confusion over responsibilities? Are the extinguishers obsolete? The recommendation must address the root cause of the problem.

Whilst audits should be planned to a schedule, there may be times when they are required with great urgency.

Following Piper Alpha, one oil major required that all UK offshore facilities be audited against the possibility that a similar event might occur in that company. There were 15 platforms and the date of delivery was eight weeks after the Piper disaster.

A team of seven experienced engineers was nominated within two days. Amongst other things, the logistics including: identifying the data required for scrutiny, data collection, hotel accommodation and helicopter availability. Each platform required the reading of an eight-centimetre thickness of pre-requested information. After reviewing the information, a set of queries was drawn up for clarification by the onshore management on the first day of the audit. Meetings were also held with the offshore team. At the end of the day a set of key observations and queries was drawn up for a platform visit on the second day of the audit. The third and final day involved writing the audit report whilst information was "fresh in the head" of the audit team. The audits took place to a very tight timetable whilst the hundred-plus deaths at Piper Alpha were still fresh in everyone's mind. This was a very tiring, stressful and emotional time for the audit team.

The target was achieved and all of the "findings" had been ranked into:

RED = Action needed
YELLOW = Review the action and determine the priority
GREEN = Probably not a priority but the action could be dealt with by a procedural change (118).

Even in well-run organisations, there is often a tendency for compliance with procedures to slip. This may not be notice by personnel closely involved in the system and period external reviews are useful.

21.7.4 Example: Simple Audit

Permit to Work procedures are discussed in Section 16.6 and an example outline for a simple PTW audit is given in Table 21.4. A plant manager who could be both the initiating manager and auditor could carry this out.

The auditor should take a sample of the permits produced in one day and progress all of them, checking that the work was being carried out correctly and as laid out in the Permit. Additionally, the auditor should sample the permits issued by each shift to satisfy that there is not a "weak link".

The auditor should issue a report containing:

1 Number of permits inspected
2 Number of permits progressed on-site
3 Deficiencies noted
4 Actions/recommendations proposed or taken to correct the deficiencies

TABLE 21.4
Outline for an Audit of a Site PTW System

Terms of Reference	To examine the PTW system to ensure that work is performed in compliance with site procedure. To deliver the results to the plant supervisor (name) by (date).
The Scope	To examine a sample of PTW for the weeks x/y. To include hot work and cold work permits but to exclude confined space entry permits.
Sample Size	The lesser of 50% of the PTWs issued or 100 PTWs
Checks	1 Were they filled in correctly and unambiguously? 2 Is the task or job clearly defined? 3 Are the isolation requirements clearly specified? 4 Are the tools required clearly specified? 5 Are the hazards clearly specified? 6 Are the site precautions clearly specified? 7 Are the emergency procedures clearly specified? 8 Are prohibited acts clearly specified? 9 Are other tasks in the area clearly specified? 10 Are authorising and sign-off signatures correct? 11 If the task was delayed were new conditions specified/

The report should be issued to the Plant Supervisor and after one day discussed with him/her. Typical faults identified during Permit to Work Audits include the following:

1. Issuer signs "on behalf of . . .".
2. Conditions for a Hot Work Permit were changed on the Permit without checking the site for the reasons or causes of the change in conditions.
3. No-one inspected the work site before the job started.
4. Incompatible work was taking place in the same work area.
5. Valves were not correctly identified (tagged) and locked shut.
6. There were more permits in force than any reasonably competent issuer could supervise and check/monitor.
7. The maintenance crew had no idea of the potential risks in the area.
8. The maintenance crew had not read the PTW
9. Signatures were missing from the PTW
10. The PTW was in use beyond the period specified
11. Supporting documentation had not been prepared or was missing

There are numerous other examples that would be familiar to experienced plant personnel.

> During an audit, the auditor was in casual conversation with a process supervisor when a message came through on his radio. One interesting comment overheard was: "I know that I set the gas limit at x% but I will alter the permit and raise the limit".
>
> This is in violation of any PTW procedure. When there is a change in conditions work must stop until the change has been fully investigated. The auditor put on his "poker face" and showed no adverse reaction. However, this incident went into the audit report as a "finding".

21.7.5 Human Factors

A successful auditor will require good interpersonal skills and, at the start of the audit demonstrate they have the knowledge and experience to conduct the audit. An auditor who does not understand how the plant works or its management systems, is unlikely to be taken seriously and is very likely to be misled. They will also need to overcome the common misconception that they are "policemen" coming to criticise. The skills and approach are similar to those required for a successful HAZOP leader (see Section 5.9). The aggressive attitude of some auditors and defensive attitude of auditees is also unhelpful. An effective audit requires a positive relationship between all the parties involved. The objective of the audit is to improve safety in the plant—not to find fault.

The auditor is eliciting information from the site and personnel. To a degree the interviewee will be defensive, more particularly if the auditor is not recognised or known to the interviewee. The first task is to put the interviewee at ease. This may include using first name terms and an informal manner.

Interviews should start informally before proceeding into the more technical aspects. Where possible verbal answers should be supported by documentation. However, avoid a tone which suggest that you want documentation because you don't believe what is being said. Often a prepared list of questions is used as an agenda. These should provide guidance rather than limit discussion. The answer to one question can lead the interviewer into relevant but previously unknown areas.

It is often useful to ask the same question more than once in different ways and to different people to ensure consistency in the answers. There is no harm in summarising and checking the interviewee's answer so the auditor can be sure they have a correct understanding. This is particularly useful when auditor and interview do not have the same native language.

A good auditor should have the following attributes:

1 Behave in a professional manner
2 Be a good listener
3 Demonstrate adequate technical knowledge (and admit where his knowledge is weak)
4 Avoid conflict and avoid belittling or embarrassing the interviewee
5 Ask questions in an easy-to-understand, unambiguous way
6 Be open-minded and non-judgemental
7 Be observant of body language during interviews and notice detail during site visits

21.8 INCIDENT REPORTING AND INVESTIGATION

21.8.1 Theory of Accidents

Accidents and other incidents may appear to occur suddenly or may appear to develop over time. In reality, all accidents have a series of precursors, which may be dormant or unrevealed. The *sudden event* such as a twisted ankle will be over in one or two seconds and the investigation will rely on visual descriptions of onlookers or narrative of the injured person to give clues. The *developing event* will be more complex and takes place over time. In a generalised model, the accident will have discreet phases.

1 Build-up: When the elements or causes are put in place.
2 Initiation: When the event is initiated.
3 Escalation: When the event grows in size.
4 Controlling: When the event is brought under control.
5 Rectification: Eliminating the causes of the event and, hopefully, similar events.

Safety procedures generally operate on the basis of *Defence in Depth* and may be systematically eroded such that an *Initiating* event starts the sequence of events, which lead to the accident. If some more defences are eroded, the event may *escalate*—if not, it will be *controlled*. This concept can be visualised by the Swiss cheese model (see Section 5.1) or a bow-tie diagram (see Section 6.4). The distinction between "incident" and "accident" is important. An accident results in damage of some form: to a person, equipment, loss of containment or damage to the environment. An incident may be an accident but could also be a situation where no harm did occur but could have. Technically these are known as a *near miss*.

The difference is easily explained by an example. If a worker walks under a ladder where someone was working overhead and is hit on the head by a falling hammer that is obviously an accident. If the hammer falls but misses the worker, it is a near miss. The situation where people can walk under a ladder where work is taking place above could be the subject of a safety observation (see Section 21.3).

21.8.2 Reporting Systems

Different industries and organisations have different reporting criteria, but general practice is to define incidents under broad classifications such as: injury, property damage, releases to the atmosphere, loss of production and, more recently, damage to the company's reputation. Each type of incident is further subdivided into levels of severity. Again, the number of levels might vary and sometimes they are aligned with the company's risk matrix (see Section 6.5). Be warned that matrices can be a very blunt tool and may focus attention on classification rather than identifying root cause.

The internet has facilitated incident reporting and the spread of information resulting in important advances. The use of common reporting systems across companies, and sometimes whole industries, allows individual companies or locations to benchmark themselves against the industry as a whole. Classification of incidents in a consistent manner will allow "accident hot spots" to be identified and corrective action taken. Finally, the internet allows the rapid and widespread dissemination of information soon after an incident occurs as well as disseminating detailed accident investigations as soon as they are available. Information spread within a company is usually rapid but there are often barriers to sharing information across companies, industries and countries. Some organisations are reluctant to admit their shortcomings, particularly if it may affect their public image. Legal action, real, perceived or threatened, often prevents information spread and the details of an event—which could prove useful to prevent a recurrence—might be unavailable for years after the event and sometimes not at all.

Often incident statistics are considered of limited value beyond inclusion in periodic reports. However, detailed analysis is often very useful in identifying if there is a common cause for several incidents of a similar type or multiple incidents in a particular location. Corrective action can then take place.

21.8.3 Investigation

It is important that all incidents are investigated to identify not just the immediate (sometimes called proximate) cause but also the *root* cause. The more serious the incident, the greater the level of investigation required. Very serious incidents are often exhaustively investigated in a process which might take years and could lead to criminal prosecutions. A number of serious UK accidents[22,23,24,25] investigated by the UK HSE illustrate the level of detail that may be necessary to identify root cause.

Investigation proceeds in several stages, the UK HSE guidance on investigation[26] suggests four stages:

1 Gathering the information
2 Analysing the information
3 Identifying risk control measures
4 Creating the action plan and its implementation

Investigations should start as soon as possible after the incident, particularly interviews with those involved and the incident location visited. Interviews should be conducted as soon as possible after the event to minimise memory fade but must be conducted with sensitivity as those involved may have suffered significant trauma. Similarly, the incident location must be surveyed in depth as quickly as possible to ensure wind, rain or souvenir hunters do not destroy or remove evidence.

There is often pressure to rush through site investigations so that repairs can be made and production restarted. This must be resisted as a hasty investigation and analysis will increase the difficulty of identifying the root cause.

More detailed approaches are possible, one six-stage approach is summarised in the following table:

1	Immediate Action	Action to make the site safe for the investigation to begin. For example, electrical supplies may need to be isolated, or buildings decontaminated. However, as far as possible, the incident location should be left untouched. When the event occurs, or even just before it occurred, "*someone*", "*somewhere*" will have heard or seen "*something*". This has to be captured at some point. These people are likely to be very active in taking control of the event and cannot be immediately disturbed. Investigation therefore requires skilled, tactful interviewing as well as technical analysis. After the event, memories of the incident may be distorted or facts may be missed so it is important that detailed visual records of the incident are kept. Digital cameras and videos will provide immediately accessible data if it is safe to use them in a flammable atmosphere. It is not a question of "*what*" should be recorded, more a case of "*anything*" and "*everything*". Most will be repetitive, but some of the views from different angles will add to a three-dimensional picture. If possible, the exact physical location and time of each shot should be noted.
2	Plan the Investigation	Many organisations will have a procedure for incident investigation with the level of investigation defined depending on seriousness. This approach will work for minor incidents but detailed planning will be necessary for serious incidents. For example: 1 Is the incident site safe to approach? (could it collapse, could hazardous materials still be present?) 2 Will special protective equipment be necessary? 3 Will special equipment be necessary? (gas detectors to warn of leakage, scaffolding for access to elevated areas, cranes/bulldozers to remove debris and so on. 4 Are there any personnel unaccounted for? (what sets the need to be in place for recovery of bodies) Numerous other factors will need to be considered including how/when witnesses will be interviewed, what laboratory investigations or experiments will be necessary, how information will be recorded and how/when the site can be returned to service. As the investigation proceeds, the plan might need modification. After the incident, nothing should be disturbed until a number of photographs of the area can be taken. If time allows, a professional photographer should be used but, failing this, high quality images, which can be enlarged, are essential. Much of the information in the photographs will be irrelevant but some will be important and no one can be sure at this point which is which. If possible, the area should be grid marked for future reference. Before anything is moved, the position of each item should be noted against the grid mark, or against a set of coordinates, and then carefully labelled. The objective should be that when pieces of data are removed for inspection they can be replaced in their original place and orientation, should the need arise. The location of each piece of evidence against fixed coordinates, its orientation and its location relative to other pieces may be critical later. For close-ups, a ruler is often included in the picture to show scale.
3	Data Collection	Physical evidence can now be removed carefully, piece by piece, and stored in a safe, secure room or workshop together with plant records and operating parameters leading up to the incident. As soon as possible after the incident eyewitnesses must be interviewed. The interviewer will be part of the investigative team and must know what facts to seek out. It is essential this is done quickly as time will start to distort the memory and also, if eyewitnesses discuss the event, the memory of a key witness may be influenced by others. The interviewer or interviewer(s) must be tactful and sympathetic—the interviews are to obtain information, not to find someone to blame. It is very likely that witnesses will be suffering from some degree of shock, may be frightened and will not respond to a pressured interview. The witnesses should be put at ease and invited to write what they saw and heard in their own words. Cross referencing should occur between witness statements and also against other documentation such as log sheets, permit to work forms and other documentation. CCTV footage may also be useful, if available. Most modern process plant is controlled by DCS which will include a historian recording process variables at the time of the incident. As the investigation continues there may be further data requirements—many incidents are caused by actions taken over weeks, months or years before the incident occurs and a trawl through historical records may be necessary. Good record-keeping from the conceptual design stage is helpful.

4 Data Analysis	Generally, a preliminary report will be required within a short time frame, typically a day or two. A final report on a complex incident could take months and, if external authorities or lawyers become involved, years The investigation can be a long and complicated process to identify the sequence of events which led to the incident. Frequently, there will be a number of ways an event could occur. For example, a gas leak could have been ignited in a variety of ways requiring investigation to discount the less likely options and, if possible, confirm the source. This may require some experimentation. Risk assessment methods such as fault and event tree diagrams (see Section 6.4), may be helpful in this respect. The immediate cause of an incident may be identified but the root cause(s) may take considerably more work. A variety of specialist techniques[27] and computer programmes have been developed for this purpose. Analysis of equipment damaged in the incident may involve destructive testing and the value of information obtained by testing needs to be assessed carefully. Where this is necessary samples must be recorded in detail before testing takes place. A simple way of determining if the root cause has been reached is the *Why Test?* Which works by looking at the cause of each step of the investigation and asking the question "why?". For example: 1 Why did the gas escape occur?—because of corrosion 2 Why did the corrosion occur? wall thinning 3 Why was the thinning not detected?—there was no regular inspection of the line 4 Why was there no regular inspection? 5 . . . 6. . . . This process should continue until the question "why?" cannot be asked any further. Generally, there will a number of contributing factors to a major incident and each will need to be explored. The investigation would then work to consider if a credible sequence of events, supported by facts has been determined. Sometime more than one series of events may be credible. To use the analogy of a jigsaw, the corners may be in place, some of the sides in place and groups two or three other pieces fitting together—some being fitted incorrectly. It is also possible that the pieces could be fitted together in a different way to give another picture. At this point, it should be possible to eliminate certain scenarios with reasonable confidence. Using a criminal analogy, there will now be the some useful "lines of enquiry" but the investigation team should still be open-minded. The team can now address the hard data obtained from the area of the incident and decide which pieces of data support the various scenarios. The physical evidence of the incident will often provide useful information. By reviewing blast damage, such as the distortion of steel work, lampposts and other structures the epicentre and approximate size of an explosion could be determined. The extent of charring or softening of a steelwork can provide information on the size and temperature of a fire. Physical examination of components might identify the mode of failure. Note that a major event may occur in stages. A relatively small blast may result in rupture of pipework leading to a large liquid spill and fire. Dust explosions are a classic example of the multistage effect. A small explosion will cause significant air turbulence often causing combustible powder to form a flammable cloud resulting in a second or larger explosion. A literature search for similar incidents will also prove useful. Eventually, one or more of the scenarios will be proven to be wrong and can be dismissed. Some of the data will require more detailed analysis. The team should also examine historic data from plant records and operating parameters. There may be other useful observations where someone saw/heard "something" before the incident and feels it may be useful. Reverting to the analogy of the jigsaw, one picture could be developing as the likely scenarios and other scenarios may be less likely—but still not capable of being rejected. Eventually (and hopefully rapidly), one incident scenario will become the probable one. The main evidence should, where possible, be supported by a second set of evidence (corroborative evidence in the forensic analogy) but there will nearly always be a small amount of unsupported evidence and data, which fits no scenario. The objectives, where possible, are either to prove this data is irrelevant or to find supporting evidence. Finally, the whole sequence of events should become clear and there will be a high degree of confidence that there is a consistent, proven story. At this point, it should be possible to predict that there should be other evidence—this may be available already but not fully recognised or it may be necessary to make further investigations.
5 Corrective Actions	Most investigations will result in recommendations to prevent a recurrence. These will generally be fully detailed in the investigation report but urgent recommendations which may have implications for safety are often made prior to issue of the full report.

6 Issue of Report	The report should detail the circumstances leading up to the event, the event itself, how the event was controlled, analysis of causation, lessons learnt plus recommendations. Note that recommendations often include actions to be taken if a similar event occurs as well as recommendations to prevent a recurrence. The report should cover both proximate and root causes. In many cases, the report will discuss all possible causes and discuss which were rejected and why. The development of the Incident will be fairly factual but will often contain comment, particularly where information is not fully available or conclusive. This section will compliment and link with the causes. The Lessons learnt come at two levels. In the first there are the conclusions, which will complement the causes and flow into the recommendations. The second level will be the messages, which may require to be cascaded down through the site/company and the industry as a whole. It would be naive to believe that all recommendations can be complete within a few days; inevitably there will be some form of risk ranking given to the recommendations which will influence the time frame. Some recommendations may be fundamental and must be implemented before restart of the plant. Others may be deferred and could be implemented later. Some may require significant capital investment, involving design, procurement and construction. This may involve a significant lead time, during which temporary measures may be necessary. A key part of the report is the summary. This is likely to be more widely read than the full report, particularly by senior managers, and it is therefore essential that it contains all the key information and recommendations. The report will usually be circulated throughout the site and other company sites. Sometimes reports, or summaries, are published more widely—often anonymously. With the increased use of company intranets, incidents and reports can be circulated companywide very quickly. The final incident investigation report may have to be submitted to the regulatory authorities. In the case of serious events, particularly if there are fatalities, they will conduct their own investigation.

There is often a temptation to shorten the investigation process when a simple solution fits the facts. However, these may be incomplete or driven by mindset.

> "An operator slipped on a staircase, twisted his ankle and was absent for 17 shifts. The staircase seemed to be in good condition and so did the man's boots."
>
> An initial reaction is probably that it is just another of those incidents that we can do nothing about—another of those occasions when "*Man told to take more care*" could have appeared in the accident report. However, on the plant concerned they were not satisfied with the easy way out. They looked into it more thoroughly. They asked the injured man why he had not used the handrails. It then came to light that the handrails were covered with plastic and that when anyone wearing insulated footwear used them, they got an electric shock when their hand touched bare metal. As he ran his hand along the plastic coating he became charged, and as soon as he touched a piece of bare metal an electric spark jumped from him to the metal. The spark, of course, was not serious enough to cause any injury, but it was unpleasant, and people tended, therefore, not to use the handrails.

A fact, sometimes forgotten, is that the primary purpose of any investigation is determine what happened to prevent a reoccurrence. If individuals are found to be negligent or deliberately cause damage disciplinary action is justified. However, the object is not to find "someone to blame", although this is not always obvious, particularly if the incident is known to the public and the press are involved.

21.9 THE BIZARRE

Safety is a serious matter and continuing attention is necessary to prevent every accident and minimise the effects of those that occur, there will be some instances which are so unusual and unlikely that no method of hazard identification will identify them in advance. In some cases, there can be no effective lessons learnt.

Hydrogen cyanide (HCN) is a hazardous material which can polymerise exothermically, it is flammable and toxic. A number of incidents involving HCN have been reported.[28] In one case, a tank of hydrogen cyanide became unstable and polymerised resulting in a minor explosion and release of HCN liquid which formed a pool under an adjacent pipe rack and ignited. By a happy coincident the initial explosion ruptured a water line running above the pool fire. Large-scale release of toxic vapour was prevented because HCN escaping from the rupture formed a pool fire and burnt (the products of combustion being harmless). The fire did minimal damage as the area was cooled by water discharged from the ruptured pipe.

An electrician needed to repair a low-voltage underground cable at a manhole. Having assessed the hazards, he arranged an insulating mat in the area he would be working and, as it looked like rain, he erected a canvas frame over the manhole cover. As he was working on the insulated mat in the dry an inquisitive dog came to see what was happening, received an electric shock and bit the electrician.

A leak of heat transfer fluid melted the insulation on a telephone cable. The plant supervisor phoned the fire brigade to report and an electric short circuit from the damaged cable ignited the leak.

A shift operator cut his hand whilst removing the metal foil from the top of a jar of coffee. The injury was minor but the company's strict accident reporting procedure required an investigation. The facts were simple and the investigation was speedily complete. However, the chair of the investigating committee was struggling to complete the section of the investigation format dealing with "actions to prevent a recurrence". In desperation, he suggested that, in future, the operator drink tea and not coffee.

At a chemical plant at Fukuyama, Japan A cat fell into a tank of carcinogenic hexavalent chromium. Staff initially noted yellow paw prints and then spotted the cat on CCTV. The cat escaped from the site and the public were warned not to approach it[29].

There have also been cases of feral cats crawling between the phases of electrical feeds and causing a short circuit. The cat burnt to death instantaneously but production was lost for a period whilst the site slowly came back on line. Problems with rodents biting through cables are also well known in the industry.

Good procedures and a good safety culture will go a long way to prevent major, and minor, incidents from occurring but eternal vigilance is also necessary.

REFERENCES

1 Rosenzweig M, Trevor Kletz Bequeaths Better Process Safety, *Chemical Processing*, 19 November 2013, www.chemicalprocessing.com/environmental-health-safety/article/11340195/trevor-kletz-bequeaths-better-process-safety

2 Kelleher Potts M & Pomares T, *Retaining Knowledge for Process Safety Management (IChemE Symposium Series 162)*, IChemE, 2017, www.icheme.org/media/15511/paper-58.pdf

3 Mathis T, Unions and Behaviour Based Safety: The 7 Deadly Sins, *ProAct Safety*, October 2009, https://proactsafety.com/articles/unions-and-behavior-based-safety-the-7-deadly-sins

4 Wallace I G, *Developing Effective Safety Systems*, IChemE, 1996

5 UK Health and Safety Executive, *Managing for Health and Safety*, HSE, 2013, www.hse.gov.uk/pubns/priced/hsg65.pdf

6 Norwegian University of Science and Technology (NTNU), *Risk Metrics*, www.ntnu.edu/ross/risk/risk-metrics#:~:text=Individual%20risk%20per%20annum%20(IRPA,activities%20a%20during%20one%20year

7 Tims C, A Guide to Tracking & Calculating Lost Time Incident Rate (LTIR), *EHSInsight*, 17 February 2023, www.ehsinsight.com/blog/a-mini-guide-to-tracking-and-calculating-lost-time-incident-rate-ltir

8 UK Health and Safety Executive, The Fire at Hickson and Welch Ltd, Tims C, A Guide to Tracking & Calculating Lost Time Incident Rate (LTIR), *EHSInsight*, 17 February 2023, HSE, 1994, www.icheme.org/media/13704/the-fire-at-hickson-and-welch-ltd.pdf

9 Norwegian University of Science and Technology (NTNU), *Risk Metrics*, www.ntnu.edu/ross/risk/risk-metrics#:~:text=Individual%20risk%20per%20annum%20(IRPA,activities%20a%20during%20one%20year

10 Lewis S, *Risk Criteria—When Is Low Enough Good Enough?*, 12 June 2007, www.traficom.fi/sites/default/files/17845-L6_Risktec.pdf

11 Henrich H W, *Industrial Accident Prevention A Scientific Approach*, McGraw-Hill, January 1959

12 Private Communication

13 IK Health and Safety Executive, *Developing Process Safety Indicators A Step-by-Step Guide for Chemical and Major Hazard Industries (HSG254)*, HSE, 2008, www.hse.gov.uk/pubns/priced/hsg254.pdf

14 Center for Chemical Process Safety (CCPS), *Process Safety Metrics Guide for Leading and Lagging Indicators*, CCPS, 2022, www.aiche.org/sites/default/files/docs/pages/ccps_process_safety_metrics_-_v4.1.pdf

15 Webb P, *Process Safety Performance Indicators—A Tool to Help Senior Managers Show Process Safety Leadership* (IChemE Symposium Series 154), IChemE, 2008, www.icheme.org/media/9696/xx-paper-01.pdf

16 Anderson M, *Behavioural Safety and Major Hazards: Magic Bullet or Shot in the Dark*, HSE, www.hse.gov.uk/humanfactors/assets/docs/magicbullet.pdf

17 Baker J A (chair), *The Report of the BP US Refineries Independent Safety Review Panel (p 134)*, January 2007

18 Stuart, *Moving from a Danger Culture to a Safety Culture*, March 2015, https://acsdchas.wordpress.com/wp-content/uploads/2015/03/moving-from-danger-to-safety-culture.pdf

19 Lardner R, Fleming M & Joyner P, *Towards a Mature Safety Culture* (IChemE Symposium Series No. 148), IChemE, 2001

20 Safety Culture Maturity Model is a registered trademark of The Keil Centre

21 The Keil Centre, *Safety Culture Maturity Model (OTO 0049)*, HMSO, 2000, https://webarchive.nationalarchives.gov.uk/ukgwa/20230103131249/www.hse.gov.uk/research/otohtm/2000/index.htm

22 Hon Lord Cullen, *The Public Inquiry into the Piper Alpha Disaster*, HMSO, November 1990, www.hse.gov.uk/offshore/piper-alpha-public-inquiry-volume1.pdf

23 Hon Lord Cullen, *The Public Inquiry into the Piper Alpha Disaster*, HMSO, November 1990, www.hse.gov.uk/offshore/piper-alpha-public-inquiry-volume2.pdf

24 Parker R J (chair), *The Flixborough Disaster, Report of the Court of Inquiry*, HMSO, www.icheme.org/media/17752/the-flixborough-disaster-report-of-the-court-of-inquiry.pdf

25 UK Health and Safety Executive, *The Fire at Hickson & Welch Ltd*, HSE, 1994, www.icheme.org/media/13704/the-fire-at-hickson-and-welch-ltd.pdf

26 UK Health and Safety Executive, *Investigating Accidents and Incidents A Workbook for Employers, Unions, Safety Representatives and Safety Professionals (HSG245)*, HSE, 2004, www.hse.gov.uk/pubns/hsg245.pdf

27 Guevara P, A Foolproof Guide to Root Cause Analysis, *Safety Culture*, 28 May 2024, https://safetyculture.com/topics/root-cause-analysis/

28 Williams C & Schwarz N, *Liquid Hydrogen Cyanide Polymerisation Hazards (IChemE Symposium Series No. 160)*, IChemE, 2015, www.icheme.org/media/8681/paper-52-hazards-25.pdf

29 Muzaffar, Cat Falls into Chemical Vat in Japan—Residents Warned to Stay Away from Animal, *Independent*, 13 March 2024, www.independent.co.uk/asia/japan/fukuyama-cat-chemical-tank-do-not-touch-b2511716.html

22 Emergency Response

22.1 INTRODUCTION

Emergency planning is aimed at identifying and responding to credible events. Credible implies a situation where there is good reason to consider an event can happen without wandering into the realm of the barely possible such as a meteor strike. A toxic release is *credible* in an industrial environment or on transport routes but not in a remote area where it is not manufactured, used or on a transit route.

Whilst every event is different, in broad terms, there are five stages leading up to and beyond an emergency:

1 Build-up—where the protective safety barriers decay (the holes in the Swiss cheese get bigger) (see Section 5.1).
2 Initiation—the single event which sets off a chain reaction
3 Escalation—this may or may not be a case but if it is there is no way of predicting the outcome
4 Control—simple for a minor event but complex for a major event.
5 Rectification—dealing with the aftermath.

Whilst a limited amount of planning for emergencies has always been done, this is becoming increasingly sophisticated. Some major emergencies, such as Chernobyl, Bhopal and Beirut appear to have been the subject of limited emergency planning. However, it is now accepted practice to develop an *Emergency Response Plan* (ERP) so that if an emergency does arise there is a framework in place to provide a response. Whilst fires, explosions and gas releases are the most likely scenarios within the process industries a comprehensive emergency plan will look at other situations such as: earthquake, storm, flood, terrorist attack or civil unrest. The list will vary depending on the installation and location. Marine and offshore facilities will need to consider vessel collision, man overboard and a diver in trouble. Secondary effects also need to be considered; a fire might generate toxic fumes, and an explosion in an old building may release asbestos into the atmosphere.

Major accident scenarios should have been identified during the project development using hazard identification techniques (see Chapter 5) and quantified for a magnitude and likelihood (see Chapter 6). The major incidents should have been characterised, the effects identified and appropriate preventative and protective systems installed. This helps to identify the key requirements of the ERP.

Most incidents are likely to be small—a fire in a wastepaper basket or a minor vehicle collision—and can easily be dealt with by onsite emergency response personnel. All incidents should, of course, be reported for investigation to prevent a recurrence (see Section 21.8). Other incidents, which may not be the worst-case credible event, still need consideration. The effects of a tank fire might be much less than a VCE but could still have significant implications in terms of hazard to personnel, damage, loss of production, pollution and damage to company reputation.

Natural perils can be assessed from various databases; unfortunately these are often only available on a subscription basis. More detailed information will be available from local meteorological data at the design stage. The effects of global warming may cause extreme events which exceed design criteria, particularly for existing plants. Terrorism and civil unrest will require a threat assessment by a security specialist. The events identified must be *credible*. Process facilities catching fire is credible, but an incident caused by impact from an asteroid would be considered *incredible*. Note that sometimes events which appear incredible are, in fact, credible.

DOI: 10.1201/9781003510680-26

The ERP is often split into onsite activities—aimed at controlling the incident—and offsite activities to ensure the safety of the surrounding population and property. The latter may involve liaison with local councils, mutual aid partners,[1] local emergency services and emergency accommodation/food/supplies/fuels and others. Ideally, the offsite exposure should be low, with planning controls limiting development around the site. Unfortunately, this is not always the case, particularly with older facilities.

22.2 EMERGENCY RESPONDERS

The provision of adequate emergency resources is a vexed subject. Emergency response managers will want a high level of both physical resources and trained personnel. However, for most of the time the equipment will not be needed and personnel will not be employed in emergency response activities. Excessive resourcing is expensive but under-resourcing will leave a company unable to effectively respond. Emergency responders are often employed on secondary tasks such as gas testing, fire watchers for hot work and inspection of firefighting equipment. The responders will also spend a significant proportion of their time training.

The most "economic" solution will vary with local conditions. In Western countries the number of professional firefighters will be limited with external resources being called upon. These will be either auxiliary responders from the site, local authority responders, mutual aid responders or a combination. Only small sites will rely exclusively on external response. In the developing world, salaries are much lower and many, relatively, small sites have full-time fire crews. As process plants operate 24/7, emergency response cover must also be continuously available.

Auxiliary responders will generally be shift workers who will abandon their normal role to provide the response to an emergency either directly or in support of full-time firefighters. In locations where staff live onsite (such as offshore platforms or desert locations) off-shift personnel should be available within a few minutes.

Local authority responders will be full-time personnel trained for domestic and industrial firefighting. However, they may lack the specialist training needed for chemical incidents or hydrocarbon fires. Without adequate hazard awareness and leadership, they may be a hazard to themselves and others. Local authority responders can be given adequately training to provide a high level of support.

> A fire occurred in a plant producing calcium carbide and the local authority fire brigade were called. There appears to have been poor communication and the hazardous properties of calcium carbide were not explained to the firefighters (calcium carbide reacts with water to produce acetylene). As often happens, the firefighters applied water to the fire. This produced acetylene which exploded resulting in several fatalities (135).

Mutual aid responders will often be professional or auxiliary responders working for companies with similar hazards. However, they may not be familiar with locations other than their own site, and familiarisation visits to other locations are desirable. Some mutual aid responders may be from dissimilar industries but provide personnel and equipment; for example airport firefighters may form part of mutual aid partnerships as they will have significant quantities of foam which would be useful for fighting tank fires.

Two important considerations which are not always considered are physical fitness and training. Emergency responders, particular firefighters will need higher levels of physical fitness than most other shift workers. For example, vertigo (fear of heights) might be acceptable for some staff but would be a severe drawback for a firefighter who is likely to climb ladders on a regular basis. Fitness requirements for UK municipal firefighter are publicly available.[2] Training is required for not only firefighting but also first aid and rescue skills. Some personnel will need additional training depending on their roles in an emergency, driving and leadership, for example.

"Practice makes perfect", so the saying goes, and this is as true in emergency response as elsewhere. Whilst theoretical training is important, practical training in conditions as close as practical to a real event is importance. Many large sites will have training facilities on-site; these are often limited and do not provide training on all the hazards that could be encountered. Lack of training for jet and running fires are common examples. There are a number of specialist training centres around the world that do have comprehensive facilities but courses can be expensive. Training needs to be supplemented by onsite drills to simulate fires and other activities such as recovery of a casualty within plant areas. Small drills are useful for training responders; they provide limited experience of a major incident and occasional large-scale simulations, involving mutual aid and municipal responders should take place. Ideally, this should be an annual occurrence but due to the difficulties of organisation many are cancelled. An intermediate form of simulation—tabletop exercise—can be used to simulate a major emergency without the need for mobilising responders. A tabletop exercise involves the emergency management team occupying the Emergency Coordination Centre (ECC) and responding to a simulation where information is provided via phone lines, messenger or other means. Emergency response management is discussed in more detail in Section 22.3.

To avoid panic, exercises are often notified in advance; this takes away an element of realism but avoids a situation where a training exercise is mistaken for a real event. Planning a major site simulation is a major exercise. Timing is important to avoid gridlock on local roads and notification to site personnel and the surrounding area will be necessary to avoid panic.

All simulations should be recorded (and ideally videoed), if this can be done safely, for post-exercise review which will identify positive actions and areas for improvement. Care is necessary to avoid a post-drill review from being too self-congratulatory.

22.3 EMERGENCY RESPONSE PLAN

Development of the ERP can be considered in three stages:

1. Planning (based on credible scenarios)
2. Development
3. Monitoring/practice

Several military sources are claimed to have said that "no plan survives first contact with the enemy",[3] and the same can be said of an ERP. To be effective, the plan must lay down guidelines and advice in key areas but not be so prescriptive that it becomes difficult to follow if the emergency is not developing in the way the ERP anticipates. The ERP therefore provides a guideline and criteria to assist the emergency management team (EMT) and other responders. For large sites, the ERP is supplemented by *pre-plans* to provide a rapid response to likely situations.

The plan should be sufficiently flexible to cope with unexpected events, for example:

1. Difficulty of emergency services or key personnel reaching the site. The latter point is generally covered by having a back-up contact for each key role in the ERT.
2. Changes in meteorology (heavy rain, changes in wind direction and other factors.).
3. Limitation in firefighting resources, particularly water and foam. (Major supplies of foam generally have emergency contact numbers and can ship foam around the world at short notice.)
4. The control of information from the site to the families who may be affected.
5. Gridlock due to peak traffic at the time of the incident. Who and how traffic is controlled offsite?

6 Public relations (including contact with the media).
7 Security to control entry to the site and deter photographers and voyeurs who try to get too close to the incident, putting themselves and others at risk.

22.3.1 Command Structure

A common source of problems is the lack of a clear command structure for the EMT. The chain of command will vary depending on the size of the incident. For minor incidents, there may be no need to mobilise the command structure but this will be essential for a major event.

The structure generally consists of three levels:

1 Strategic who will control the overall response to the emergency. This might well be remote from the site and located at the site's corporate office and be responsible for high-level decision-making and communication with the press and government agencies.

As in military situations, decisions in an emergency cannot efficiently be made by committee. At the strategic and, to a lesser extent, tactical levels some collaboration may be possible but at the operational level decisions need to be made quickly. It is therefore essential that responsibilities are clearly defined in the emergency procedures.

At a site level, the key members of the ERT will largely be drawn from senior members of the staff and will include:

1 Emergency Manager	Generally, the site manager
2 Liaison Manager	To deal with the press and outside agencies as well as the families of those who may be involved in the incident.
3 Facilities Manager	Often the site engineering or maintenance manager who has responsibility for ensuring utility requirements (water, power, fuel for vehicles) are available.
4 Health and Safety Manager	To advise on safety and environmental issues which could affect the site
5 Logistics Manager	To control the vehicle movements and ensure resources are available. One of the duties may be to control traffic flow around blocked roads or near smoke or toxic releases.

Additional team members may be necessary depending on the individual site. For example, a marine coordinator will be necessary for incidents offshore or at a jetty.

6 Tactical which will manage the emergency at the site. Generally, the site EMT will act as the tactical level. They will decide how to fight the fire, how to allocate resources and how to initiate search and rescue. For an extended emergency, they will ensure that all personnel have adequate rest and sustenance.

7 Operational this is the "front line" of emergency response: the personnel directly involved in firefighting and rescue operations. Often this will be site personnel, particularly operators, as well as emergency responders. Problems can occur at the operational level. Local authority firefighters will not have the same level of technical knowledge as the plant operators and, may put themselves and others in danger. For this reason, it is common practice for the shift supervisor to act as the *forward controller* and liaise with external firefighters The situation has improved in recent years with the development of HAZCHEM and similar systems (see Section 3.7).

In the UK these three levels are often called *gold command*, *silver command* and *bronze command*.

It is useful for key members of the emergency response function to be readily identified. External responders will not know site personnel and even company staff may not know who performs what role in the emergency organisation. For this reason, it is useful for key individuals to wear high-visibility clothing and colour-coded helmets with their position in the emergency organisation clearly identified.

The involvement of the local or national government personnel varies greatly between countries. In some countries, the local authority fire chief, has responsibility for firefighting which can lead to difficulties if the fire fighters are inadequately trained or under-resourced for the type of incidents that could occur.

On 25 November 1990[4] a major fire ignited at Stapleton tank farm at Denver airport. Heat from the original fire ruptured several fuel tanks releasing significant amounts of fuel. Whilst flying into/out of the airport was still possible, there were major difficulties in supplying fuel and difficult, costly and time-consuming measures were necessary to keep flights operating. Firefighters fought the blaze but were hampered by a number of factors including poor weather conditions. Firefighting operations continued for around two days without success. Eventually a specialist hydrocarbon fire team arrived and smothered the fire with large amounts of foam, extinguishing it within around 30 minutes.

In most cases good collaboration between the site and the local authority personnel is achieved but cases can arise where local authority personnel take control but have little understanding of the risks and appropriate firefighting response. Good communication and interpersonal relationships are essential and regular visits to the site by local authority personnel and involvement in drills help in this respect.

22.3.2 Planning

The first stage in developing an ERP is to identify the types of situations which may occur and what the response should be:

1. What can go wrong and where might it occur? Scenarios and consequences should have been identified during hazard identification (see Chapter 5), and risk assessments (see Chapter 6).
2. Who and what might be affected (both off- and onsite)?
3. What are the objectives and limitations of the plan?
4. What onsite and offsite resources (both personnel and equipment) will be required? (see Section 22.4).
5. How can escalation to a larger event be prevented?

These questions are indicative of the requirements for planning and can be expanded and modified depending on individual circumstances. For example, an offshore platform or a remote production site will often have to rely on its own resources with help being unavailable for a rapid response but being essential for site evacuation. As noted in Section 22.1, the scenarios to be considered will usually go beyond process incidents and involve numerous non-process related situations.

Whilst the greatest emphasis is placed on control of the incident, there will often be a separate specialist group responsible for evacuation of site personnel and rescue of anyone who is trapped.

Once the emergency scenarios have been identified, the ERP can be more fully developed. Most plans have a number of essential common features:

1. Call-out
2. Evacuation

3 Command structure
4 Event classification
5 Roles and responsibilities of the EMT
6 External resources (local authority and mutual aid responders; see Section 22.4).
7 Procedure for response to specific scenarios

Each is discussed in more detail in the following sections.

22.3.2.1 Call-Out

The ERP should specify how the alarm should be raised and the first actions to be taken. Generally, an incident will be reported to the emergency personnel. On a large site with dedicated emergency responders, this is likely to be the site fire station; on a smaller site this is likely to be the process control room. Once the alarm is raised, the onsite emergency resources will be mobilised and if necessary external resources can be asked to provide support.

The members of the ERT will have been pre appointed and their details (phone numbers and car registration numbers) noted at the site entrance to ensure they can easily enter the site. Only authorised personnel, including emergency responders, should be allowed entry. Security personnel must be especially vigilant to prevent trespassers (reporters, looters, spectators and others) gaining access, hindering rescue efforts and possibly putting themselves at risk.

The normal advice to the person reporting a fire is to only fight the fire *if it is safe to do*. In all other situations, the area should be evacuated.

An emergency situation will generally be communicated around the site by an alarm. There are generally two or three tones/durations:

1 One tone/duration to warn of a fire or other event
2 A separate tone to indicate a gas escape
3 Finally, a tone/duration to indicate the "all clear"

In noisy areas, the audible tone may be supplemented by flashing lights.

22.3.2.2 Evacuation

The names and approximate locations of personnel on-site, particularly in process, storage and utility areas, should be known at all times (see Section 16.4). This will assist in the location of personnel (including visitors/contractors) who have not responded to the roll call following an alarm. The time that all emergency responders enter and leave the incident area will be recorded and regular updates on their exact location maintained. The use of modern radio systems and GPS tracking will assist this. However, the length of time responders are within the incident area must be tracked—particularly when Breathing Apparatus (BA) is being used.

It is standard practice to provide muster (also called assembly) points where a head count will be taken and, if necessary, search and rescue operations initiated. Whilst they should be close to the area being evacuated, they must also be sufficiently remote that personnel will not be affected by the event taking place. An exception to this general rule occurs if there is a significant gas release, particularly at sites handling toxic or noxious gases. In this case, personnel must travel cross wind to a safe location. Wind direction indicators should be installed so that wind direction can be easily determined. Typically, these are "windsocks" which are large enough and brightly coloured making them easily visible.

In some instances, escape masks or short duration BA sets may be necessary to escape from affected areas such as poorly located control rooms and *safe refuges*. Personnel working in these areas should have ready access to escape equipment and often carry them during all site work. Note that these are designed for protection when evacuating from a safe area and not for involvement in rescue operations when full BA is required.

If personnel are unaccounted for after a head count, search and rescue operations will take place—if it is safe to do so. In some cases, search patterns will have been prepared in advance to allow a rapid and thorough search to be undertaken. Search and rescue should form part of regular emergency drills. Unless this is done, there is a danger that the search will be random with some areas being searched several times whilst others are not visited.

22.3.2.3 Event Classification

Most ERPs categorise incidents into three tiers:

1. Minor events which are easily controlled with the minimum of resources (for example a wastepaper bin fire controlled with a single extinguisher)
2. A significant event but which can be controlled by the site's internal emergency resources
3. A major event which will require the involvement of external resources (this will include any event with offsite effects)

22.3.2.4 Emergency Management Team

Details of the EMT have been discussed earlier. The ERP should list the members of the team and their responsibilities as well as alternatives, to allow for people who are unavailable for any reason.

22.3.2.5 External Resources

A major event will require the involvement of local authority emergency services and, if available, mutual aid partners.

The role of local authority firefighters has previously been discussed but medical resources may be needed involving ambulances and external first aiders entering the site with evacuation to hospitals for the more seriously injured. The police will have a number of roles including the coordination of activities off site (evacuation of the area around the site, closing roads to keep sightseers away, ensuring access is available for emergency vehicles to reach the site, evacuation of casualties and so on).

The police will often be involved in post incident investigation, particularly if there are fatalities or sabotage/terrorism is suspected.

22.3.3 Emergency Coordination Centre

A centre of operations for the strategic response will be required. Whilst a few large organisations have a dedicated room, most companies equip a meeting room which can be converted to an ECC if required. Note that sometimes the term "Emergency Control Centre" is used—this can lead to confusion with the plant control room and Emergency Coordination Centre is a better designation. At one point in time, it was considered desirable for the emergency to be coordinated with the safe shut down of the plant in the process control room. However, this is now considered undesirable for two reasons:

1. Many process control rooms are still located close to process areas and may be damaged or unapproachable in an emergency situation. Even if the control room is blast resistance and pressurised access to/from the control room could involve travel through dangerous areas. The ECC should, ideally, be located remotely from hazardous areas.
2. The process operators will need to concentrate on the safe control and shut down of the process facilities and may be distracted by the large number of radio messages and visitors associated with ECC activities.

Equipment required in the ECC will include the following:

1 Communication Equipment	Whilst communications are more speedily achieved by radio or telephone, there might be occasions where documents need to be transferred or radio/telephones are unavailable. It would be prudent to have *runners* at the ECC who can transfer messages if all else fails.
2 Radios	For contact with site personnel and emergency responders
3 Telephones	This should include phones for emergency use only whose numbers are not known to any but essential personnel. The published numbers for the site are likely to be overwhelmed by press enquiries and the families of on-duty staff calling. Chargers for phones and radios must be available and any devices kept in the ECC should be fully charged.
4 Plot Plans of the Site	Overall and large-scale plans will be required
5 Map(s) of the Off-Site Area	Particularly if evacuation is required. This should show evacuation routes and any refuge centres such as schools and community centres.
6 Flip Charts and Pens	To record in visible format what is intended and how it is progressing. Timelines may also be included.
7 Clock	For multinational companies several clocks set to critical time zones may be appropriate
8 Appropriate Documentation, including (a) P&IDs (b) PFDs (c) Local plot plan (d) Under-surface drains including effluent drains (e) Suitable hard standings for emergency services (f) Location of key isolation points	Documentation is increasingly available in electronic format and accessed via a screen. However, a major event might cut off power to some or all parts of the site. Even if an Uninterruptible Power Supply (UPS) is available, it would be prudent to have up-to-date paper documentation available.
9 ERP (including pre-plans)	
10 Hazard Studies	
11 Mutual Aid Agreement (if one exists)	
12 Details of Fire Protection Systems	Including stocks of firefighting chemicals
13 Details of Other Emergency Equipment	Pollution control equipment, portable generators, pick-up trucks designated for emergency use
14 Material Safety Data Sheets (MSDS)	How to deal with the releases and the toxicology
15 Details of Any Current Impairments	See Section 22.5.
16 Call-Out List for Key Personnel	This will need to be updated at frequent intervals and should include contact details
17 Telephone Numbers for Emergency Services (local hospitals, government agencies, meteorological centre and so on)	Diggle[5] gives a more comprehensive list of contacts. As there may be different types of injury, it is important to know both the capacity of nearby hospitals and their resources and skills (such as the capability of dealing with serious burns and the availability of ventilators for casualties affected by toxic gases).
18 Recording Equipment	For post-incident investigation. In some cases, this may be a dedicated member of the ERT. Increasingly audio and video recordings are used.
19 Basic Catering Facilities	

22.3.4 Procedure for Response to Specific Scenarios

The response to each type of incident will differ but will follow a similar BASIC pattern. In some cases, actions may be taken in parallel rather than in series. Actions likely to be common to likely scenarios are listed in the following table.

1 Raise the alarm	
2 Activate the emergency response and set up the ECC.	This may involve calling off-duty staff
3 Evacuate all non-essential personnel	Some personnel may need to remain in the hazard area to ensure a safe plant shut down or assist emergency responders In some cases, personnel may need to shelter in a safe location
4 Shut down the plant.	The extent of shut down will vary depending on the situation. A fire in one small area of the site might lead to a local shut down with the rest of the site remaining in operation. Clearly the sectional isolation of the area causing concern is a priority. In a major fire the whole plant must be isolated and made safe.
5 Conduct a "head count" and initiate a search pattern for any missing personnel	It is important that there is a continuous and up-to-date record of who is on the site. This includes emergency personnel who enter the site as well as staff unaccounted for after evacuation.
6 Record conditions at the site, take photographs and take witness statements	In many cases, there will be a requirement to inform the authorities after all but a minor incident. The collection of evidence will assist the investigation and reduce the period when repairs can be undertaken and production restarted. It is essential that witness statements are taken within a few hours of the event as memory becomes distorted with time. Photographs and other site investigations will have to wait until confirmation that the site is safe to enter. Even then special measures, such as the use of BA may be necessary.

The response to larger incidents will be similar in terms of the actions taken but significantly different in scale. Management of the incident will be much more complex.

As well as front line activities such as rescue and firefighting, the logistical and planning requirements will be significant and include the following:

1 Communication with the media
2 Communication with local authority emergency authorities (to arrange evacuation of casualties by ambulance, ensuring roads are kept clear for emergency vehicles and evacuation of personnel outside the site if necessary)
3 Coordination of mutual aid partners
4 Communication with the families of on shift personnel (who may be injured or worse)
5 Ensuring that adequate firefighting resources (equipment, personnel, foam concentrate and water) are available
6 Ensuring that emergency utility requirements are available (temporary power supplies, fuel for fire appliances, additional foam stocks and so on).
7 Preparation of staging areas for emergency vehicles and personnel arriving at the site.
8 Monitoring of discharges from the site (for example disposal of contaminated fire water or plumes of smoke containing toxic products of combustion).
9 Provision of food and water to emergency responders and, for long duration events, temporary accommodation.

22.3.5 Pre-Plans

Whilst it is essential to keep the ERT to a framework for emergency response, it is common practice to develop *pre-plans* for specific fire situations, especially large storage tanks. Pre-plans may be appropriate for other incidents such as toxic releases. Whilst a rapid response to a fire will generally shorten its duration and limit damage, this will not be the case if the initial fire attack is inadequate or cannot be sustained. Inadequate planning can also put personnel in danger. A couple of examples will illustrate the point.

1 If a manual response to a fire in an electrical switch room is required, the response of some firefighters will be to enter the room and apply some form of fire extinguishment. However, unless, breathing apparatus is worn and the power supply isolated before entry, the firefighters will be at risk. These precautions may seem obvious, but they have sometimes been omitted.
2 There is a relatively standard approach to fighting fires in floating roof tanks. If possible, the tank contents are pumped out and foam solution is applied to form a "blanket" over the surface of the tank. It is important that the foam falls as a continuous stream in one place and then spreads over the tank surface. This is often achieved by aiming the foam jet at the inner tank wall and letting it slide down to the burning liquid surface. For this to work the foam application must be from a safe location and continue until the tank surface is completely covered with water/foam solution. Application generally continues for a further period to ensure that foam breakdown is matched with application of fresh foam. Sometimes this approach fails because the foam application is interrupted. This could happen for several reasons but is commonly due to the stock of available foam concentrate being exhausted. It is therefore necessary for firefighters, in advance of the fire, knowing where to position the foam nozzles to ensure the stream will reach the tank without exposing equipment or personnel to excessive levels of thermal radiation. This, in turn, will depend on the location of fixed fire water hydrants. Additionally, the water and especially the foam requirements needed to fight the fire need to be known to ensure a fire attack will not run out of foam. An example of a tank fire pre-plan is shown in Figures 22.1 and 22.2.

Figure 22.1 shows the areas where firefighters might require protection or may be unable to access due to high levels of thermal radiation. Note that the contours differ from those discussed in Section 9.5 which are influenced by wind direction. Figure 22.2 shows contours based on wind from any direction and represents a worst case. Some areas within the contours may be approachable at some points in time but will become hazardous if the wind direction changes.

Floating roof storage tanks provide a good example of what constitutes a credible event. The most common type of fire on a floating roof tank is in the rim seal between the walls and the roof. Flame size is low and these fires are relatively easily extinguished. Less common are fires involving the whole roof of the tank, generally after the roof has sunk for some reasons, Figures 22.1 and 22.2 are based on this scenario. Even less common is a fire which involves a number of tanks but these still occur if the tanks are located close together.

Pre-plans may include additional details, for example the routes to be taken by fire trucks. Roads may be affected by fumes or thermal radiation which may make them impassable. However, the greater the level of detail, the more likely that something will go wrong and the plan will become unworkable.

Similar detailed levels of planning can be performed for other scenarios.

ERP- 7 EMERGENCY RESPONSE PLAN FOR:	ALL FLOATING ROOF TANKS GENERIC FULL SURFACE FIRE		Doc. ERP-7 **Approved: XX** Rev. 0. Date: November 2015.
STRATEGY Site ESD - Halt ship loading operations - Deploy water monitors for cooling any adjacent tank or tanks affected by radiant heat – Allow tank burn out whilst cooling adjacent exposures – pump out of crude if possible - If foam attack decision is made, ensure all resources in place before attempting foam attack – If foam attack fails all personnel to be evacuated from site and await boilover.			
IMMEDIATE RESPONSE	**ACTIONS**	**EQUIPMENT / RESOURCES**	**INFO / COMMENTS**
[ ] Control Room/Technicians	Investigate and confirm fire event.	Affected tank MOV's and Site ESD Valves	Any available technicians to act as Local Evacuation Officers (LEO's) and direct evacuees to a safe muster point.
[ ] Control Room/Technicians	Stop tank and ship operations and Site ESD – isolate tank		
[ ] Control Room	Actuate site alarm and start fire pump.	Site alarm switch and fire pump start in CR	
[ ] Technicians	If safe and possible, ensure plant evacuation.		
[ ] Control Room	Inform LAFB (Fire Brigade) and Refinery Fire Station	Telephone 999, Refinery Fire Station xxxx	Team Leader to act as initial OSC until relieved.
[ ] Control Room	Inform Refinery to halt pumping and advise Berth Supervisor		
[ ] Control Room	Alert duty on-call, Asset Manager and Ops Engineer		
[ ] Control Room	Check ballast rundown is in commission for firewater supply		
[ ] OSC	Ensure OSC "jacket" is on for recognition.	Telephone/ Radio	
[ ] OSC	Ensure this section actions are checked done.	Telephone/ pager/radio	
[ ] OSC	Make contact with arriving LAFB officer and handover ERP	This ERP	
2nd RESPONDERS	**ACTIONS**	**EQUIPMENT / RESOURCES**	**INFO / COMMENTS**
[ ] Authorised Person/Security	Establish evacuation status/any missing persons.	Access Control Read Out/Print Out	
[ ] Security Guard	Handover this ERP LAFB Fire Officer and direct all fire vehicles to incident scene.	This ERP	
	Respond with site fire vehicle	Site fire vehicle.	
[ ] Technicians	Deploy water monitors for cooling adjacent tank.	2 x 2500lpm portable water monitors, 24 x 70mm delivery hose.	
[ ] Technicians	Establish a fire control point for oncoming fire brigade units to report.	Control point to be at nearest/safest site telephone.	
[ ] Technicians/OSC	Line up available tanks to take affected tank pump out.	Tank MOV's	Inform LAFB Fire Officer that all listed actions have been carried out up to this point.
	Check all actions in this section are marked and give ERP to LAFB Officer on arrival.		
[ ] Technicians/OSC			
[ ] OSC			
3rd RESPONDERS	**ACTIONS**	**EQUIPMENT / RESOURCES**	**INFO / COMMENTS**
[ ] LAFB Fire Officer	Report to incident control point for incident information.	Control Point	If extinguishment decision taken, site fire pumps can only provide 26,500lpm approx. so other supply arrangements are required.
[ ] LAFB Fire Officer	Deploy water monitors for cooling adjacent tank.	4 x 2500 lpm portable water monitors, 48 x 70mm delivery hose.	
[ ] LAFB Fire Officer	If decision is to attempt extinguishment, set-up/actuate foam monitors onto full surface fire. Delegate officers as Safety Observers to give advance warning of boilover event.	4 x 15000 lpm foam monitors, 216,000 litres of 3% fluoroprotein foam, 240 x 70mm delivery fire hose.	Foam concentrate based on 3% Fluoroprotein at a rate of 10.4 lpm/m² for 120 mins. Total foam stock on site is 20,000 litres, so additional foam supplies are required if foam attack is to be made.
[ ] LAFB Fire Officer	Maintain foam application until fire is extinguished and thereafter until a secure foam blanket established.	Total water demand for foam and water monitors is 75,000 lpm	
INCIDENT POTENTIAL HAZARDS Crude oil full surface fires in tanks, if left to burn freely, create a heatwave which sinks into the tank and on contacting a water layer causes a "boilover" with sometimes violent or severe ejection of burning crude. This event, in worst cases, may lead to burning crude flowing over an area of 300 metres diameter. Personnel should not work inside the tank bund after the initial stages of the fire due to boilover hazard. A boilover may occur at any time after the first few hours of the incident and no reliance should be placed on anticipating the heatwave reaching the base of tank before a boilover will occur. It is prudent to expect a boilover anytime after the first 2/3 hours of the incident.			
OTHER CONCERNS Foam should not be applied to a crude tank fire until there is sufficient freeboard in the tank since foam application without space in the tank to accommodate will lead to slopovers and foam wastage. This is most important where overfill has caused incident and where crude oil is at or very near to the tank shell height. It may be that the fire needs to burn for an hour or more before it is safe to apply foam. Firewater and foam solution in bund must be monitored and controlled. Equipment for pump out and location for		RADIOACTIVE (RA) HAZARDS / ASBESTOS HAZARDS / TOXIC HAZARDS 1 RA Instrument/Gauges on Seal Water Drum V-511 on MVR Unit. It is Cs-137 material with 2627 MBq. No asbestos is present. No toxic materials.	

FIGURE 22.1 Typical Tank Fire Pre-plan (reproduced courtesy of ENRg Consultants)

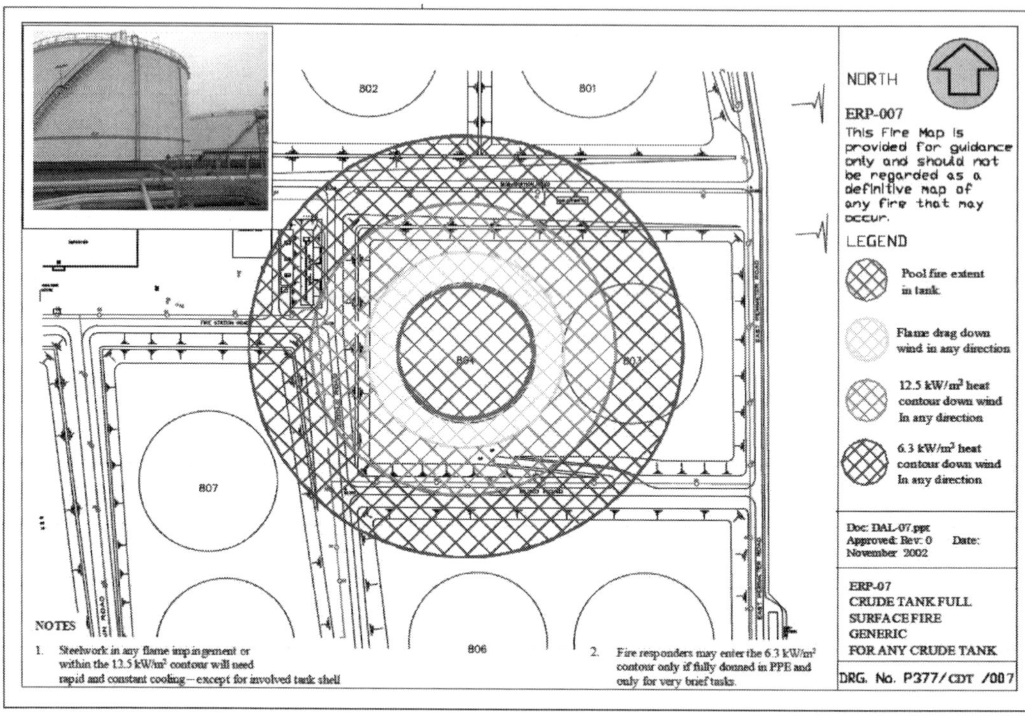

FIGURE 22.2 Typical Tank Fire Pre-plan ((reproduced courtesy of ENRg Consultants)

22.4 RESOURCES

A fire station or other store for emergency equipment will be necessary and must, be located away from areas where it might be damaged by an incident. Similarly, a first aid facility in a safe location will be required. In most cases, this will provide limited treatment and will effectively be a *casualty clearing station* following triage where minor injuries can be treated and the seriously injured are evacuated to hospital. First aid facilities in remote areas or offshore platforms may have more extensive facilities.

The only other structure likely to be needed might be a safe refuge (see Section 8.7). These can become hot and carbon dioxide levels can build up whilst oxygen levels fall. A person in the refuge begins to sweat and the pulse and breathing rates increase (due to the higher carbon dioxide concentration). Physiological symptoms can result in panic. The adequacy of the refuges needs to be determined, and breathing air cylinders may need to be provided particularly when evacuating the refuge to another safer zone. Whilst safe refuges are generally for protection against fire and gas, they may be installed for other reasons. Ships and offshore platforms in areas prone to piracy generally have safe refuges where the crew can barricade themselves inside until a rescue force arrives. Communication between the EEC and the safe refuge is important so that head counts can be carried out and time until evacuation communicated.

Equipment types required for firefighting is discussed in Section 9.12; the development of preplans (see Section 22.3) will determine the resources needed for specific scenarios. The requirement for additional equipment will vary from site to site but will include medical supplies, hospital capability and pollution control equipment. In locations where there is a concentration of process plants, each mutual aid partner may contribute to a "pool" of equipment suitable for the worst case. It is important to ensure that all equipment is to the same specification and interchangeable between the various mutual aid partners. Hose couplings, in particular, may vary between different companies.

Other resources may also be needed and will vary from industry to industry. Emergency showers and eye baths will be necessary where chemical spills could occur—these are relatively common incidents at loading and offloading locations. Pollution control requirements will vary with oil spills requiring the use of booms and dispersants. Spills of chemicals might require the use of neutralising agents and chemical adsorbents.

22.5 IMPAIRMENT PROCEDURE

Whilst emergency responders at all levels should be aware of the resources available, it is also important that they know of any defective equipment. In some cases, mitigation for equipment which is out of service may be required. For example, flame detectors will usually be isolated in areas where hot work is taking place to avoid false alarms; a team of firefighters may be required during this period. At other times, no special action will be necessary beyond ensuring that responders are informed. For example, if a water deluge system is unavailable, firefighters will need to develop an alternative means of fire attack. Even for the short periods when equipment is out of service for maintenance, emergency responders must be informed. It is common practice to write a specific procedure to control this situation.[6] Impairments should be identified as part of the risk/job safety assessment which takes place before any work activity takes place.

The need to prevent inadvertent isolation of fixed protection, such as sprinkler and deluge systems is recognised and it is common practice to lock isolation valves open to prevent them from "being closed to prevent leakage".

In addition, emergency responders must be kept informed of road closures so that alternative means of access to all parts of the site can be determined. Fire appliances may have large turning circles and some road junctions might be too "tight" for access—it is better to know this in advance rather than have an appliance become "stuck" in an emergency.

REFERENCES

1 Mutual aid partners are personnel from other industrial companies
2 UK Fire Service, *Physical Tests*, UK Fire Service, www.fireservice.co.uk/recruitment/physical/
3 Anon, *No Plan Survives First Contact with the Enemy*, https://quoteinvestigator.com/2021/05/04/no-plan/
4 US National Transportation Safety Board (NTSB), *Aircraft Accident Report Fuel Farm Fire at Stapleton International Airport, Denver Colorado, November 1990*, US Department of Commerce, National Technical Information Service, 1 October 1991, www.ntsb.gov/investigations/AccidentReports/Reports/AAR9107.pdf
5 Diggle W M, *Major Emergencies in a Petrochemical Complex: Planning Action for Emergency Services. Process Industry Hazards (Symposium Series 47)*, IChemE, 1976 we should check this one
6 Swiss Reinsurance Corporation (Swiss Re), *Fire Protection System Impairment*, https://corporatesolutions.swissre.com/insights/knowledge/risk-engineering-services-fire-protection-system-impairment.html

Part E

The Future

23 The Future

23.1 INTRODUCTION

The most accurate prediction of the future is probably that whatever is predicted will be incorrect. Despite this somewhat cynical viewpoint, some form of prediction is necessary for everything from balancing a household budget to determining production quantities and costs for determining the viability of a new project—which loops back to Chapter 12. A common assumption is to look at the past and extrapolate to the future and this works to an extent but sooner or later a discontinuity will arise. In 1894, the UK newspaper *The Times* predicted that within 50 years every street in London would be under nine feet of (horse) manure; this was labelled the Great Manure Crisis.[1] The development and widespread use of electric trams and motor vehicles prevented this unpleasant prospect (assuming that the forecast was accurate). Similar predictions of societal or technological change based purely on extrapolating past trends are frequently inaccurate, if not farcical.

There are many situations which cause significant change in society. The development of computing, and its spin-off is probably the biggest single example since the end of the Second World War. There have, of course, been many events which have caused a degree of change, ranging from the development of more lethal weapons (both nuclear and otherwise), natural catastrophes and, central to the purpose of this book, industrial accidents.

Whilst technological change may be slow and steady, there are periods of rapid change. War and pandemic are examples within living memory. In 1939 fighter planes could fly at around 560 km/hr. In 1945 the German Messerschmitt 262 could fly at nearly 880 km/hr. Rockets as a major weapon of war appeared for the first time.

In this short chapter, two separate approaches have been used to consider how the process industries might develop over the next few decades.

Extrapolation	The dangers of using the past to predict the future have been discussed earlier. However, there are existing, "heritage" situations which must be in the resolved in the future, these are discussed in Section 23.2.
Blue Sky Thinking	There are a number of industries, industrial processes and methodologies which either are in their infancy or have still not reached significant commercial acceptance. These are discussed briefly in Section 23.4.

23.2 HERITAGE ISSUES

In one form or another the chemical and allied industries have been around for millennia. Some sites dating from the 19th-century are still in operation and the legacy of other sites is in evidence in the form of contaminated land and dumps of chemical waste.

> Love Canal in New York State[2] was developed in the 1890s but in the 1920s became a dump site for municipal waste. In the 1940s it was purchased by Hooker Chemical Company which used the site to dump around 20,000 tonnes of waste from the manufacture of dyes, perfumes and solvents for rubber and synthetic resins. In 1953 the land was sold to a local school district. Over the next 30 years, there were numerous cases of ill health and eventually the community was demolished. The situation at Love Canal led to the setting up of a Superfund to clean up the area.

DOI: 10.1201/9781003510680-28

The location and contents of dumps may not always be known and, when discovered, ownership and responsibility for clean-up can be difficult to determine. Decontamination raises significant economic and technical problems but in the long-term pressure for the development of undeveloped land (greenfield or brownfield) will increase.

There will always be a need for basic industrial chemicals ranging from acids to chlorine and caustic soda. Whilst manufacturing processes are well established, future improvements cannot be ruled out. Sulphuric acid production changed from the Lead Chamber process to the more efficient Contact process[3] and chlorine manufacture, after many years moved from mercury to diaphragm cells.[4]

Recovery/recycling of materials is another process industry with potential. Fires in recycling facilities are frustratingly common and this is one area, at least, where improved attention to safety is desirable now. Mine tailings have been reworked for secondary recovery of gold[5] and the potential for reworking of other abandoned or discarded wastes may be practical with both economic and environmental benefits.

The provision of energy will be a key challenge. The oil, gas and petrochemical industries will decline as the move to *net zero* accelerates although they will still be economically important for several decades. Of increasing importance will be the hydrogen economy and large-scale activities are currently planned. Hydrogen production, transport and storage are well understood technologies but currently on a scale far smaller than will be required to replace gasoline and diesel for road transport. The concept of carbon capture is well known but not yet widely practised. Vastly increased use of hydrogen will see storage/dispensing facilities on a scale not dissimilar to current vehicle service stations. Hydrogen has a very high energy density making it a dangerous material with the public currently having limited knowledge of safety issues. As hydrogen is difficult to liquefy and expensive to store as a gas, the transport of ammonia with conversion to hydrogen at the point of use is being explored. Ammonia is flammable, although less so than hydrogen but is toxic. There is plenty of scope for the ingenuity of chemical engineers as the energy industries change.

Electrical power is likely to be increasingly important but will require major infrastructure developments. Whilst "clean energy" (wind, wave and solar) to produce electricity has increased massively since the start of the 21st century, much research is still needed, particularly in the field of energy storage. The importance of nuclear power is uncertain. Nuclear power plants have nearly always been delivered late and over budget. Current plans for smaller cheaper modular reactors are underway but are not yet in operation. There are encouraging results from nuclear fusion research, but commercial development is still a long way off.

Governments (and intra-government bodies) will need to smooth the path from fossil fuel to renewables. An approach modelled on "management of change" (see Section 16.9) would be useful but governments do not always act in the best interests of their citizens or the planet—always having one eye on the next election.

There has always been a wide range of non-energy industries which employ chemical engineers: fine chemicals, dyes and colourings, pharmaceuticals, foodstuffs and even brewing. With the advent of new technology, the role of chemical engineers will expand to take in industries such as nanotechnology, biotechnology and computer chip production. The plethora of new routes for chemical manufacture will also require chemical engineers as production moves from the laboratory and pilot plant to full-scale manufacture. Risk assessment and regulation may struggle to keep up with these developments and future unwelcome surprises are likely. The recovery of "rare earths" used in electronics and magnets is fundamental to civilisation and processes to extract and recycles these materials will be fundamental.

The boom in chemical manufacture after the Second World War, up to the 1970s was relatively unconstrained by regulation—almost all of it related to occupational safety. This has now changed in many parts of the world. The regulatory authorities, in some countries at least, have spent considerable time and effort investigating the hazards of new technology. In the UK for instance, there have been numerous studies into risks associated with the hydrogen economy.[6]

Whilst the future development of the process industry cannot be mapped with certainty, it is likely that there will be a much higher level of risk assessment and regulation of new technology going forward than has previously been the case.

Another trend which is not hard to predict is the increased use of computerisation and automation. Many of the activities that were undertaken manually at the dawn of the process industries are now automated. Advances in control and instrumentation during a working lifetime are phenomenal but have not occurred without problems, alarm flooding (see Section 13.9) being perhaps the most obvious. A more ominous threat is cyber-attack, particularly where remote monitoring of equipment is taking place. One of the first and still the most famous examples of a cyber-attack in the process industries was the Stuxnet attack on Iran's nuclear fuel enrichment programme.[7] Cyber-attack is a growing problem. The danger is recognised on both sides of the Atlantic; both the UK HSE[8] and CCPS in the USA have published guidance.[9] Cybersecurity is likely to be an area of increasing interest as attackers seek new methods of attack and protectors look for new methods of defence.

The use of pen and paper is starting to be replaced with computer terminals in control rooms and tablets in the field. This is considered to have two dangers. First, many computer formats constrain the information that can be entered (as anyone filling in an online form will know). Second, if not adequately designed logbooks and other recording systems may be open to later alteration—making accident investigation difficult. One trend is the use of QR labels on field instruments. The field operator scans the code and then enters to field instrument reading manually. Hopefully, the operator will realise if a reading is abnormal and investigate. It is only one stage further to automate the system and have a field reading scanned in to a table directly or signalled back to the control room. The latter solution is likely to appeal to managers as a way of reducing the size of operating team. Field operators as well as taking reading are also very effective fire and gas detectors and all-round site monitors. Any attempt to reduce their numbers must be critically reviewed.

23.3 AGEING PLANT

One issue which should surprise no one is the problem of ageing plant. Management have increasingly tried to "sweat the assets" over the years and poor (or lack of) maintenance is a serious concern. Some forms of ageing, such as corrosion are well known and covered earlier (see Chapter 20). However, the problems are likely to increase as plants are run beyond their design life—particularly in the oil, gas and petrochemical industries where the transition away from fossil may not justify construction of new plant before the industry starts to decline.

The UK HSE has conducted research[10] and published guidance[11] on ageing and detailed information is available from numerous other sources. Techniques for estimating the remaining life of critical equipment are available but are often highly technical and the preserve of a few specialist companies. In future the consideration of ageing needs to progress beyond the potential for corrosion and obsolete equipment to consider new ageing mechanisms which may not be related to the physical aspects of the plant. Will there, for example, be an adequate number of trained staff to run the plant? Younger operators are unlikely to want to work on plants which will shut down within a few years leaving them at risk of redundancy. Older operators, if they do not want to retire may become unfit for work. The effort required to upgrade plant control systems to cope with serious cybersecurity issues might be excessive.

Finally a reassessment of exporting manufacture (along with pollution) to the developing world is being rethought. The desirability of having critical manufacturing facilities thousands of miles away is being reconsidered. If there is a rush to build new facilities "at home", the problems of cost and speed of construction may lead to familiar issues of "corner cutting". Political issues also affect siting of facilities with new plant and pipelines avoiding politically sensitive areas even at increased cost.

23.4 "BLUE SKY" THINKING

Engineers are generally pragmatic beings who don't pack a crystal ball along with their laptop and safety boots. It is difficult to "back winners" from the various new products and processes which may develop in the next decade or so. Advances in medical science, including the development of artificial blood and the growing of organs might be possible. Deep-sea mining of minerals remains a controversial topic but is likely to go ahead in some form at some point in time. Biological processes for chemical manufacture are also likely to be commercially viable in the future.

For the last 75 years the process industries have generally worked on the principle that "bigger is better" with plant sizes ever increasing. Plants for methanol, for example, can now reach a capacity of 5,000 tonnes/day.[12] A change to modular units which can be easily shipped to location and minimise travel costs may be an option for large plants with huge capital costs which may be underutilised in their early years of operation. New processes might produce more customised products with batch size limiting capacity. Increasingly universities are developing *research hubs* with cross-fertilisation by sharing ideas across disciplines. Engineering and Scientific disciplines will become more homogenous and less separated.

Having damaged the earth so badly since the dawn of the industrial age, extreme measure might be needed to reverse climate change. Planting trees to absorb carbon dioxide is often put forward as a solution but increasing heating with no real sign of a reduction in fossil fuel usage, may make more extreme measures necessary—*Terraforming* to use a phrase beloved by science fiction fans. The development of colonies (requiring highly reliable life support systems) on other planets is also likely during the lifetime of people currently training to be engineers.

In parallel with new products and processes, will be new or modernised equipment and methodologies. Process intensification might, finally, start to utilise Hi-Gee technology (see Section 11.2). Some of these are beginning to appear such as online corrosion probes and Dissolved Gas Analysis (DGA). Computer systems are increasingly sophisticated with Advanced Process Control (APC) being installed in most new plants (as well as cyber protection). The concept of a *digital twin*[13] is now becoming established.

New assessment methods, or enhancements to existing methods are likely, probably using some form of software. This opens up a whole now range of problems and safety experts and regulatory authorities will need a lot of persuasion that any form of computerised system will be as effective as the, largely manual, systems, such as HAZOP, currently employed. This does not mean that automation cannot assist some parts of the risk and hazard assessment but the idea of relying on a "black box" will be an anathema to most process safety specialists at this point in time.

An area which is ripe for exploration is the development of tools to assess the safety culture of organisations at the highest level. Directors and senior managers often make decisions without adequately considering the implications at the "shop floor level". Large cuts in maintenance budgets are likely to lead to future failures. "Efficiency savings" are often an illusion and will translate into negative features like reduced training, reduction in manpower or the cancellation of improvements (generally safety not production related). As one commentator said, only half in jest, "accountants are the biggest cause of accidents".

Increasing concern is being paid to ethical concerns and environmental, social, governance (ESG) reviews are now being performed by some companies although there must be suspicion that some, at lease, have an element of *greenwashing*. The concepts of sustainability (which could be considered as a subset of environmental concerns) and resilience are achieving increased importance and show a movement towards a more holistic view of hazard and risk, encompassing process facilities within the environment and business supply chain. Whilst this would appear a welcome development, it creates a much more complex atmosphere for management, communication and effective governance. Safety, along with environment, will remain important considerations—but probably will not be treated by politicians or industrialists with the importance they deserve. It is possible, in some cases, that previously hard one safety or environmental improvements could be put into reverse.

23.5 BOLDLY GO—BUT SAFELY

The process industries, in one form or another, will continue to thrive. In most cases, the driver for development will be profit; there may be exceptions in time of war, pandemic or other national/international emergencies.

A selection of possible technologies and techniques has been discussed in Sections 23.2 and 23.4 but are likely, at best, to be only partially correct. Whatever processes enter commercial production, the need to ensure safe conditions from "cradle to grave" will be necessary. The topics covered in this book will remain valid.

A rather utopian possibility is the ability to predict when accidents are likely to occur, rather like the film Minority Report predicts crimes that are about to happen. Concepts that may facilitate this, such as Rasmussen's *Drift to Danger* model[14] could assist but need considerable refinement and are likely to meet resistance at both plant and management levels.

To paraphrase Gene Roddenberry of Star Trek fame, the process industries will boldly develop to supply products to society (and profit for their owners). But regardless of whether a process is novel or well established, the price of safety is eternal vigilance.

REFERENCES

1 Utrecht University, *The Great Manure Crisis*, www.uu.nl/en/research/urban-futures-studio/initiatives/mixed-classroom-techniques-of-futuring/mobility-museum-2050/the-great-manure-crisis#:~:text=In%201894%20the%20Times%20predicted,solution%20was%20found%20for%20years

2 Abhari R, Returning to Love Canal After Forty Years, *Loss Prevention Bulletin (No. 280)*, August 2021, www.icheme.org/media/16584/lpb280_pg17.pdf

3 Wainwright R, A World of Sulfur: Part 2, *World Sulphur Magazine*, 27 December 2018, www.worldfertilizer.com/special-reports/27122018/a-world-of-sulfur-part-2/

4 Schmittinger P (editor), *Chlorine Principles and Industrial Practice*, Wiley-VCH, 2000

5 Weyer C, Scyphers J & Gabelman J, Secondary Gold Recovery & Reclamation of Historic Properties, *21st Mine Design Operations & Closure Conference*, May 2013, www.mtech.edu/mwtp/presentations/2013_presentations/Charmaine%20Weyer_Joe%20Scyphers.pdf

6 Stolecks-Antczak K, Hazards of Hydrogen Transport in the Existing Natural Gas Pipeline Network, *Journal of Power Technologies*, 98(4), 2018

7 Kushner D, *The Real Story of Stuxnet*, IEEE Spectrum, 24 May 2024, https://spectrum.ieee.org/the-real-story-of-stuxnet

8 UK Health and Safety Executive, *Cyber Security*, latest edition, www.hse.gov.uk/eci/cyber-security.htm

9 Center for Chemical Process Safety (CCPS), *Managing Cybersecurity in the Process Industries: A Risk-based Approach*, Wiley, April 2022

10 Wintle J, Moore P, Henry N, Smalley A & Amphlett G, *Plant Ageing Management of Equipment Containing Hazardous Fluids or Pressure (RR509)*, 2006, www.hse.gov.uk/Research/rrpdf/rr509.pdf

11 Horrocks P, Mansfield D, Parker K, Thomson J, Atkinson T & Worsley J, *Managing Ageing Plant a Summary Guide*, HSE, www.hse.gov.uk/research/rrpdf/rr823-summary-guide.pdf

12 Williams C, Trinidad and Tobago Building 5,000 Tonne/Day Methanol Plant, *Oil & Gas Journal*, 8 August 2003, www.ogj.com/refining-processing/article/17255516/trinidad-and-tobago-building-5000-tonne-day-methanol-plant
13 Anon, Digital Twin Improves Plant Design and Operational Performance, *Control Engineering*, 30 March 2022, www.controleng.com/articles/digital-twin-improves-plant-design-and-operational-performance/
14 Rasmussen J, Risk Management in a Dynamic Society: A Modelling Problem, *Safety Science*, 27(2/3), 1997

Appendix A: Simple HAZOP Study

The Process and Instrumentation Diagram (P&ID) shows a new-build house and the domestic water system to be installed (see Figure A.1). An attempt has been made to avoid "line crosses" (a key feature of a P&ID). Line numbers are given the prefix "L", tanks the prefix "T" and valves (taps) the prefix "V". There is a ball cock "BC" (float valve) on the head tank T1 and a temperature control on the immersion heater "TC". The piping is all 25 mm nominal diameter and is fully pressure rated to the maximum supply pressure from the service provider. This will not necessarily be the case for real studies. Each line will have a designator for the fluid handled, a unique line number, a nominal diameter, materials of construction reference, a rated pressure reference and a thermal insulation reference (Section 14.4). This can be a long string of letters and numbers all of which will be unique to that line!

The *word picture* or description of the system shown in the P&ID is:

> *Water (clean and germ free) from the service supplier is piped from the water main in the street through a new section of piping and Valve Vi at a pressure of 3 barg and 15°C. V1 is inside the house. The line splits to supply cold water to a sink through V2 or to a head tank T1. The level in the head tank is controlled by BC 1 (a float valve). From T1 the water flows into a second tank T2 where it is heated to 60°C by an immersion heater on "on/off control". Hot water flows to the sink through V3. The sink has a drain closed by a plug. The flow rate is to <u>each outlet</u> will not exceed a maximum of 0.5 l/s but can be anywhere between 0.0 l/s and 0.5 l/s depending on the opening of taps (valves). A line from L4 (not shown) feeds the toilet cistern.*

The range of flows means that the study has to consider "some flow when it should be no flow" and "some flow when it should be a peak flow".

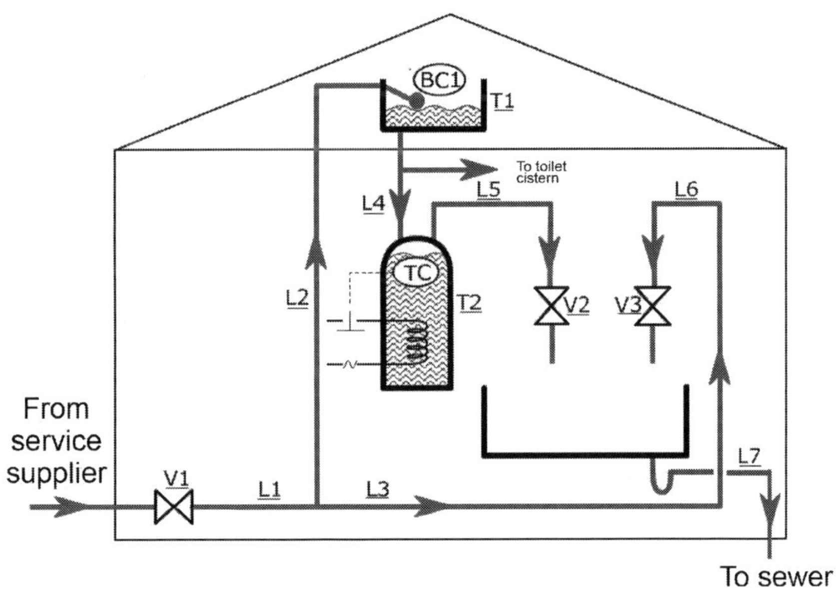

FIGURE A.1 Simplified P&ID of Domestic Water System

NOTE

In a major project the process description may take around 30 minutes to read, and of, course there will be many interlinking P&IDs.

It is often forgotten that the study may have to start within the upstream process and end within the downstream process. A failure in the upstream may affect the process under study and a failure in the process under study may affect the downstream process.

It is necessary to understand the structure of a HAZOP. First it is necessary to have a *parameter* and *guideword matrix*. The parameters are physical process conditions that can be measured and the guideword is a variation of that parameter or a deviation from the normal. Not all combinations of parameters and guidewords are relevant; only those which are included in the matrix. If the whole combination (meaningful or not) is used the study will be twice as long! For the water supply example, the applicable guidewords are shown in the following table.

Parameter (to Right) Guideword (Below)	**Flow**	**Temperature**	**Pressure**	**Level**
High	X	X	X	X
Higher	X	?	X	X
Low	X	X	X	X
Lower	X	—	X	X
Reverse	X	—	—	—
As well as	X	—	—	—
Other than	?	—	—	—

Other guidewords may be appropriate, depending on the specific location. In this case, the following need to be considered:

1 Access
2 Maintenance

There is also a "disguised node" in the form of Tanks 1 and 2. Sometimes these can be treated as part of a node (as has been done, in part, in this study; BC is inside the tank boundary). It is often worth treating units such as tanks and pumps and others with their own particular parameters/guidewords.

It is necessary now to select "nodes" or lines where there is a clear "intent". This should be marked as the lead line. When complete the line should be clearly identified.

Node	**Intent**
1 V1 to tank T1—L1—L2	Flow of 0 to 0.5 l/s of clean water at an inlet pressure of 300kPa and 15°C into T1 controlled by a ball cock BC in T1.
2 V1 to sink and V2—L1/3/6	Flow of 0 to 0.5 l/s of clean water at an inlet pressure of 300kPa and 15°C (as earlier).
3 T1 to T2—L4	Flow of 0 to 0.5 l/s of clean water to the cold tap and the toilet at minimum pressure (hydrostatic head) 3 barg and 15°C
4 T2 to sink and V3—L5	Flow of 0 to 0.5 l/s of clean water and hydrostatic head and 60°C
5 T1 Ideally a new node	To act as a head tank/reservoir for water feed to the immersion heater and T2

The order of the nodes clearly follows that of the process flow and the final node (the tank, T1) should in an ideal world be treated last as the analysis of the "flows", "pressures" and "temperatures" and "as well as" are less applicable to the tank, and, if put as node 2 or 3 it could destroy some of the fluidity of the questioning and analysis.

Having defined the parameters, guideword matrix and nodes, the facilitator must now introduce him/herself to the team giving name and background or affiliation. The Facilitator could be independent of the team. He/she must explain what is required of them and also any "ground rules". For example, mobile phones must be switched off. The HAZOP secretary then introduces themselves followed by the rest of the team giving name, discipline and affiliation. The facilitator must ensure that the team is at ease and explain that there are no silly questions all options must be clearly and clinically examined.

The facilitator will have already defined the start and end points as well as likely nodes.

One of the teams will then describe the word picture of the diagram. The facilitator will define the study start and end points and then select a node for study. Clearly the start point is at the junction from the water supply main but the end point could be the drain from the sink or the entry to the sewer itself.

NOTE

In a real HAZOP the study should end at the downstream point, in this case it would be the main effluent sewer.

It will be noted that the sequence of the "interrogation" does not follow the matrix. This is deliberate—by following the matrix rigidly the HAZOP becomes sterile and quite destructive to creative questioning. If the team has identified a possible cause it is better to take that to a conclusion and to, temporarily, break out of the matrix. The facilitator will be ensuring that the HAZOP process is being correctly followed. Too many changes in the logic flow can result in issues being forgotten or overlooked.

The following is illustrative of how the questioning in a HAZOP study might develop in an undergraduate setting.

Facilitator *"Can I start with a peak flow of 0.5 l/s and examine how it could exceed 0.5 l/s?"*

Clearly the facilitator is looking at **high flow** and to do this they will provide a "clue". It can be tedious with no creativity if the parameter and guideword are trotted out without any structure. Additionally, the facilitator has already noted that the flow is controlled by a ball cock (float valve). However, the facilitator is encouraging the team to consider a line rupture, possibly caused by low-temperature or poor-quality fittings.

Team *"This could be caused by a line break"*

Facilitator *"What might cause this?"*

Team *"Poor fittings, sprung union or even frost damage in the attic!"*

Facilitator *"Good! Now what should be the actions—there may be more than 1"*

Facilitator *"How can you isolate this leakage?"*

Team *"Where is the master isolation? It is shown inside the house but is it accessible?"*

Facilitator *"Who will take this as an action to the builder?"*

Fred raises his hand.

"Thanks Fred. May I also suggest that the orientation is checked such that it can be closed by hand?"

The team has reached the answer that the facilitator wanted. Identifying the orientation is also important but could become a bit laboured. This comment shows the team that the facilitator has a good understanding of the design.

Facilitator *"What might be done to stop the fittings leaking?"*

Team *"The lines could be pressure tested before entry and where possible joints should be soldered. Further the lines L1 and L 2 should be thermally insulated inside the house—particularly the attic area".*

Facilitator *"Thank you—who will take this action back to the builder?"*
Jane raises her hand.
"Thanks Jane". "Are there any other causes?"
There is a pause in the conversation.

There seems to be a pause in the thinking process; the Facilitator knows from experience that there is another issue and has to prompt the team without giving them the answer.

Facilitator *"What controls the flow?"*
Team *"The ball cock".*
Facilitator *"What happens if the ball cock is damaged and not water tight?*
Again the word issue is high flow—when there should be no flow.
Team *"The tank will fill up!"*
Facilitator *"And then what happens?"*
Team *"The tank will overflow"*
Facilitator *"Then what happens, what is the effect of this?"*

The question contains a clue:—is there an overflow? This question is now encouraging lateral thought process. The second part of this question could be left out of a play act so as to test the team. The Facilitator may have to prompt the team if they miss it.

Team "The tank T 1 will overflow into the attic. An overflow should be installed on T1 to drain outside the house"
If the need for an overflow has been missed it will be necessary to prompt a reply by a new question: *"OK, but what are the implications and how might they be negated?"*
There is a secondary issue which could be introduced by questioning. "Does the overflow self drain or is there a low point trap which could leak?
Facilitator *"Who will take this action?"*
John raises his hand.
"Thanks, John."

So far questioning has covered **high flow**. What about **no flow**? The ball cock stuck closed, for example, due to damage or ice in T1. The issue of ice will be repeated later. Low temperature will also produce a comment on ice. In general, about 90–95% of all comments and actions are repeated under different combinations of parameter and guideword.

The HAZOP secretary will be making a continuous record of the comments in a spreadsheet with the main actions and person responsible for them.

It would be possible to carry out a full HAZOP which would take some time to record but the key issues can be recorded such that a lecturer could "play act" the study with students. The weakness with one discipline in a classroom environment, as already indicated, is that there is no cross fertilisation of ideas or skills.

It is essential that the Facilitator acts as a facilitator and not try to conduct the study single-handedly or ignore input from the team. The Facilitator has to ask pointed questions and try to explore unlikely possibilities. No firm guidance can be given as how this can be achieved but there are some useful questions:

1 "What happens then?" (I do not know! Do you"?)
2 "Sorry could you explain that to me?"
3 Direct the team to a detail in the P&ID and ask them to analyse the failure modes. This is asking the team "To think more deeply". "What happens if . . . ?"

The outcomes of the discussion must be recorded accurately, and in sufficient detail, in the HAZOP recording sheets.

Ref.	Parameter and Guideword	Question	Cause	Possible Actions
1.1	High flow	What could cause a flow of greater than 0.5 l/s?	1 Line break, 2 Leaking joints 3 Rupture due to ice	1.1.1 Should all lines be pressure tested and use soldered joints? 1.1.2 Check access to and orientation of V1 1.1.3 Thermally insulate L1 and L2, particularly in the attic
1.2	No Flow	What else other than ice could cause "no flow?"	1 Ball cock stuck Shut 2 Ice plug	1.2.1 Is there adequate access and isolation for removal (Access and maintenance)? 1.2.2 See earlier actions
1.3	High Flow (when there should be "no flow")	No flow has already been considered. The expectation is that there would be no flow in this line for most of the time. How might it be more than 0.0l/s?	1 Ball cock leaking/weeping due to wear or age resulting in overflow of T 1 2 Toilet flushed at the same time as running a hot tap	1.3.1 Fit an overflow to outside the house. 1.3.2 See 1.1.4 Confirmation of issue 1.3.3 Check that ball cock valve can handle both flows
1.4	Pressure/No	1.4 What might create No pressure in L 1—L 2	1.4.1 Line rupture 1.4.2 Loss of water supply to T1 supply due to a BWM 1.4.3 Loss of all water to the house	1.4.1 See Action 1.1.2 1.4.2 Size the Tank T1 to have adequate reserves for flushing toilets and washing hands. 1.4.3 Fit a second COLD water supply to the cold tap. Is the direct supply ideal?
1.5	Temperature/ Low	1.5 What might be the cause and effect of low temperature	1.5 Frost conditions leading to line rupture	1.5.1 See Action 1.1.2 1.5.2 See action 1.1.2 & 1.1.3
1.6	Pressure/ High	1.6 Is this credible other than ice conditions (already considered in 1.1	1.6.1 Pressure is controlled at source 1.6.2 Ice plug expands and leads to the rupture of the pipe 1.6.3 Water Hammer	1.6.1 No action required—noted See 1.5.1 1.6.2 Verify Water hammer has been considered in the piping spec
1.7	Temperature/ High	1.7.1 Is this credible other than summer conditions (higher) or a fire? 1.7.2 Should all piping be "non-plastic"?	1.7 The main cause might be a fire.	1.7 Noted. No action
1.8	Flow of water "as well as"	1.8 What else may come with water? Debris? Germs? What damage might it create? Can it score the ball cock such that it leaks?	1.8 New build!	1.8.1 Flush the line clean and carry out a biological test before entry to the house! 1.8.2 See also cause 1.3
	The study will cover all of the parameters and guidewords and finally look at the "others"—access and maintenance			
1.9	Level high/low	Record that it was not appropriate in node 1and that no deviations could be found		
1.10	Access and Maintenance	Can BC be accessed and maintained?	1.10.1 Is there a roof access hatch? 1.10.2 Is the BV accessible in the T1?	1.10.1 Verify that there will be a roof access hatch 1.10.2 Verify that there is clearance for removal of BV between the roof and the T.
	Note: Isolation of BV to be covered under the Tank T 1 node bring forward at that time			
	Etc.	Etc.	Etc.	Etc.

It will be noted that *many* actions have confirmation under other combinations of Parameter and Guideword. This is quite normal, but it is essential that they are cross referenced in case of a future audit. Likewise, the need to insert "no action required" to show that the combination was considered but no actions were noted needs to be recorded.

This listing is appropriate for L1 and L2

Tank T1 COULD have debris in it (as well as) from construction! How is it removed?
Can T1 be drained? NO! Is a drain line required?
In the case of T1 the inflow rate is defined by the BC and the outflow rate is defined by either the size of V2 or the toilet cistern. Is it worth asking about the relative capacities?

Some of the comments in 1 could apply to L3.

T2 is a repeat of T1 it does not need an overflow but it does require a drain! However it requires a drain to maintain the immersion heater.

The exit line from T2 (L5) is shown coming from the top of the tank. The parameter and guideword "low temperature" should spot a failure of the immersion heater (control of heater itself), power failure (trivial) as well as the "short circuit" of L4 to L5. L4 should enter the bottom of the tank T2 such as to create a reservoir of hot water. T2 does not have a drain so the removal of the immersion heater is problematical.

The discussion can be turned onto the effluent line. The guidewords "as well as" and "reverse" (flow) can be revealing. Where is the rodding eye to clear a choke in the effluent? Is the "U" bend in the sink deep enough to prevent reverse flow?

Once complete the line should be fully marked off as "complete". When all lines are fully marked off the study on that P&ID is complete. On a complex line or P&ID where there are many interconnections it will be necessary to check each connection in detail!

Project Intent: As Described Earlier

The final worksheets would look something similar to that shown in the following list as a total record:
<u>Worksheet for first node of study</u>
<u>Facilitator: Tutor, Scribe: Joe</u>
<u>Attendees: Fred, Jane, Jim, Fred and Jack</u>
P & ID attached
Date X/Y/ZZ
Node 1 V1 to tank V1—L1—L2.
Intent: Flow of 0 to 0.5 l/s of clean water at an inlet pressure of 300kPa and 15°C into T1 controlled by a ball cock BC in T1.

Ref No	Parameter	Guideword	Cause	Effect	Protective systems	Action	On
1.1	Flow	Higher	Line break, sprung joints rupture due to ice	Leakage into house or attic	None	1 Should all lines be pressure tested and all soldered joints? 2 Check the access to and orientation of V1. 1.1.3 Consider lagging the lines L1 and L2 particularly in the attic	Fred Jane Jane

Ref No	Parameter	Guideword	Cause	Effect	Protective systems	Action	On
1.2	Flow	No	Ball cock stuck shut due to damage or ice	T1 Overflows and water enters attic	None	1.2.1 Is there adequate access and isolation for removal? (Access to the attic and under the roof trusses for maintenance) See 1.1.2 1.2.2 Consider the need for lagging T1 as well as the piping	Jim Jim
1.3	Flow Flow	Higher when it should be when it should be no High	Ball cock weeping due to wear and tear or age resulting in overflow of T1 Cistern flushes when sink is filled**	Damage to the ceilings under the tank T1 drained	None (at present None obvious	1.3 1 Consider the need for an overflow to outside the house. 1.3.2 See 1.2.1 Confirmation of issue 1.3.3 Check the size of BC—can it supply the two flows?	Fred Fred
1.4	Pressure	No	Line rupture Loss of supply due to a BWM	See 1.1 No water for flushing toilets, washing hands and cups of tea!	None	1.4.1 Noted 1.4.2 Consider sizing the Tank T1 to have adequate reserves for flushing toilets and washing hands. 1.4.3 Consider fitting a second COLD water supply to the cold tap from T1. 1.4.4 Review the direct supply to the cold tap, is it ideal?	Jane Jane Jane
1.5	Temperature	Low	Frost	Frost conditions leading to line rupture	None	1.5.1 See Action 1.1.2 1.5.2 See action 1.1.2 & 1.1.3	Noted
1.6	Pressure	High	Ice plug formation Overpressure at Supply Water hammer*** caused by sudden closure of BC resulting in pipe rupture	Line ruptures leading to internal flooding when the line thaws out. As earlier	Probably pressure controlled at source	1.6.1 See 1.5.1 1.6.3 Verify with the builder that a full hydraulic surge analysis has been carried out	Noted Noted Jack
1.7	Temperature	High	Fire Very hot day	Piping springs	None	1.7.1 Noted 1.7.2 Builder to consider the need for piping expansion	Jack
1.8	Water flow	As well as	New piping. Muck and germs in the feed line	Potential illness Damage to BV	None	1.8.1 Builder to consider flushing the line clean and carry out a biological test before entry to the house! 1.8.2 See Cause 1.3	Jack

Ref No	Parameter	Guideword	Cause	Effect	Protective systems	Action	On
1.9	Access Maintenance		Unable to access V1 and BC	Unable to stop flow Unable to maintain BC	May be OK	1.9.1 Builder to verify that there will be a roof access hatch 1.9.2 Builder to verify that there is clearance for removal of BV between the roof and the T1 1.9.1 Is there a master isolation between the main supply and V1?	Fred Fred Fred
1.10	Flow	Reverse	Siphon out of T1 if the service line is severed	Loss of back-up water	V1?	1.10 Verify that there is a siphon breaker at BC in T1	Jack
1.11	Level	High				See 1.3	See 1.3

It will be noted that the matrix and others have all been considered. The line can now be marked off in a highlighter pen as complete.

** *Was put in to illustrate the fact that the "cause" could be out with the boundary of the node under review but the effect will be within the node! (And vice versa!)*

*** *Will only be spotted with the correct team spectrum.*

Now move to the next line which is best L1, L3 and L6 as it is almost a repeat on L1 and L2. This can be examined "by exception" but again the actions MUST be recorded even if it is "cut and paste".

The results for the high/low level in T1 would be:

Ref No	Parameter	Guideword	Cause	Effect	Protective systems	Action	On
5.1	Level	High	In flow greater than out flow	Tank overflows. Damage to the ceilings under the tank	None	5.1.1 Consider the need for an overflow to outside the house. 5.1.2 See 1.2.1 Confirmation of issue Confirmation of issue)	Fred
5.2	Level	Lower	Inflow less than out flow	Tank Empties	None	5.2 Consider the need for lagging T 2 as well as the piping See 1.2 where this is discussed	Fred Jim
5.3		Maintenance/ Access	Lack of isolation valves in the system	5.3.1 The only isolation for BC is the master isolation V1 5.3.2 T 1 can be drained through T2 but is it adequate? 5.3.3 No "master isolation" between the supplier and the house		5.3.1 Considerer an isolation in L2 close to BC 5.3.2 Consider the need for a drain on T1 to the outside 5.3.3 Verify that there is a master isolation in the water supply to the house	Jack Jack Fred

Other combinations of parameter and guideword were not appropriate to this node. Record that no deviations could be found.

Do not underestimate the amount of recording! If there is a future upset the results of the HAZOP will be given very close attention.

Note that this simple study assumed a new site using existing water supplies. However there was a warning at the beginning that in reality the start and end points may have to be somewhat remote from the immediate process under study. If this had been done the presence of Cryptosporidium would be flushed out (sorry for the pun!)

Note that there are some useful devices that can be used.

1 Splitting the nodes can add some more definition or analysis particularly with equipment (tanks, pumps, heat exchangers and other equipment).
2 A number of P & IDS should be linked to show the process flow and assist the identification of causes and effects external to that specific node/P & ID. It could be some P & IDs beyond!
3 It is often useful to make a note that a particular issue should be covered in a later node so as to maintain the "flow of the discussion". The action is recorded so it will not be forgotten but it will allow the logic to flow without interruption.
4 This will also apply to an interface between services (cooling water and the like) and the process fluid. It is better to keep the process and service discussion separate so as to keep the logical flow of the discussion and a record made to review the interface at a later stage. (also make a note on the P & ID of some of the potential issues as a back-up.)
5 Even if a combination of parameter and guideword produces "No Action Required" it is important that this is recorded as it shows that it was considered. Otherwise, it could be revealed as a "missing item" in a future audit—not an unusual situation.

This study is simple but it does show the rigour required in a HAZOP study and (unfortunately) if it is done without any attempt to "add colour" by asking "awkward questions" it will become completely boring!! The skill is to make it a challenge!

Appendix B: Example Operating Procedure

Title: Start-up of Cooling Water Pump No 123
SOP 321
Prepared by: J. Bloggs
Authorised by: A.N. Other
Copies to: Records
All Plant B Standing Instruction Dossiers
All Plant B Supervisors
Last Issue: 1 March 2014

OBJECTIVE OF THIS STANDING INSTRUCTION

This Standing Instruction gives a list of all of the contacts to be made when starting up Motor Driven Pump P–123. It also gives a list of the pre-start-up checks, preparation tests and the final start-up process. Within the introduction are found the operating envelope and the prohibited operating zone.

PRE-START-UP CHECKS

1 Ensure that the pump's Denko lubricator is full of grade "xyz" oil and there are no oil leaks. In the event of any leaks contact the maintenance department before starting.
2 Ensure the coupling guard is securely in place.
3 Ensure there are no slip plates in the process lines and start-up blanks are removed.
4 Ensure all valves are closed-shown as a C for status—(other than any steam trap isolation valves if fitted.
5 Check the pump rotation and if necessary, change the electrical connections.

(Note—the three-phase system has Red (R, Blue (B) and Yellow (Y) connections. If the pump turns in the wrong direction the vector diagrams inside the pump are in error and it is likely that no forward flow is possible. The solution is to reverse two phases in the motor RBY rotates one way and RYB rotates the other)

PRE-START-UP CONTACTS

1 Contact all control rooms on the plant (telephone Nos. 124, 125 and 126) and inform them that new cooling water pump is to be brought online in one to two hours time and that operators may have adjust cooling water flows.

PREPARATION TESTS

Reference should be made to piping and instrument line diagrams (P&ID No 10, Figure B.1).

1 Notify all control rooms on the plant (telephones 124, 125 and 126 that the extra cooling water pump will be brought on line in 15 minutes.

2 Prime the pump—open pump P-123 suction valve CW123 valve about two turns then open the body vent valve 10 to clear (drive out) any air or gas. Close valve V10 when liquids are detected.
3 Open valve V1 fully open the cooling water pump minimum flow valve CW123, valve 3.
4 Start the pump.
5 Slowly open valve V2 in CW456 to allow any control systems to function properly and slowly close the recycle line V3
6 Check the pump including the seal for leaks
7 Notify all users that the pump is now running.

POST-START-UP

1 Check the lubricator levels.
2 Check the pump-bearing temperatures are not more than 50°C.
3 Check the vibration levels on the Pump are acceptable (less than 2 mm/sec velocity).
4 Check pump 123 for evidence of cavitation; if any is detected fully vent gases through valve V10 and then close valve 10.

Note

This is a trivial example which illustrates the need for detail in each step. In some cases, there may not be a minimum flow recycle and then after venting the pump the discharge valve V2 should be opened one turn and the pump started. Then the valve should be slowly opened fully as controls are established.

When there is no minimum flow by-pass the NRV is a critical item and must supply a reasonably secure seal against revere flow.

It is possibly worth noting that some SOPs refer to handling emergencies and upsets. For example, what is the procedure following loss of cooling water or power?

Whilst this example is trivial the startup of a turbo-compressor could have the following operating instructions

1 Process circuit preparation and alignment
2 Functional testing of the protective systems
3 Lube and seal preparation
4 Cooling water preparation
5 Notification of sudden power demand at startup
6 The actual steps during the startup
7 The actual steps following shut down
8 Further instructions will be required for a steam-driven unit, warming up the steam mains, sealing steam, condenser sets and ejectors.

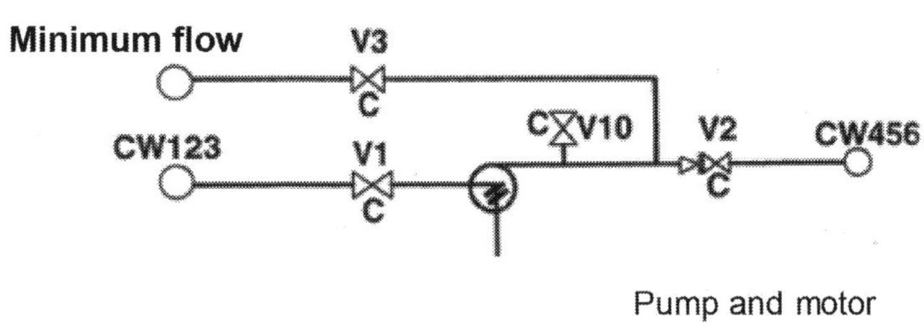

FIGURE B.1 Simplified P&ID No 10

Each operating instruction should be around 5 to 10 pages in length. For a major item of equipment such as a compressor, instructions take a lot of writing and training, particularly the dynamics of change during the three or four minutes of stabilisation.

Index

Note: Page numbers in *italics* indicate figures, and page numbers in **bold** indicate tables in the text.

T

U

V

Made in the USA
Las Vegas, NV
19 September 2026

37532dd9-6e06-4dce-a018-0aa16aae3932R01